T0140221

Massive MIMO Detection Algorithm and VLSI Architecture

Leibo Liu · Guiqiang Peng ·
Shaojun Wei

Massive MIMO Detection Algorithm and VLSI Architecture

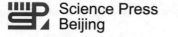 Science Press
Beijing

 Springer

Leibo Liu
Institute of Microelectronics
Tsinghua University
Beijing, China

Guiqiang Peng
Institute of Microelectronics
Tsinghua University
Beijing, China

Shaojun Wei
Institute of Microelectronics
Tsinghua University
Beijing, China

ISBN 978-981-13-6364-1 ISBN 978-981-13-6362-7 (eBook)
https://doi.org/10.1007/978-981-13-6362-7

Jointly published with Science Press, Beijing, China
The print edition is not for sale in China Mainland. Customers from China Mainland please order the print book from: Science Press.
ISBN of the China Mainland edition: 978-7-03-060210-7

Library of Congress Control Number: 2019930265

Translation from the Chinese language edition: 大规模MIMO检测算法VLSI架构— —专用电路及动态重构实现 by Leibo Liu, Guiqiang Peng, Shaojun Wei, © Science Press 2019. Published by Science Press. All Rights Reserved.

© Springer Nature Singapore Pte Ltd. and Science Press, Beijing, China 2019
This work is subject to copyright. All rights are reserved by the Publishers, whether the whole or part of the material is concerned, specifically the rights of translation, reprinting, reuse of illustrations, recitation, broadcasting, reproduction on microfilms or in any other physical way, and transmission or information storage and retrieval, electronic adaptation, computer software, or by similar or dissimilar methodology now known or hereafter developed.
The use of general descriptive names, registered names, trademarks, service marks, etc. in this publication does not imply, even in the absence of a specific statement, that such names are exempt from the relevant protective laws and regulations and therefore free for general use.
The publishers, the authors, and the editors are safe to assume that the advice and information in this book are believed to be true and accurate at the date of publication. Neither the publishers nor the authors or the editors give a warranty, express or implied, with respect to the material contained herein or for any errors or omissions that may have been made. The publishers remain neutral with regard to jurisdictional claims in published maps and institutional affiliations.

This Springer imprint is published by the registered company Springer Nature Singapore Pte Ltd.
The registered company address is: 152 Beach Road, #21-01/04 Gateway East, Singapore 189721, Singapore

Preface

As one of the core technologies for future mobile communications, the massive MIMO technology can effectively improve the network capacity, enhance the network robustness, and reduce the communication latency. However, the complexity of baseband processing increases sharply as the number of antennas increases. Therefore, the design of high-performance massive MIMO baseband processing chips, especially the design of massive MIMO detection chips featuring low complexity and high parallelism, has become a technical bottleneck that restricts the broad application of the massive MIMO technology in communications systems.

This book first introduces the process of the team's research on efficient massive MIMO detection algorithms and circuit architectures. On the basis of the analysis on the existing massive MIMO detection algorithms, the team has optimized the algorithms from different aspects such as computation complexity and parallelism, conducted mathematical theoretical analyses and proven that the massive MIMO detection optimization algorithms proposed by the team have the advantages of low complexity and high parallelism and can fully satisfy the requirements for detection accuracy. Finally, by using the ASIC as a carrier, the team has verified that the chips based on the proposed massive MIMO detection algorithms feature high energy efficiency, high area efficiency, and low detection error.

In the process of designing the massive MIMO detection chip, we learned that the massive MIMO detection chips based on the ASIC are suitable only for application scenarios with very high requirements for the processing speed; however, some application scenarios require massive MIMO detection chips to have certain flexibility and scalability so that the massive MIMO detection chips can support different standards, algorithms, and antenna sizes and adapt to the evolution of standards and algorithms. After we conducted certain analyses, we believe that the reconfigurable computing architecture is a very promising solution. On the basis of the analyses on and common feature extraction from a large number of existing massive MIMO detection algorithms, the team has designed the data channels and configuration channels that are applicable to massive MIMO detection algorithms, involving PEs, interconnections, storage mechanisms, context formats, and configuration methods. Thus, the team has completed the design of a massive MIMO detection reconfigurable processor.

The massive MIMO detection reconfigurable processor may also be applicable to future wireless communications systems such as Beyond 5G. There are three main reasons: First, wireless communication algorithms are now developed in the repeated iteration and optimization processes. In the process of solving the limitation problem of commercial algorithms, the update of an algorithm, no matter whether it is an optimized algorithm or a newly designed algorithm, has a strong logical continuation relationship, which provides an internal logical basis for design of the reconfigurable processor architecture. Second, the design for PEs and PEAs of the massive MIMO detection reconfigurable processors fully considers the requirements for flexibility and scalability so that the PEs and PEAs can meet the hardware requirements and foreseeable future needs of various algorithms at present. Third, the design methodology is applicable to all the massive MIMO detection reconfigurable processors. Therefore, the hardware implementation requirements for future algorithms can be met. Hence, after corresponding algorithm analyses are conducted, the optimization and design of the reconfigurable processor architecture based on the design methodology will become a universal process.

This book consists of seven chapters. Chapter 1 introduces the development trend of wireless communication technologies including the development and research status of the massive MIMO technology and the MIMO detection technology, analyzes the advantages and disadvantages of the MIMO detection chip based on the ASIC and instruction-level architecture processor in aspects such as performance, power consumption and flexibility, proposes the dynamic reconfigurable chip technology for MIMO detection, and analyzes the feasibility for implementing the proposed technology. Chapters 2 and 3 introduce the linear massive MIMO detection algorithm and the corresponding circuit architecture, respectively, and analyze the advantages of the linear detection optimization algorithm proposed by the team from different aspects such as algorithm convergence, computation complexity, and detection performance. The experimental results have shown that the circuit designed on the basis of the algorithm proposed by the team has higher energy efficiency and area efficiency, and thus verified that the optimization algorithm proposed by the team is more suitable for hardware implementation. Chapters 4 and 5 introduce the nonlinear massive MIMO detection algorithm with high detection accuracy and the corresponding circuit architecture, respectively, and compare the nonlinear massive MIMO detection algorithm proposed by the team with other algorithms from different aspects such as algorithm convergence, computation complexity, detection performance, and experimental results. The results have shown that the complexity of the algorithm proposed by the team is within the acceptable range while the algorithm implements high detection accuracy. Chapter 6 provides detailed information on the dynamic reconfigurable chip for massive MIMO detection. First, the chapter uses the reconfigurable computing architecture as the target hardware platform to analyze mainstream massive MIMO detection algorithms at present, including common logic extraction from algorithms, feature extraction of data types and parallelism analysis on algorithms. Then, the chapter provides a detailed analysis on the hardware architecture design of the dynamic reconfigurable chip for massive MIMO detection from different aspects of data channels and configuration channels, and introduces the design method for the hardware architecture specific to the massive MIMO detection algorithm. Chapter 7 provides an outlook on application of the VLSI architecture for massive MIMO detection on the server, mobile terminal and edge computing sides.

This book embodies the nearly 6-year collective wisdom of the wireless communication baseband processor team from the Institute of Microelectronics of Tsinghua University. Thanks to the classmates and colleagues of the team members including Peng Guiqiang, Wang Junjun, Zhang Peng, Wei Qiushi, Tan Yingran, Yang Haichang, Wang Pan, Wu Yibo, Zhu Yihong, Xue Yang, Li Zhaoshi, Yang Xiao, Ding Ziyu, and Wang Hanning for their participation. Thanks to our engineers Wang Yao, Ying Yijie, Kong Jia, Chen Yingjie, Wang Guangbin, Wang Lei, Li Zhengdong, Luo Senpin, Jin Yu, et al. for their participation. Thanks to Prof. Wei Shaojun for his support for and guidance to the preparation of this book. Thanks to Editor Zhao Yanchun from Science Press for her suggestions on the publication of this book. Finally, I give thanks to my wife and children for their understanding and tolerance of my work. Without their support, it is hard to imagine how I could finish this work. They are also an important impetus for my future efforts and progress.

Beijing, China   Leibo Liu
August 2018

Contents

Abbreviations

ACC	Accumulator
ADPLL	All digital phase-locked loop
AHB	Advanced high-performance bus
ALU	Arithmetic logical unit
AMBA	Advanced microcontroller bus architecture
AR	Augmented reality
ARM	Advanced RISC machine
ASIC	Application-specific integrated circuit
ASIP	Application-specific instruction set processor
AU	Arithmetic unit
BB	Branch and bound
BLER	Block error rate
BPSK	Binary phase-shift keying
BTS	Base transceiver station
CBU	Column-broadcast unit
CC	Convolutional coding
CDMA	Code-division multiple access
CG algorithm	Conjugate gradient algorithm
CGLS	Conjugate gradient least square
CGRA	Coarse-grained reconfigurable array
CHEST	Channel estimation
CHOSLAR	Cholesky sorted QR decomposition and partial iterative lattice reduction
CM	Complex multiplication
CORDIC	Coordinate rotation digital computer
CoREP	Common Reports
CP	Cyclic prefix
CPA	Control program assist
CPLD	Complex programmable logic device
CPU	Central processing unit

CSG	Closed subscriber group
CSI	Channel state information
CSIR	Receiver channel state information
DDR	Double data rate
DMA	Direct memory access
DSP	Digital signal processor
DVFS	Dynamic voltage and frequency scaling
ELP	Energy latency product
EMI	Electromagnetic interference
EPD	Expectation propagation detection
FBMC	Filter bank based multicarrier modulation
FBS	Forward–backward splitting
FDD	Frequency-division duplexing
FEC	Forward error correction
FER	Frame error rate
FFT	Fast Fourier transform
FIR	Finite impulse response
FPGA	Field-programmable gate array
FSM	Finite-state machine
GI	Guard interval
GPP	General purpose processor
GPU	Graphics processing unit
GR	Givens rotation
GSM	Global System for Mobile communication
HART	Highway addressable remote transducer
HDL	Hardware description language
HEVC	High Efficiency Video Codec
HMD	Head-mounted display
HT	Householder transformation
i.i.d.	Independent identically distributed
I/O	Input/output
IaaS	Infrastructure as a Service
IASP	Instruction set architecture processor
IFFT	Inverse fast Fourier transform
IIC	Intra-iterative interference cancellation
IoT	Internet of things
ISI	Intersymbol interference
ISP	Internet Service Provider
JED	Joint channel estimation and data detection
JTAG	Joint Test Action Group
LBC	Lower bound of cost
LDPC	Low-density parity-check code
LLC	Last level cache
LLR	Log likelihood ratio
LPF	Low-pass filter

LR	Lattice reduction
LTE	Long-term evolution
LUD	LU decomposition
LUT	Lookup table
M2M	Machine to machine
MAC	Multiply and accumulate
MDA	Multimode detection architecture
MEC	Mobile edging computing
MF	Matched filtering
MIMO	Massive multiple-input multiple-output
ML algorithm	Machine language algorithm
MMSE	Minimum mean square error
MMSE-SIC	Minimum mean square error-successive interference cancelation
MPD	Message passing detector
MWD	Multi-window display
NI	Network interface
NoC	Network on chip
NP problem	Nondeterministic polynomial problem
NSA algorithm	Neumann series approximation algorithm
NTL	Network topology link
OCD	Optimized coordinate descent
OFDM	Orthogonal frequency-division multiplexing
OFDMA	Orthogonal frequency-division multiple access
opcode	Operation code
OSG	Open subscriber group
PaaS	Platform as a Service
PAR	Peak-to-average ratio
PARSEC	Princeton Application Repository for Shared-Memory Computers
PCBB	Priority and compensation factor oriented branch and bound
PDA	Probabilistic data association
PE	Processing element
PEA	Processing element array
PILR	Partial iterative lattice reduction
PIP	Picture in picture
PLL	Phase-locked loop
PSD	Positive semidefinite
QAM	Quadrature amplitude modulation
QPSK	Quadrature phase-shift keying
RADD	Real-valued addition
RAM	Random-access memory
RBU	Row-broadcast unit
RC	Reliability benefits
RCM	Reliability cost model
REM	Reliability efficiency model

RISC	Reduced instruction set computer
RMUL	Real-valued multiplication
RSN	Resource node
RTL	Resistor transistor logic
SA	Simulated annealing
SaaS	Software as a Service
SC-FDMA	Single-carrier frequency-division multiple access
SD	Sphere decoding
SD algorithm	Standard deviation algorithm
SDP	Semidefinite program
SDR	Software-defined radio
SER	Symbol error rate
SIMD	Single instruction, multiple data
SINR	Signal-to-interference-plus-noise-ratio
SNR	Signal-to-noise ratio
SoC	System on chip
SRAM	Static random-access memory
TASER	Triangular approximate semidefinite relaxation
TDD	Time-division duplexing
TDMA	Time-division multiple access
TGFF	Task graph for free
TSMC	Taiwan Semiconductor Manufacturing Company
UBC	Upper bound of cost
UMTS	Universal Mobile Telecommunications System
VLIW	Very long instruction word
VLSI	Very-large-scale integration
VOPD	Video object plane decoder
VR	Virtual reality
WiMAX	Worldwide Interoperability for Microwave Access
WLAN	Wireless local area network
ZF	Zero frequency
ZF-DF	Zero-forcing decision feedback

Chapter 1
Introduction

With the rapid development of the people's demands for mobile communication in their daily life, the complex data communication and processing will become an important challenge to the future mobile communication. As the key part of the developing mobile communication technology, massive multiple-input multiple-output (MIMO) technology can improve the network capacity, enhance the network robustness and reduce the communication delay. However, as the number of antennas increases, so does the baseband processing complexity dramatically. The very large scale integration (VLSI) chip is the carrier of the massive antenna detection algorithm. The design of the massive MIMO baseband processing chip will become one of the bottlenecks in the real application of this technology, especially the design of massive MIMO detection chip with high complexity and low parallelism.

In order to meet the data transmission requirements of wireless communication in future and address the relevant power consumption issues, the massive MIMO detection chip needs to achieve high data throughput rate, high energy efficiency and low delay. In the meantime to support different standards, algorithms, antenna scales, etc., the massive MIMO detection chip needs to be flexible. Furthermore to adapt to the evolution of future standards and algorithms, the massive MIMO detection chip needs to be scalable. The traditional MIMO detection processors, including the instruction set architecture processor (ISAP) and the application specific integrated circuit (ASIC), cannot simultaneously satisfy the three requirements: energy efficiency, flexibility and scalability. Though the ASIC can meet the rapidly growing computing power requirements of massive MIMO detection chips and achieve the high data throughput rate, the high energy efficiency and the low delay, the standards, transmission performance requirements, MIMO scale, algorithms, etc. will usually differ in order to provide personalized and customized services as the communication technologies, standards and transmission performance requirements develop. Supporting multiple standards and multiple protocols will become one of the key considerations in the hardware circuit design. In addition, the hardware circuit design also needs scalability to cope with the rapid development of baseband processing algorithms and to ensure the reliable and seamless connection of the algorithm evolution. Therefore, the application of ASIC will be significantly limited. What's more,

© Springer Nature Singapore Pte Ltd. and Science Press, Beijing, China 2019
L. Liu et al., *Massive MIMO Detection Algorithm and VLSI Architecture*,
https://doi.org/10.1007/978-981-13-6362-7_1

although the ISAP can meet the requirements of flexibility and scalability, the application of such processors will be significantly limited because the ISAP cannot meet the requirements of the processing rate and power consumption of the future mobile communication. The reconfigurable processor, as a new implementation method, not only achieves a high data throughput rate, low energy consumption and low delay in the MIMO detection, but also boasts unique advantages in terms of flexibility and scalability. Benefiting from the hardware reconfigurability, this architecture may possibly update the system and fix the bugs while the system is running. This feature will extend the service life of the product and ensure its advantages in the time-to-market aspect. In summary, the reconfigurable processor with the MIMO detection function can properly balance the requirements applied in such aspects as energy efficiency, flexibility and scalability, and it will be an important and promising development direction in the future.

1.1   Application Requirements

Digital technology makes continuous innovation possible for different industries. Information and communication technology (ICT), media, finance and insurance industries are leading the way in the current digital transformation process [1–4]. At the same time, digitalization in the areas of retail, automotive, oil, gas, chemicals, healthcare, mining and agriculture is accelerating [5–8]. Key technologies supporting digitalization include software-defined devices, big data [9, 10], cloud computing [11, 12], blockchain [13, 14], network security [15, 16], virtual reality (VR) [17, 18] and augmented reality (AR) [19, 20]. As the quality of life improves, a variety of more advanced and complex applications are coming or appearing in people's daily lives. The conceiving of future life will be smarter, more convenient and more effective. Cloud virtualization, AR, autopilot, intelligent manufacturing, wireless electronic healthcare and other applications are driving the development of communication technologies. Communication networks are the key of all connections.

1.1.1   Typical Applications in Future

1.1.1.1   Cloud Virtualization and AR

The effective work of VR/AR requires very good bandwidth because most VR/AR applications are very data intensive [17]. Although the average data throughput rate of existing 4th-generation (4G) mobile communication networks may reach 100 Mbit/s, some advanced VR/AR applications will require higher speed and lower delay (Fig. 1.1). For example, VR and AR are revolutionary technological innovations in the consumer industry. VR/AR demands a great amount of data transmission, storage, and computing. As a result, these data and compute-intensive tasks will be

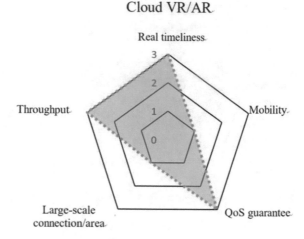

Fig. 1.1 System requirements of cloud virtualization and AR

moved to the cloud which provides rich data storage and necessary high-speed computing capability [21].

1.1.1.2 Autopilot and Other Mobile Revolutions

The key technology driving the mobile revolution—autopilot requires safe, reliable, low-delay and high-bandwidth connections [22], which are essential in the high-speed mobile and highly-dense urban environments. In the era of autopilot, comprehensive wireless connection will allow additional services to be embedded in the vehicles. The reduction of human intervention is based on the frequent exchange of information between the vehicle control system and the cloud-based backend system. For remote driving, a vehicle is driven from a distance, not the person in the vehicle. The vehicle is controlled manually instead of automatically. This technology may be used to provide quality concierge services, for example, to enable someone to work on the go, to help a driver without a driver's license, or to help a driver who is sick, drunk or not suitable for driving. Figure 1.2 shows the system requirements for autopilot and remote driving. These two technologies require high-reliability wireless transmission and less than 10 ms round-trip time (RTT). Only robust wireless communication technologies can meet the strict connection requirements. Fifth-generation (5G) and beyond 5G may become the unified connection technology to meet the future requirements for connection, sharing, remote operation, etc. [23].

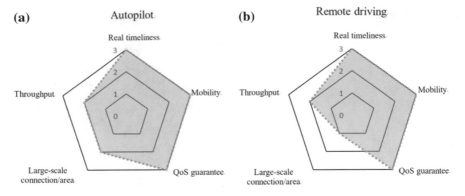

Fig. 1.2   a System requirements of autopilot, **b** remote driving

1.1.1.3   Intelligent Manufacturing

The basic business philosophy of implementing intelligent manufacturing is to bring higher quality products to market through more flexible and efficient production systems [24]. Innovation is the core of the manufacturing industry. The main development directions include sophisticated production, digitalization, and more flexible workflow and production [25]. The main advantages of intelligent manufacturing include:

(1)   Increase productivity with collaborative robots and AR smart glasses and help employees increase their production efficiency throughout the assembly process. The collaborative robots exchange analyses to complete the synchronous and collaborative automation process, and the AR smart glasses enable employees to get their work done faster and more accurately.
(2)   Accurately predict the future performance, optimize the maintenance plans, and automatically order the parts used for replacement through the state-based monitoring, machine learning, physics-based digital simulation, etc., thereby reducing the downtime and maintenance costs.
(3)   Reduce the inventory and logistics costs by optimizing the accessibility and transparency of suppliers' internal and external data.

Before the development of wireless communication technologies, manufacturers relied on wired technologies to connect the application programs. With the development of wireless solutions such as wireless fidelity (Wi-Fi), Bluetooth and highway addressable remote transducer (HART), more and more intelligent and wireless devices have emerged in manufacturing workplaces. But these wireless solutions are limited in terms of security and reliable bandwidth. Cutting-edge application connections require flexible, mobile, high-bandwidth, ultra-reliable, and low-delay communications as the basis (Fig. 1.3).

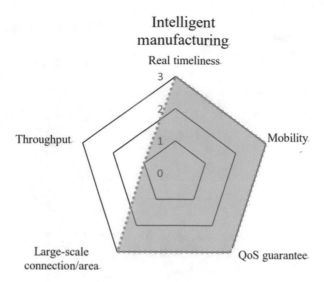

Fig. 1.3 System requirements of intelligent manufacturing

1.1.1.4   Wireless Electronic Healthcare

In Western countries and Asian countries, the population aging speed is accelerating. From 2012 to 2017, wireless networks were increasingly applied in medical devices. Healthcare professionals have begun integrating solutions such as remote audio and video diagnostics, tele-surgery, resource databases, etc., and using wearable devices and portable devices for remote health monitoring. The healthcare industry probably sees a fully personalized medical consulting service that will enable the physician's AI medical system through a 5G network connection. These intelligent medical systems can be embedded in large hospitals, family doctors, local doctors' clinics, and even medical clinics that lack on-site medical staff. Wireless electronic medical tasks include:

(1) Implement real-time health management and track patients' medical records, recommend treatment procedures and appropriate medications, and schedule follow-up visits.
(2) Provide prospective monitoring for patients through the AI model to advise on treatment options.

Other advanced application scenarios include medical robots, medical cognition, etc. These high-end applications require uninterrupted data connections, such as biotelemetry, VR-based medical training, ambulance plane, bioinformatics, and bioreal-time data transmission.

Telecom operators can work with the medical industry to become medical system integrators. They can create a good ecosystem for society and provide connection, communication and related services, such as analyzing medical data and cloud ser-

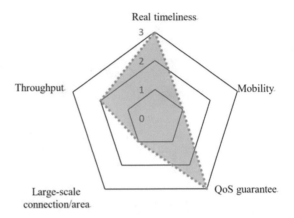

Fig. 1.4 System requirements of wireless electronic healthcare

vices, while supporting the deployment of various technologies. The telemedicine diagnostic process relies in particular on the low-delay and high-quality data services provided by 5G networks (Fig. 1.4) [26].

1.1.2 Communication System Requirements

In the era of rapid update and development of people's daily life applications [27, 28], according to the needs of different applications, communication systems and circuits will have the following points to be considered.

1.1.2.1 Data Throughput Rate

The pursuit of faster communication networks has been the driving force for the development of communication technologies and the main driving force for the development of next-generation mobile communication technologies. According to the industry's opinion, the peak data transmission rate required by next-generation wireless communication technologies is on the order of tens of Gigabit per second (Gbit/s), which is equivalent to about 1000 times the data transmission rate of 4G networks. At the same time, mobile communication technologies put higher requirements on network delay. In the 4G network era, the data transmission delay (including sending and receiving paths) is about 15 ms, which is acceptable for most current services but is unacceptable for some new applications, such as VR and AR. Next-generation mobile communication technologies are expected to achieve a delay of approximately 1 ms. The requirements for the data transmission rate and network delay are one of the major challenges in the development of communication technologies. In the future, communication networks need to be able to support a huge number of communication devices and gigantic amount of data information [2]. Due to the emergence and development of the Internet of Things (IoT) technology [29, 30] and machine to machine (M2M) technology [31, 32], communication systems will connect to a large number of other types of communication devices that access the network, in addition to personal mobile communication devices. According to forecasts, the number of these devices may reach the order of tens of billions or even hundreds of billions, which will increase the density of devices in some areas [33, 34]. For some applications that require high data rates, such as real-time data transmission and video sharing, the increased device density will have a negative impact on the system performance. In addition to the huge number of communication devices, the communication systems will have to process more communication data. According to current estimates, the network usage of mobile devices will experience a huge growth in the next few years. The ever-increasing number of devices and the need for data traffic pose a challenge to current communication systems [31, 33]. The hardware circuit design layer requires higher processing rates and shorter delays; at the same time, the mobile applications require lower power consumption and less area overhead. Therefore, chip design is also an important challenge.

1.1.2.2 Power Consumption, Area, and Energy Efficiency

In recent years, with the emphasis on environmental protection, people are increasingly pursuing a low-carbon lifestyle, which also imposes power requirements on communication systems. From the perspective of logistics, cost and battery technology, constantly increasing power consumption is also unacceptable [35, 36]. In communication systems and baseband processing circuits, the energy efficiency is measured in terms of Joules per bit or bits per joule. Therefore, the power consumption is increased by the same order of magnitude as the data transmission rate. To reduce the power consumption increase order is to maintain or improve the energy efficiency. The increase in energy efficiency is critical to the applications of IoT because most IoT communication devices are battery-powered and need to work for long periods of time without human intervention. To meet this requirement, the battery life cycle should be 10 years or even longer generally. For the IoT and M2M communication systems, in addition to energy efficiency improvement, a series of energy management technologies are needed to save energy. What's more, renewable energy can be used to power devices, such as solar cells. How to improve the energy efficiency of communication systems and circuits and prolong the life cycle of device batteries is an urgent problem to be solved in the next-generation communication technologies. In communication baseband circuit design, reducing areas will decrease the power consumption and costs to some extent. So, how to reduce the chip area is also an urgent problem to be solved.

1.1.2.3   Flexibility and Scalability

Scalability and flexibility are also the concerns of next-generation communication systems [37, 38]. The flexibility of the IoT means that the communication systems can satisfy different needs and different applications. In the future, there will be various unique needs of applications, and how to meet the needs of the different applications will be an urgent problem to be solved [27, 28]. For the same applications, different scenarios, algorithms and even performance standards will affect the selection of communication technologies. The communication systems will meet as many needs as possible. The circuit design also needs certain flexibility to meet the data processing requirements in different situations. Scalability refers to ensuring that existing service quality, etc. are not affected while new and heterogeneous devices, applications and functions are introduced based on the user needs. The application extension includes the application updates, iterations, and improvements. The technical extension includes the technological evolution and algorithm evolution. The communication circuits and systems can support extensions to different directions. Scalability is proposed based on the assumption of distribution of high-density communication circuits and devices in future, so managing the status information of a huge number of connected devices is also a problem to be considered.

1.1.2.4   Coverage

A sufficiently high network coverage is fundamental to providing stable and reliable communication services [4, 7]. For many consumer-oriented IoT applications, the IoT devices need to exchange information with mobile users, thus ensuring that users can connect to networks anywhere and be served while they are on the move is an important prerequisite for the IoT applications. For indoor applications and other IoT applications, such as smart meters and elevators that are installed in basements with low network coverage, the extended coverage is a major design direction for next-generation communication systems. The ultimate goal of this type of IoT network deployment is to provide a higher indoor coverage, therefore creating an effect equivalent to signals crossing walls and floors to support large-scale deployment of the IoT applications. The major challenge of improving the coverage is to minimize the total deployment costs.

1.1.2.5   Security

The need for security and privacy is another design requirement for communication system applications. For the M2 M applications, the M2 M network security is extremely critical to network information sharing, as the neighboring M2 M nodes can share sensitive information related to user identity and other personal information, and use such personal information for illegal activities. For the IoT applications, security and privacy are also main problems to be considered. The true identity of mobile IoT users should be protected from infringement, and location information is also important because it can reveal the physical location of the IoT devices [39, 40]. In addition to information disclosure, how to deal with human interference is also a problem that needs to be solved in the next-generation communication systems. Unlike unauthorized theft of information, human interference is the deliberate transmission of interference signals by an illegal node to disrupt the normal communication process, and even illegal blocking of authorized users from accessing wireless resources. The human illegal attack on the communication network in the communication process also challenges the security of the communication systems. During such an attack, the attacker can control the communication channels of the legitimate users, and thus intercept, modify, and even replace the normal communication information between the users. This type of attack affects the confidentiality, integrity, and availability of data and is currently the most common attack that poses a threat to the security of communications systems. The issues that threaten the security of communications systems and user privacy require adequate attention in future communication technologies.

1.2 Mobile Communication and MIMO Detection

1.2.1 Development of Communication Technologies

The first generation of mobile communication technology appeared after the theory of cellular system was proposed, and it mainly met the needs of people's wireless mobile communication. With the development and maturity of digital cellular technology, people introduced the second generation of cellular mobile communication system to implement the digital voice service, which further improved the quality of mobile communication. At the end of the twentieth century, the rapid development of Internet protocol (IP) and Internet technologies changed the way people communicated, and the appeal of traditional voice communication declined; people expected wireless mobile networks to provide Internet services, therefore the third-generation (3G) mobile communication systems emerged and were capable of providing data services. In the twenty-first century, the rapidly developing information technologies provided people with more mobile communication services, which challenged the service capabilities of 3G systems, so the 4G mobile communication systems were introduced to implement wireless network broadband services. The 4G network is an all-IP network, which mainly provides data services. Its uplink rate of data transmission can reach 20 Mbit/s, and its downlink rate can reach 100 Mbit/s, which can basically meet the needs of various mobile communication services [2, 4]. However, the rapid development of mobile Internet technology and the IoT technology have almost subverted the traditional mobile communication mode, and new mobile communication services, such as social networks, mobile cloud computing, and the

Internet of Vehicles, have proposed new demands for the development of mobile communication networks.

In 2012, the European Union (EU) officially launched the mobile and wireless communications enables for the 2020 information society (METIS) project [41] to conduct research on the 5G mobile communication networks. In addition to METIS, the EU launched a larger research project 5G infrastructure public private partnership (5G-PPP), aiming to accelerate the 5G mobile communication research and innovation of EU, and establish the EU's guiding position in the field of 5G mobile communications. The UK government set up a 5G mobile communication R&D center with a plurality of enterprises at Surrey University, dedicated to 5G research [41, 42]. In Asia, South Korea launched the "GIGA Korea" 5G mobile communication project in 2013, and the China International Mobile Telecommunications (IMT)-2020 promotion group was also established in the same year to unite the 5G research strength in Asia to jointly promote the development of 5G technology standards [43, 44].

In 2015, the International Telecommunication Union (ITU) officially named 5G as IMT-2020, and defined mobile broadband, large-scale machine communication, and high-reliability low-delay communication as the main application scenarios of 5G. Figure 1.5 shows the technical requirements of different application scenarios of 5G [4, 37]. 5G does not simply emphasize the peak rate, and it takes 8 technical indexes into consideration: peak rate, user experienced data rate, spectral efficiency, mobility, delay, connection density, network energy efficiency, and traffic. On the 5G networks, multiple types of existing or future wireless access transmission technologies and functional networks are converged, including traditional cellular networks, large-scale multi-antenna networks, cognitive wireless networks, wireless local area networks, wireless sensor networks, small base stations, visible light communications and device direct-connection communications, and they are managed through a unified core network to provide ultra-high-speed and ultra-low-delay user experience and consistent seamless connection services for multiple scenarios.

In summary, the development of 5G technologies presents new features, as described below [37, 45]:

(1) The 5G research will focus more on user experience while advancing technological changes. The average throughput rate, transmission delay, and the capabilities of supporting emerging mobile services such as VR, 3D, and interactive games will become key indexes for measuring the performance of 5G systems.

(2) Different from the traditional mobile communication systems which focus on the typical technologies such as point-to-point physical-layer transmission and channel coding and decoding, 5G system research will attach importance to multi-point, multi-user, multi-antenna, and multi-cell collaborative networking, aiming at achieving a significant increase in the system performance from the aspect of the architecture.

(3) The indoor mobile communication service has occupied the dominant position of applications, so indoor wireless coverage performance and service supporting

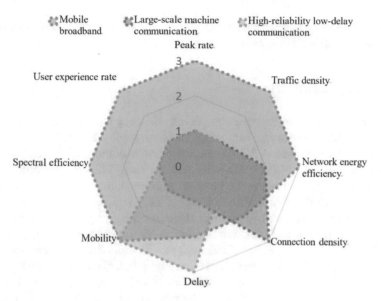

Fig. 1.5 Key technical indexes of 5G

capability will be the prior design goals of 5G systems, which changes the design concept of the traditional mobile communication: "large-scale coverage as the primary concern and indoor coverage as the secondary concern".

(4) High-band spectral resources will be more applied to 5G mobile communication systems, but wireless and wired convergence, radio-over-fiber (ROF) networking and other technologies will be more applied due to the limitation of high-band radio wave's penetration capability.

(5) 5G wireless networks that support soft configuration will be the main research direction in the future. Operators can adjust network resources in real time according to the dynamic changes of service traffic, effectively reducing network OPEX and energy consumption.

1.2.2 Key 5G Technologies

In order to improve business support capabilities, breakthroughs will be made in wireless transmission technologies and network technologies of 5G [34, 37]. Regarding wireless transmission, technologies that can further improve spectral efficiency and spectral potential are introduced, such as advanced multiple access technology, multi-antenna technology, code modulation technology, and new waveform design technology. As for the wireless network, more flexible and intelligent network architectures and networking technologies will be employed, such as software-defined wireless network architecture where control is separated from data forwarding, unified self-organizing network, and heterogeneous ultra-dense deployment. The following will introduce critical technologies of 5G mobile communication.

1.2.2.1 Massive MIMO

As an effective means to improve the spectral efficiency and transmission reliability of a system, the multi-antenna technology has been applied to various wireless communication systems, such as 3G system, long term evolution (LTE), LTE-advanced (LTE-A), wireless LAN (wireless local area network, WLAN). According to the information theory, increasing the number of antennas improves the spectral efficiency and reliability significantly. In particular, when transmitting antennas and receiving antennas increase by hundreds, the channel capacity of MIMO system will increase linearly with the minimum number of transmitting or receiving antennas. Therefore, using a large number of antennas provides an effective solution to greatly expand the capacity of the system. In the current wireless communication systems, due to the technical limits such as occupied space and implementation complexity in a multi-antenna system, the number of antennas configured on the TX/RX end is limited. For example, in the LTE system, a maximum of four antennas are used; and in the LTE-A system, a maximum of eight antennas are used. However, a MIMO system provides huge capacity and reliability gain when it is equipped with a large number of antennas, so relevant technologies have attracted the attention of researchers, including the research on the multi-user MIMO systems where base stations are equipped with a large number of antennas far more than that of mobile users under single-cell circumstance [46, 47]. In 2010, Thomas Marzetta from Bell Laboratory studied the multi-user MIMO technology which enables users to configure each base station with an unlimited number of antennas in the time division duplexing (TDD) under multi-cell circumstance, and found some features which are distinct from the single-cell circumstance where a base station is equipped with a limited number of antennas, thereby proposing the concept of massive MIMO (or large scale MIMO) [48, 49]. Based on the concept, many researchers have been devoted to the study of the base stations equipped with a limited number of antennas [50–52].

In a massive MIMO system, base stations are equipped with a huge number of antennas (usually ranging from tens to hundreds of antennas, which is one or two orders of magnitude of the number of antennas in the existing system) which serve multiple users simultaneously on the same time-frequency resource. Regarding the configuration method, antennas can be deployed as a centralized massive MIMO or distributed massive MIMO. In a centralized massive MIMO, a large number of antennas are centrally deployed on one base station. On a 5G radio access network, massive MIMO will be applied to "macro-assisted small cells": in macro cells, the lower-frequency band is used to provide comprehensive control plane services; and in small cells, the millimeter waves are applied to the highly-oriented massive MIMO beams to carry user plane services. On the 5G band, it is possible to include antennas of hundreds of orders of magnitude in an array. Such a large number of antennas can be used to generate very narrow high-energy beams to offset the high path loss of millimeter waves, making advanced multi-user MIMO (MU-MIMO) possible and improving the capacity of small cell systems. Another application of the massive MIMO technology is distributed massive MIMO where multiple beams are transmitted simultaneously from different base stations to the same mobile device, thereby reducing the correlation between antenna panels and improving throughput rate. In addition, the reflection of nearby obstacles can minimize the correlation of the beam combinations along different mobile device traces as the mobile device moves. Therefore, when beams are selected based on the channel state information (CSI) transmitted by the mobile device to the base station, not the beam power, most cells can obtain higher throughput rate. In short, if the massive MIMO technology is configured on a higher frequency band (such as millimeter waves), when beams are transmitted to a specific mobile device from base stations in different locations, the reflection of buildings reduces the correlation of the beams, thereby improving the performance of the communication systems [49, 53].

Figure 1.6 shows the main application scenarios of the massive MIMO technologies in 5G communication systems [49, 54].

Cells fall into macro cells and micro cells which can be deployed on a homogeneous network or heterogeneous network under indoor or outdoor scenarios. According to the relevant test literature, 70% of the communication of the land mobile communication system is generated from indoor environment. Therefore, the channels of massive MIMO can be divided into macro cell base stations for outdoor users & indoor users, and micro cell base stations for outdoor users & indoor users. Micro cells can be used as relay base stations as well with channels pointing from macro cell base stations to micro cell base stations. The number of antennas for the base stations is unlimited, and the number of antennas for users can be increased.

The massive MIMO system boasts the following advantages [46, 55]:

① The spatial resolution of massive MIMO technology is significantly enhanced compared with the existing MIMO technology. With the massive MIMO tech-

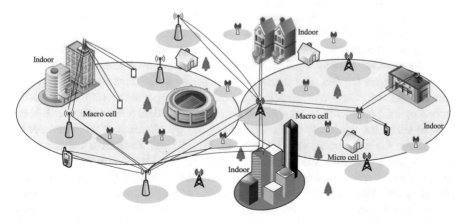

Fig. 1.6 Application scenarios of the massive MIMO technologies

nology, spatial resources can be deeply exploited, enabling multiple users on the same network to communicate with a base station concurrently on the same time-frequency resources based on the spatial freedom provided by the massive MIMO technology. The spectral efficiency can be greatly improved without increasing the density and bandwidth of the base stations.

② The massive MIMO technology enables beams to be concentrated in a narrow range, thereby dramatically reducing interference.

③ The massive MIMO technology enables the transmission power to be greatly reduced, thereby improving power efficiency.

④ When the number of antennas is sufficient, the simplest linear pre-coding and linear detectors tend to be optimal, and both noise and uncorrelated interference are negligible.

1.2.2.2   Co-time Co-frequency Full Duplex Technology

The co-time co-frequency full duplex (CCFD) technology refers to the bidirectional communication at the same time on the same frequency [44]. In the wireless communication system, the inherent transmitting signals of the network side and device side have self-interference to the receiving signals. Due to the technical limitation, the existing wireless communication system cannot achieve co-time co-frequency bidirectional communication, and the two-way links are distinguished by time and frequency, which correspond to the TDD and frequency division duplexing (FDD), respectively.

In theory, the CCFD technology can double the spectral efficiency compared with the traditional TDD or FDD mode, and effectively reduce the end-to-end transmission delay and signaling overhead [56]. When the CCFD technology adopts independent antennas for transmitting and receiving, since the transmitting and receiving antennas are close to each other and the power of the transmitting and receiving signals is greatly different, the co-time co-frequency signal (self-interference) will have a strong interference to the receiving signal at the receiving antenna. Thereby, the core problem of CCFD is to suppress and eliminate the strong self-interference effectively. In recent years, researchers have developed different technologies to offset self-interference, including digital interference offset, analog interference offset, mixed interference offset, and interference offset by an antenna installed in a specific position. Most self-interference can be offset under specific scenarios using these technologies. Meanwhile, the researchers have developed relevant experimental systems to verify the feasibility of the CCFD technology. Under specific experimental conditions, the experiment result can reach 90% of the theoretical capacity of a CCFD system. However, the experimental system considers a simple interference model where a small number of antennas are deployed in a single base station and providing small broadband services. Theoretical analysis and systematic experimental verification have not been conducted for a CCFD system employing a complex interference model where a plurality of antennas are configured for multiple cells and providing large broadband services. Therefore, an in-depth analysis should be conducted to develop more practical self-interference cancellation technologies under a complex interference model where multiple antennas are configured for multiple cells to provide large broadband services [51, 53].

In addition to the self-interference cancellation technology, the research on the CCFD technology includes designing low-complexity physical layer interference cancellation algorithms and studying power control and energy consumption control problem of the CCFD system [57]. Apply the CCFD technology to cognitive wireless networks to reduce collisions between secondary nodes and improve the performance of the cognitive wireless networks [58]. Apply the CCFD technology to heterogeneous networks to solve wireless backhaul problems [59]. Combine the CCFD technology with relay technology to solve hidden device problems, congestion-induced throughput problems, and end-to-end delay problems in the current network [60, 61]. Combine the CCFD relay with MIMO technology and beamforming optimization technology to improve end-to-end performance and anti-interference capability of the system [62].

In order to improve the application of the CCFD technology to the wireless networks in future, we still have a lot of researches to do [63]. In addition to an in-depth research on the self-interference cancellation problem of the CCFD technology, we need more comprehensive consideration of the opportunities and challenges faced by the CCFD technology, including designing low-power, low-cost, miniaturized antennas to eliminate self-interference; solving coding, modulation, power allocation, beamforming, channel estimation, equalization, and decoding problems of the CCFD system physical layer; designing protocols of the medium access layer and higher layers; confirming interference coordination strategy, network resource management and CCFD frame structure in the CCFD system; and conducting effective combination of the CCFD technology with large-scale multi-antenna technology and system performance analysis.

1.2.2.3 Ultra-Dense Heterogeneous Network

Following the diversification, integration and intelligence development requirements of the 5G network and the popularization of intelligent devices, data traffic will present an explosive growth gradually. Measures such as reducing the radius of a cell and increasing the number of low-power nodes will become one of the core technologies to meet the 5G development needs and support the network growth mentioned in the vision. An ultra-dense network will assume the ever-growing the data traffic of 5G networks [64, 65].

Since the 5G system includes both new wireless transmission technologies and subsequent evolution of various existing wireless access technologies, the 5G network must be a multi-layer and multi-wireless access heterogeneous network where a plurality of wireless access technologies (such as 5G, 4G, LTE, universal mobile telecommunications system (UMTS), and Wi-Fi) coexist and macro stations responsible for basic coverage and low-power small stations that cover hotspots (such as Micro, Pico, Relay, and Femto) are configured [66]. Among these large number of low-power nodes, some are carrier-deployed and planned low-power macro nodes, others are user-deployed and unplanned low-power nodes, and the user-deployed low-power nodes may be of the open subscriber group (OSG) type or the closed subscriber group (CSG) type, which makes the network topology and features extremely complex.

According to the relevant statistics, in the 50 years from 1950 to 2000, the improvement of voice coding technology and multiple access channel and modulation technology caused the resource efficiency to increase by less than 10 times and the adoption of wider bandwidth caused the transmission rate to increase by tens of times, while the reduction of the cell radius resulted in the spatial multiplexing rate of the spectral resources to increase by more than 2700 times [67]. Therefore, reducing the cell radius and increasing the spatial multiplexing rate of spectral resources to improve the transmission capability per unit area is the core technology to support the 1000-times growth of traffic in the future. In the conventional wireless communication system, the reduction of the cell radius is completed by dividing a cell. As the coverage of a cell is decreased, the optimal site location cannot be obtained, which increases the difficulty of further cell division. The system capacity is improved by increasing the number of low-power nodes, which means an increase in site deployment density. According to predictions, in the coverage area of a macro station in a future wireless network, the deployment density of various low-power nodes employing different wireless transmission technologies will reach 10 times the density of the existing station deployment, and the distance between the stations is shortened to 10 m or even smaller [68, 69]. The number of users per square kilometer can be up to 25,000 [70], and even the ratio of activated users to stations in the future can reach 1:1, that is, each activated user will have a service node. By that time, an ultra-density heterogeneous network is formed.

Although the ultra-density heterogeneous network architecture of 5G has great development prospects, the reduction of distance between nodes and the increasingly dense network deployment will make the network topology more complicated, which leads to an incompatibility with existing mobile communication systems. On 5G mobile communication networks, interference is an inevitable problem. The interferences on the networks mainly include co-frequency interference, shared spectral resource interference, and interference between different coverage levels [71]. In the existing communication system, the interference coordination algorithms can only solve the single interference source problem. On the 5G network, the transmission loss of the adjacent nodes generally has little difference, so the strength of multiple interference sources is approximate, which further deteriorates the network performance and makes the coordination algorithms difficult to deal with the interference. In addition, due to the large difference in service and user demands for quality of service (QoS), 5G network requirements are met by using a series of measures to ensure system performance, mainly: the realization of different services on the network [72], coordination schemes between nodes, network selection [73], and energy-saving configuration methods [74].

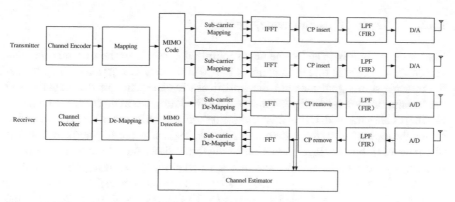

Fig. 1.7 Baseband algorithm processing flow of MIMO-OFDM system

1.2.3 MIMI Baseband Processing

The MIMO technology is combined with the orthogonal frequency division multi-plexing (OFDM) technology and they are widely used in current mainstream commu-nication protocols, including global system for mobile communication (GSM), code division multiple access (CDMA), LTE, worldwide interoperability for microwave access (WiMAX), and Wi-Fi. The OFDM technology helps improve the bandwidth usage of the system significantly, and the MIMO technology helps double the trans-mission rate and reliability. Figure 1.7 shows a typical MIMO-OFDM baseband algo-rithm processing flow which can be simplified as a combination of multiple one-way ODFM baseband signals [75]. In the ODFM communication systems, signals will be modulated and mapped after they are encoded and interwoven in a channel. The signal amplitude and phase represent digital information. Then, after serial/parallel conversion, the signals are changed to parallel data streams, mapped to sub-carriers, and added to empty sub-carriers. Using the inverse fast Fourier transform (IFFT), data are modulated to multiple orthogonal sub-carriers and changed to data streams after some parallel/serial conversion. In order to suppress multipath interference and ensure the orthogonality between different sub-carriers, the signals will be cyclic prefix (CP) extended. After passing the low pass filter (LPF), the signals are changed to analog signals by digital-analog conversion, and modulated to carriers and sent to the channels for transmission. After the channel transmission, the original data are restored from the orthogonal carrier vector by fast FFT at the receiving end using the opposite process. On the basis of OFDM system operations such as channel encod-ing, modulation and Fourier conversion, the MIMO technology adds low-density parity check code (LDPC) options to channel encoding, and multi-antenna channel estimation and signal detection.

The following analyzes the channel encoding and decoding, the signal modulation and demodulation, the MIMO signal detection, the FFT & IFFT, and the finite impulse response (FIR) filter in details.

1.2.3.1   Channel Encoding and Decoding

Channel encoding is to add some new supervised symbols to the transmission data according to a certain rule to implement the error or error correction coding, so that the transmission signal matches the statistical characteristics of the channel, and the communication reliability is improved. Convolutional coding (CC) is a common channel coding method. Using this method, a sequence of information is converted to a code through a linear, finite state shift register. Typically, the shift register consists of a K-level (k bits per level) and n linear algebraic function generators. The binary data are shifted into the encoder and moves k bits along the shift register each time. Each k-bit input sequence corresponds to an n-bit output sequence [76], and its coding efficiency (code rate) is defined as $R_c = k/n$. The parameter K is referred to as the constraint length of the CC, and the CC conforming to the above parameters is simply referred to as the $(n, k, K + 1)$ convolutional code. In some high-rate protocols, in addition to CC, Reed-Solomon code (RS code), Turbo code, LDPC, or a combination of these codes, such as RS-CC and Turbo-CC, are also used as an alternative.

If CC is used at the transmitting end, the Viterbi decoding algorithm is used for decoding at the receiving end. The Viterbi decoding algorithm is a maximum likelihood decoding algorithm. Instead of calculating the metric of each path, it receives and compares segments one by one and then selects a possible decoding branch. In the actual decoding process, a very long code stream is usually divided into a number of small segments (length L) and then decoded respectively, thus avoiding huge hardware overhead caused due to over-long code stream. Finally, the decoding results of each segment are concatenated to obtain the maximum likelihood decoding of the entire sequence. According to the experiment (computer simulation) results, when $L \geq 5 K$, the performance reduction of the Viterbi algorithm after segmentation is negligible compared with the optimal algorithm performance [77].

1.2.3.2   Signal Modulation and Demodulation

In the OFDM communication systems, the sub-carrier signals will be modulated and mapped after they are encoded and interwoven in a channel. By changing the amplitude, phase or frequency of the signal carriers to transmit baseband signals, modulation is implemented. If the digital information volume represented by the modulated signal is increased, corresponding data transmission rate is improved. However, a high data transmission rate may result in an increase in the bit error rate (BER) after demodulation. Therefore, different modulation modes are used in different channels to implement bit stream to complex conversion. The modulation modes supported by wireless communication standards such as LTE, WLAN, and WiMAX include binary phase shift keying (BPSK), quadrature phase shift keying (QPSK), and quadrature amplitude modulation (QAM).

At the receiving end, the process of converting complex data into a bit stream is a constellation demodulation process. Demodulation is the inverse process of modulation. In the demodulation process of extracting the input bits from the received signal, the received complex data has a certain difference from the original value due to the interference of the channel noise. Therefore, decision conditions need to be specified in the demodulation process. The farther the received signal is away from the decision boundary, the higher the decision accuracy. Table 1.1 shows the demodulation boundary conditions in different modulation systems.

Table 1.1 Demodulation boundary conditions in different modulation systems

Demodulation type	Decision boundary	Decision and constellation mapping
BPSK	$I = 0$	Q, b_0, +1, $\bullet$ 0, -1, +1, $Q=0$, -1
QPSK	$I = 0\ (m = 0)$ $Q = 0\ (m = 1)$	Q, $b_0 b_1$, $\bullet$ 10, +1, $\bullet$ 00, -1, +1, $Q=0$, $\bullet$ 11, -1, $\bullet$ 01

(continued)

Table 1.1 (continued)

Demodulation type	Decision boundary	Decision and constellation mapping
16QAM	$I = 0 \ (m = 0)$ $Q = 0 \ (m = 1)$ $Q = 2/\sqrt{10}$ $I = 2/\sqrt{10}$	
64QAM	$I = 0 \ (m = 0)$ $Q = 0 \ (m = 1)$ $I = \pm 4/\sqrt{42} (m = 2)$ $Q = \pm 4/\sqrt{42} (m = 3)$ $I = \pm 2/\sqrt{42},$ $\quad I = \pm 6/\sqrt{42} (m = 4)$ $Q = \pm 2/\sqrt{42},$ $\quad Q = \pm 6/\sqrt{42} (m = 5)$	

1.2.3.3 MIMO Signal Detection

In an MIMO communication system, the signals obtained at the receiving end are the linear superposition of independently transmitted signal symbols, so these symbols need to be separated at the receiving end.

Figure 1.8 shows a MIMO system which is equipped with N_t transmitting antennas and N_r receiving antennas. The data streams are divided into N_t data substreams, which are mapped through the constellations and then sent to the transmitting antennas.

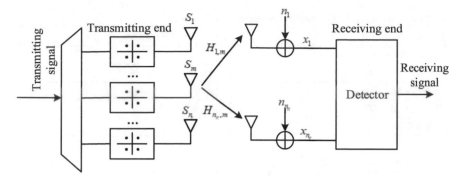

Fig. 1.8 MIMO system

At the receiving end, each antenna receives the signals sent by corresponding transmitting antenna. The symbols received by all the receiving antennas are represented by a vector $y \in \mathbb{C}^{N_r}$, and the relationship shown by Formula (1.1) is established.

$$y = Hs + n \qquad (1.1)$$

where, $s \in \mathcal{O}^{N_t}$ is a transmitted signal vector containing all user data symbols ($\mathcal{O}$ represents a set of constellation points), and $H \in \mathbb{C}^{N_r \times N_t}$ is a Rayleigh flat fading channel matrix whose elements $h_{j,i}$ are channel gains from the transmitting antenna $i(i = 1, 2, \ldots, N_t)$ to the receiving antenna $j(j = 1, 2, \ldots, N_r)$. $n \in \mathbb{C}^{N_r}$ are additive Gaussian white noise vectors whose components are independent and subject to the $N(0, \sigma^2)$ distribution. Signal detection is the process of eliminating noise interference based on the known reception vector y and the estimated channel matrix H and then calculating and determining the sent vector s. The focus of MIMO detection algorithm research is to make a trade-off between detection performance and computing complexity. Detection performance is usually measured by using the BER.

1.2.3.4 FFT and IFFT

In the OFDM system, the IFFT technology is used to implement serial-to-parallel conversion for the modulated sub-carriers and frequency domain-to-time domain conversion. The orthogonal spectrum is formed between the sub-bands for transmission. The pilot and guard symbols are also imported in some fixed sub-bands for data frame synchronization and spectrum estimation. After the IFFT, zero padding or guard intervals (GI) need to be inserted before and after the OFDM symbols based on the specific protocol to eliminate the inter symbol interference (ISI). At the receiving end, the FFT technology is used for demodulation accordingly. Regarding the algorithm structure, IFFT shares similarity with FFT, so only FFT is exemplified below. Table 1.2 lists the number of FFT points used by some OFDM systems.

Table 1.2 Number of FFT points used by some OFDM systems

OFDM system	Number of FFT points
WLAN (IEEE 802.11 series)	64
DAB	2048, 1024, 512, 256
UWB (IEEE 802.15.3)	128
WiMAX (IEEE 802.16e)	2048, 1024, 512, 128
IEEE 802.22 (CR)	2048

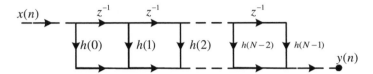

Fig. 1.9 Structure of direct FIR filter

1.2.3.5   FIR Filtering

The FIR filter changes the time domain or frequency domain property of a signal through a certain operation process, and finally outputs the signal in a sequence form. Based on structure, FIR filters fall into straight-line filters, cascaded filters, and linear phase filters. The following content introduces a straight-line filter. A straight-line filter is also referred to as a horizontal FIR filter, and it mainly converts a sample point $x(n)$ to a required $y(n)$ based on the unit impulse response $h(n)$ with length N using the multiply-accumulate centralized method. The unit impulse response $h(n)$ is also referred to as a tap coefficient. Formula (1.2) shows the straight-line FIR filter, and Formula (1.3) shows the system input/output relationship. If Formula (1.3) is expanded using a network structure, its structure is shown as Fig. 1.9. Based on a specific response function, the FIR filter can implement low-pass, high-pass, and band-pass functions. In the baseband signal processing, a low-pass filter is selected generally to process data in transmission and receiving. The order of the FIR filter affects the processing effect of the data. Generally, a 32-order FIR filter can achieve better results.

$$H(z) = \sum_{n=0}^{N-1} h(n)z^{-n} \tag{1.2}$$

$$y(n) = \sum_{m=0}^{N-1} h(m)x(n-m) \tag{1.3}$$

1.2.4   Difficulties in Massive MIMO Detection

Compared with the traditional MIMO technology, the massive MIMO technology is still an emerging technology, and researchers have encountered many challenges in the technology commercialization process.

1.2.4.1   Channel Estimation

First, the massive MIMO system has large-scale antenna arrays, and corresponding channel responses comply with certain Law of Large Numbers (LLN).

Second, the massive MIMO uses the TDD technology currently. This technology is different from the FDD technology and boasting channel reciprocity characteristics. The study on the TDD technology is still challenging. Last, the pilot pollution problem of the massive MIMO system is still not resolved. This problem will appear when the orthogonal pilot sequence is used in a cell and the same pilot sequence group is used between cells [4, 7]. The main reason for this problem is that when users use the same set of training sequence or non-orthogonal training sequence, the training sequence sent by the users of the neighboring cells is non-orthogonal. As a result, the channel estimated by the base station is not the channel used between the local users and the base station, but the channel polluted by the training sequence sent by the users of other cells.

1.2.4.2   Channel Modeling

In the massive MIMO system, base stations are equipped with a large number of antennas, so the spatial resolution of the MIMO transmission is significantly improved. The wireless transmission channels boast new characteristics, and channel models applicable to the massive MIMO system need to be systematically discussed. Under the given channel model and transmitting power, accurately characterize the maximum transmission rate supported by the channel (namely the channel capacity), and thus reveal the influence of various channel characteristics on the channel capacity, to provide important basis for optimal design of the transmission system, and performance assessments such as spectral efficiency and energy efficiency.

1.2.4.3   Signal Detector

Signal detection technology in the massive MIMO system has a critical impact on overall system performance. Compared with existing MIMO systems, base stations in the massive MIMO system are equipped with a large number of antennas, so a massive amount of data will be generated, which puts higher requirements on RF and baseband processing algorithms. The MIMO detection algorithms are expected to be practical, balance between low complexity and high parallelism, and boast hardware achievability and low power consumption.

1.2.4.4   CSI Obtainment

Under the high reliability and low latency requirements of 5G, the estimation of CSI must be real-time and accurate [78]. The CSI plays a supporting and guarantee role in the later channel modeling and communication. If the CSI cannot be captured quickly and accurately, the transmission process will be severely interfered and restricted [79]. According to available research result, if a fast fading module is introduced to the massive MIMO system, the system CSI will change slowly with time. Besides, the number of concurrent users served by the system is irrelevant with the number of base station antennas, and is limited by the CSI obtainment capability of the system.

1.2.4.5   Device Design of Large-Scale Antenna Array

It is well known that too-small space between antennas will result in mutual inter-ference, so how to effectively deploy a large number of antennas in a limited space becomes a new challenge.

There are many challenges in the research of the above problems. As the research is deepened, researchers give high hope to the application of the massive MIMO technology in 5G [54]. It is foreseeable that massive MIMO technology will become one of the core technologies that distinguish 5G from existing systems.

1.3   Status Quo of MIMO Detection Chip Research

The MIMO detection chips fall into ISAP and ASIC according to its system archi-tecture. The typical architecture of ISAP includes the general purpose processor (GPP), the digital signal processor (DSP) and the application specific instruction set processor (ASIP). The following content introduces the existing MIMO detection chips.

1.3.1   ISAP-Based MIMO Detection Chip

The GPP, DSP, and graphics processing unit (GPU) of ISAPs boast universality. Usually, MIMO detection algorithms are applied to these processor architectures, and no special architecture is designed for MIMO detection. For example, in the literature [80, 81], the MIMO detection algorithms are optimized and the optimized algorithms are mapped to the GPU for hardware implementation. This book focuses on ISAPDEbased MIMO detection chips. While maintaining versatility, ASIP supports specific architecture optimization design for an algorithm, so that related operations can be completed more efficiently and in a targeted manner. The following content is a brief analysis and introduction of the relevant architecture design.

In the literature [82], an efficient lightweight software-defined radio (SDR) ASIP is designed. The increase in its efficiency is contributed to: carefully selected instruction sets, optimized data access techniques which efficiently utilize function units (FUs), and flexible floating-point operations using runtime-adaptive numerical precision. In the literature [82], a concept processor (napCore) is presented to demonstrate the impact of these techniques on the processor performance, and its potentials and limitations are discussed compared to an ASIC solution. In the literature [82], the author also introduces this processor prototype napCore as a fully programmable floating-point processor core that can support these efficiency-support measures. In the literature [82], this processor prototype napCore applies to vector-based operation algorithms. The napCore is a fully programmable single instruction multiple data (SIMD) processor core designed for vector arithmetic. In the literature [82], linear MIMO detection is used as a typical application because linear MIMO detection is widely applied and practical. Similar results can be obtained for other vector algorithms (such as linear channel estimation and interpolation). According to the literature [82], a well-designed lightweight ASIP can provide better flexibility than a non-programmable ASIC does, while guaranteeing energy efficiency as high as that of the ASIC.

Figure 1.10 shows the pipeline structure of the SIMD core. The instruction word is requested and obtained from the program memory in the prefetching standard unit, received in the instruction fetch phase after one cycle, and then interpreted in the decoding phase. In this phase, operations of all subsequent phases will be configured. In the literature [82], the following four arithmetic levels (EX1, EX2, RED1, RED2) are designed to match the processing scheme of standard vector arithmetic operations. This is a combined computational logic design for multiplication and subsequent addition. In the EX1 and EX2 phases, a complex multiplication operation is mainly performed, in which EX1 performs a real-valued multiplication operation, and EX2 performs an accumulation to form a complex value result. The Newton iteration unit for scalar reciprocal is also located in EX1. In the following RED1 and RED2 phases, the results of the previous multiplication unit can be further processed by additions. For example, these additions can be configured as an adder tree to implement the algorithm requirements. Besides, the PrepOp-EX2 unit in EX2 can read an extra vector operand from the vector register as an input of RED1 for multiply-accumulate operations. After processing in the RED2 phase, the result is written back to the vector memory or scalar/vector register file.

For the programmable architecture of SIMD or the very long instruction word (VLIW), etc., the processor has a programmable architecture with inherent parallelism and the effective operand acquisition mechanism is a challenging task. In order to accomplish this task, a very different data access pattern must be implemented, which also caused the complex operand acquisition architecture shown in Fig. 1.11

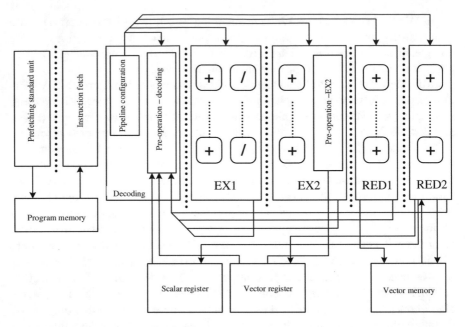

Fig. 1.10 Pipeline structure of SIMD core

to be proposed for the first operand. © [2016] IEEE. Reprinted, with permission, from Ref. [82]

The literature [82] describes a series of measures to optimize the architecture so as to calculate the complex vector algorithm flexibly and efficiently. A versatile instruction set for complex vector arithmetic increases data throughput rate. Optimized operand acquisition schemes, including intelligent bypass and vector arithmetic affine permutation units, further increase the data throughput rate of the architecture and thus achieve high area efficiency. Energy efficiency can be optimized by numerical change of floating-point operations, which allows the programmer to adjust the numerical accuracy at runtime based on the application requirements, thereby reducing switching activity and energy consumption. In the 90 nm process, the area efficiency reaches 47.1 vec/s/GE (vec represents the number of signal vectors; GE represents the number of logic gates), and the energy efficiency reaches 0.031 vec/nJ.

In the literature [83], the detection and decoding operations are implemented by connecting with a sphere decoding (SD) and a forward error correction (FEC) core in a system on chip (SoC). The network-on-chip (NoC) flexibility enables the SD and FEC to be used as stand-alone units or as an integrated detection and decoding chain. Figure 1.12 shows the structure of the SoC. A phase-locked loop (PLL) provides each unit with a separate clock with a range of 83–667 MHz. This allows each unit to be adjusted to the optimum operating point to achieve the required data throughput rate

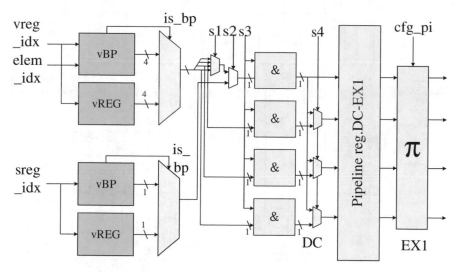

Fig. 1.11 Data acquisition schematic for the first operand. © [2018] IEEE. Reprinted, with permission, from Ref. [82]

with the minimal power consumption. The field programmable gate array (FPGA) interface operates at 500 MHz and provides 8 Gbit/s data streams in each direction.

The SD core consists of an ASIP that includes a control path and a vector data path to support SIMD vectorization (e.g., for OFDM systems). The data path is divided into several FUs, as shown in Fig. 1.13. Since the pipeline cannot be directly applied to the data path that is based on the SD feedback loop, the author proposes a 5-level pipeline for independent MIMO symbol detection in order to improve the data throughput rate. Through the output port of the caching FU, data generated by one FU can be directly operated by another connected FU, thereby avoiding the storage of intermediate data. The memory interface is designed to support simultaneous access to channel and symbol data to avoid data throughput degradation. Access to the conditional memory is assisted by the flow control unit in the control path.

The flexible FEC module contains a programmable multi-core ASIP that is capable of decoding CCs, Turbo codes, and LDPC codes. The FEC module consists of three identical independently-programmable processor cores and they are connected to the local memory via a connected network, as shown in Fig. 1.14. In the architecture, any number of cores can work together on code blocks, and different codes can be decoded simultaneously on separate clusters. This makes dynamic core and multi-mode operations possible. Each core contains a control path and a SIMD data path. The data path includes four processing elements (PEs) that take advantage of the similarity of key algorithms in the basic operations of the decoding algorithms in an isomorphic form. The internal 16 PEs allow data to be processed in parallel in a grid form and then for Viterbi and Turbo decoding, or for processing updates of the 8 LDPC nodes in parallel. The connected network may be configured as an inherent

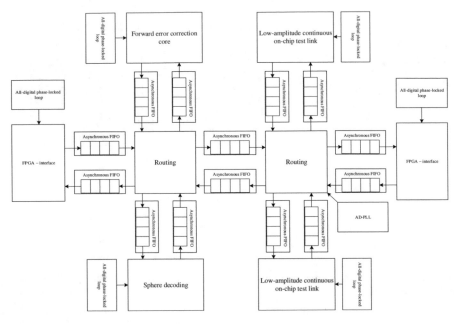

Fig. 1.12 SoC structure. © [2018] IEEE. Reprinted, with permission, from Ref. [83]

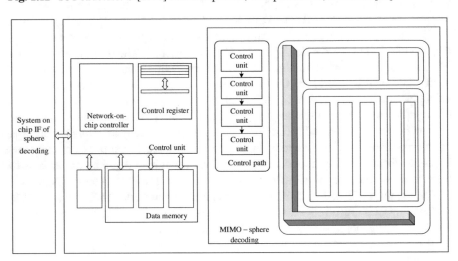

Fig. 1.13 SD module containing 5-level pipeline. © [2018] IEEE. Reprinted, with permission, from Ref. [83]

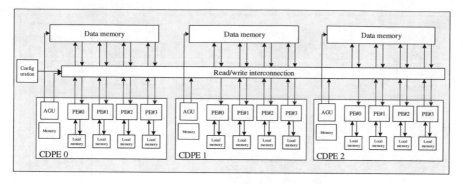

Fig. 1.14 Flexible FEC architecture module. © [2018] IEEE. Reprinted, with permission, from Ref. [83]

random permutation network element to perform Turbo decoding or a bucket shift required to replace the sub-matrix of the LDPC parity check matrix.

The SoC is fabricated in a TSMC 65 nm CMOS process. The chip area is 1.875 mm × 3.750 mm = 7.03125 mm^2, which includes all 84 input/output (I/O) units. The MIMO detector unit supports 64-QAM and 4 × 4 mm MIMO transmission. The core power supply works under 1.2 V and occupies 0.31 mm^2, including 2.75 KB static random access memory (SRAM). Its average power consumption is 36 mW at 1.2 V and 333 MHz. The compromise between MIMO detection data throughput rate and signal-to-noise ratio (SNR) is adjustable, ranging from 14.1 dB SNR and 296 Mbit/s data throughput rate to 15.55 dB SNR and 807 Mbit/s data throughput rate. Moreover, the MIMO detector unit can be configured to perform a minimum mean square error-successive interference cancellation (MMSE-SIC) detection algorithm that can achieve a data throughput rate of 2 Gbit/s.

In the literature [84], a heterogeneous SoC platform boasting runtime scheduling and fine-grained hierarchical power management is proposed. This solution can adapt to dynamically changing workloads and semi-deterministic behavior in modern concurrent wireless applications. The proposed dynamic scheduler can be implemented by software on a GPP or a specific application hardware unit. Obviously, the software provides the highest flexibility, but it can become a performance bottleneck for complex applications. In the article, the possible performance bottleneck caused by the flexibility is overcome by implementing the dynamic scheduler on the ASIP.

In the literature [84], the SoC consists of 20 heterogeneous cores (8 of which are Duo-PE) which are connected by a hierarchical packet-switched star-grid NoC, as shown in Fig. 1.15. The Duo-PE consists of a vector DSP and a reduced instruction set computer (RISC) core and is connected to the local shared memory. Such setup improves regional efficiency and data locality. Each Duo-PE is equipped with a direct memory access (DMA) for simultaneous data prefetching and task execution. To support fine-grained fast power management, each Duo-PE is equipped with a dynamic voltage and frequency scaling (DVFS) unit. The NoC works at 500 MHz clock frequency and 80 Gbit/s high-speed serial link, which forms a compact top

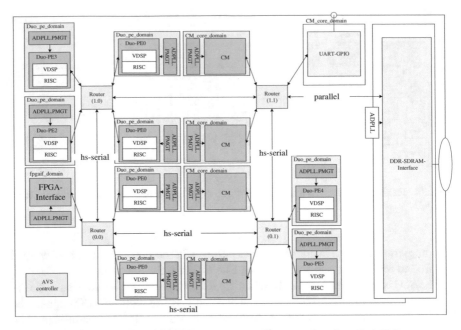

Fig. 1.15 SoC structure. © [2018] IEEE. Reprinted, with permission, from Ref. [84]

floor layout. The all-digital phase-locked loop (ADPLL) is connected to each unit and allows clock frequency adjustment in the range of 83–666 MHz. The DDR2 (double data rate 2) interface that connects to two 128 MB global memory devices provides 12.8 Gbit/s data transmission rate. The FPGA I/O interface provides 10 Gbit/s data transmission rate. The application processor is a Tensilica 570T RISC core with 16 KB data and 16 KB instruction cache. It executes the application control code and sends a task scheduling request to the dynamic scheduler. Based on the Tensilica LX4 core, the dynamic scheduler efficiently implements adaptive power management and dynamic task scheduling (including resource allocation, data dependency check, and data management). The dynamic scheduler analyzes the scheduling request at runtime and configures the dynamic voltage and frequency of the PE according to the current system load, priority, and deadline to optimize task scheduling and allocation at maximum.

The SoC is fabricated by using the TSMC 65 nm low power-complementary metal oxide semiconductor (LP-CMOS) technology. It integrates 10.2 M logic gates and occupies a size of 36 mm^2 (6 mm x 6 mm). The MIMO iterative detection and decoding part occupies a size of 1.68 mm^2, including 93 KB SRAM. Each Duo-PE has an area of 1.36 mm^2, of which 0.8 mm^2 is for two dual-port 32 KB memory. The RISC core works on the maximum frequency 445 MHz at 1.2 V. The dynamic scheduler occupies 1.36 mm^2, including 64 KB data memory and 32 KB instruction memory. It works on the maximum frequency 445 MHz at 1.2 V, implementing 1.1 Mbit/s data throughput rate and consuming 69.2 mW power. At the PE level, the ultra-high-speed DVFS follows the dynamic adaptive control of the dynamic scheduler to further improve energy efficiency. Flexible iterative and multi-mode processing units improve area performance and increase energy efficiency by three times compared with that of related study results.

1.3.2   ASIC-Based MIMO Detection Chip

The ASIC is an integrated circuit that is specifically designed and manufactured according to specific user requirements and specific electronic systems. Compared with ISAP, ASIC is characterized by small area, low power consumption, fast processing speed, high reliability and low cost. ASIC-based MIMO detectors focus on not only detection accuracy but also on chip performance, aiming to achieve a good compromise between the two. At present, ASIC design is a hot topic for both traditional MIMO detection and massive MIMO detection. The following content describes the existing ASIC-based MIMO detectors.

1.3.2.1   ASIC Design for Traditional MIMO Detection

In the literature [85], an MMSE multi-domain LDPC code iterative detection decoder is proposed for 4×4, 256-QAM, MIMO systems to achieve excellent detection accuracy. To minimize delay and increase data throughput rate in the iterative loop, the MMSE detector is divided into four task-based pipeline levels so that all pipeline levels can run in parallel. Both the pipeline level number and delay of the detector are minimized, and the long critical paths are interwoven and placed in the slow clock domain to support high data transmission rates. The data throughput rate of the MMSE detector has doubled. To reduce power consumption, automatic clock gating is applied to the phase boundaries and cache registers to save 53% detector power and 61% decoder power.

The MMSE detector consists of four parallel pipeline levels, as shown in Fig. 1.16. The channel information of the decoder and the log likelihood ratio (LLR) of the priori symbols are preprocessed in the first stage to generate an MMSE matrix. The matrix is then MMSE filtered using the LU decomposition (LUD) in the second and third phases. In this process, interference cancellation is done in parallel. In the final phase, the SNR and symbol LLR are calculated and used as the input of the multi-domain LDPC decoder. The LUD in the second phase contains critical paths and demands long delays, making the pipeline and data throughput rate hitting a bottleneck. Since the Newton iterative reciprocal solution unit determines the internal loop delay of the LUD, a parallel reciprocal computation structure is reconstructed in the literature [85], which can shorten the second phase from 18 cycles to 12 cycles. In order to relax the timing constraints on the critical paths of the second and third phases, a double slow clock domain is created in the literature [85] for these two phases to reduce hardware resource overhead. As a result, the number of logic gates is reduced

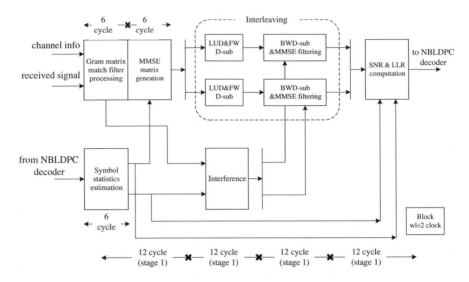

Fig. 1.16 MMSE detector modules. © [2018] IEEE. Reprinted, with permission, from Ref. [85]

and the data throughput rate is increased by 38%. In the final stage, SNR calculation is simplified using the characteristics of the algorithms. By the optimization, the final chip area is reduced by 50% and power consumption by 46%. In this architecture, a total of 70.9 KB registers are used to cache data between the detector and the decoder and data at various levels. The registers are used in place of memory arrays to support high access bandwidth and accommodate small memory blocks. Due to the pipeline structure, most of the registers used in the design should not be updated frequently. Otherwise, the power consumption overhead is reduced. In the literature [85], the access mode is optimized, power consumption of the detector is reduced by 53% by enabling register clock gating at idle.

In the 65 nm CMOS process, the final MMSE detector works at 1.39 Gbit/s on the maximum frequency 517 MHz, with the area reduced to 0.7 mm^2 and the power consumption decreased to 26.5 mW. The MMSE detector also achieves 19.2 pJ/bit energy efficiency.

The literature [86] describes the ASIC implementation of a soft-input soft-output detector for iterative MIMO decoding, proposes a parallel interference cancellation algorithm based on low complexity MMSE, and designs a proper VLSI structure. By reducing the number of required matrix inversions, the computational complexity is reduced without performance loss. A corresponding VLSI top-level architecture is designed, which includes all necessary channel matrix preprocessing circuits. This architecture uses LUD-based matrix inversions, so it is superior to the matrix inversion circuits of other MIMO systems in terms of area and data throughput rate. The key to achieving high data throughput rate is to use cells that are based on custom Newton iterations. To achieve high data throughput rate, the author divides

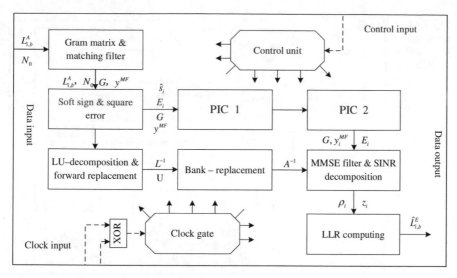

Fig. 1.17 VLSI's top-level architecture. © [2018] IEEE. Reprinted, with permission, from Ref. [86]

the algorithms into eight subtasks which are executed in parallel pipeline. Figure 1.17 shows the top-level architecture of VLSI and the division of related computations. The architecture consists of eight processing units, where the six processing steps of the algorithms are mapped onto the processing units. This architecture boasts the following two advantages:

① Achieve consistently a high data throughput rate.
② Each processing unit can be designed, optimized and verified separately, so the development and verification time is reduced.

All FUs share the same basic units and perform assigned tasks in a time-sharing manner. The basic unit architecture is shown in Fig. 1.18. It is composed of a finite state machine (FSM) that controls the data memory, an arithmetic unit (AU) for a set of specific tasks, and a connected network (all memory is allocated to all AUs in parallel). In order to maximize the clock frequency and minimize the circuit area, the detector uses fixed point algorithms. The internal word length of the AU and memory are optimized with the support of numerical simulation. The feed-through function available in an FU allows the data of all memory to be transferred from the FU to its subsequent FU in parallel within the exchange cycle. Moreover, the feedback path enables the AU to use the computational results immediately in subsequent processing cycles. By inserting a pipeline register at the input end of each AU, the critical path length can be reduced by 1/3. Moreover, some AUs also pass the computational results to the next FU within the exchange cycle to reduce the number of idle AUs.

In the literature [86], LUD requires an accurate calculation of 18 clock cycles. Therefore, the reciprocal unit consumes a maximum of three clock cycles per recip-

Fig. 1.18 Basic unit
architecture. © [2018] IEEE.
Reprinted, with permission,
from Ref. [86]

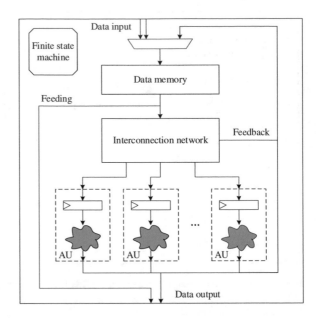

rocal. In addition, according to the simulations, 15-bit accuracy is sufficient to enable
the loss of detection performance to be neglected. During evaluation of the underlying
architecture, two solutions meeting the given constraints are obtained, and they are
described in Fig. 1.19. In the 90 nm CMOS process implementation, the sequential
architecture requires a 4-bit lookup-table (LUT) to perform two Newton iterations.
The pipeline architecture requires an 8-bit LUT and an additional pipeline register
to perform a single iteration, and the final area is 2.5 times that of the sequential
architecture. Since the design goal is to maximize the clock frequency of the entire
detector, the pipeline architecture is used.

In the 90 nm CMOS process, the chip area is 1.5 mm^2 and the data throughput rate
is 757 Mbit/s. Compared with other MIMO detectors, performance gets significantly
improved with this design. The power consumption of the ASIC is 189.1 mW and
the energy efficiency is 0.25 nJ/bit per iteration.

1.3.2.2   ASIC Design for Massive MIMO Detection

In recent years, the massive MIMO technology has become a hot research. In the
massive MIMO system, the previous detector design of traditional MIMO encounters
performance bottle due to the dramatic increase of the computational complexity.
Massive MIMO detectors can reduce the computational complexity while ensuring
detection accuracy, so as to improve data throughput rate and reduce power consumption per unit area, so they are applied more and more.

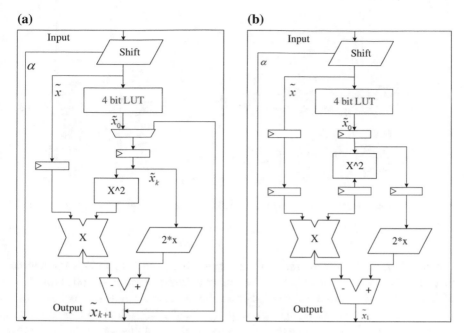

Fig. 1.19   a Reciprocal unit design of the sequential architecture; **b** pipeline architecture. © [2018] IEEE. Reprinted, with permission, from Ref. [86]

In the literature [87], a massive MIMO detector with 2.0 mm^2 and 128 × 16 bits is proposed, which provides 21 dB array gain and 16 × multiplexing gain at the system level. The detector can implement an iterative expectation propagation detection (EPD) algorithm up to 256-QAM modulation. Figure 1.20 shows the EPD algorithm architecture. It contains the input data memory, the MMSE-PIC module, the approximate time matching module, the symbol estimation memory module and so on. The Gram matrix and matched filtering (MF) vector (y^{MF}) are cached in memory, and the memory can be reconfigured to achieve flexible access modes. The MMSE parallel interference cancellation algorithm optimizes detection performance by eliminating interference between uplink users. The constellation point matching unit improves the estimation of the transmitted symbols by combining the constellation information. The detection control unit dynamically adjusts the calculation operations and iteration times per iteration process. To support vector calculations of different lengths, the architecture scale is configurable to implement the dynamic dimensionality reduction. When a batch of estimated signals are determined as reliable, their subsequent calculations will be frozen and removed. Dynamic dimensionality reduction enables the complexity to be reduced by 40–90%. With appropriate threshold selection, the possibility of early freezing of subsequent calculations is minimized, thereby reducing the loss of SNR, even neglecting the loss of SNR. During silicon-level circuit design, combine this adaptive architecture with coarse-grained clock gating saves 49.3% of the power consumption.

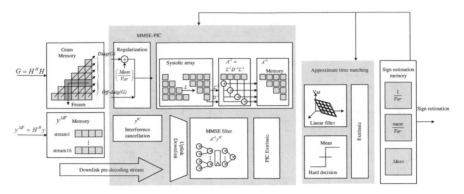

Fig. 1.20 EDP circuit architecture design. © [2018] IEEE. Reprinted, with permission, from Ref. [87]

One of the most compute-intensive and most critical parts of EPD is the matrix inversion module in the MMSE-PIC filter. In the existing research, systolic arrays are often used to implement LDL decomposition, thus achieve accurate matrix inversion. Systolic arrays are characterized by the highly-unified architecture, efficient routing, and simple control. However, due to the need for zero padding input, the hardware utilization of a systolic array architecture is only 33.3%. In the literature [87], a thin LDL systolic array is implemented and it combines underutilized PE circuits into a 16 × 16 array. With this design, the hardware utilization is increased to 90% and interconnection overhead is reduced by more than 70%. As shown in Fig. 1.21, the PE in the conventional systolic array performs division (PE0), multiplication (PE1) or multiple and accumulate (MAC) (PE2 and PE3) operations and passes its output to the adjacent PE. In a thin systolic array, every three PEs in a row are combined. This method shortens the data transmission of the systolic array. The thin array uses buffers to limit data movement, and thus maximizes data reuse. The reuse of data is particularly advantageous in design because the basic processing unit requires a relatively long 28-bit data width to support different channel conditions. Compared with conventional systolic arrays, the thin array architecture has its silicon area reduced by 62%. In addition, with the thin array, data transmission delay is reduced so that a larger portion of the time can be spent on data processing.

EPD chip is manufactured by 28 nm process, occupying 2.0 mm². Under 1 V voltage, the EPD chip runs at 1.6 Gbit/s on 512 MHz. Under 0.4 V positive bias voltage, the EPD chip can work at 1.8 Gbit/s on 569 MHz, an 11% increase in the data throughput rate. Corresponding core power consumption is 127 mW, and energy efficiency is 70.6 pJ/bit. For low-power applications, the EPD chip can work under 0.2 V negative bias voltage at 754 Mbit/s. In this case, the power consumption is reduced to 23.4 mW. In the article, the EPD chip provides flexibility in modulation and channel adaption, supports uplink/downlink processing, and implements high energy efficiency and area efficiency.

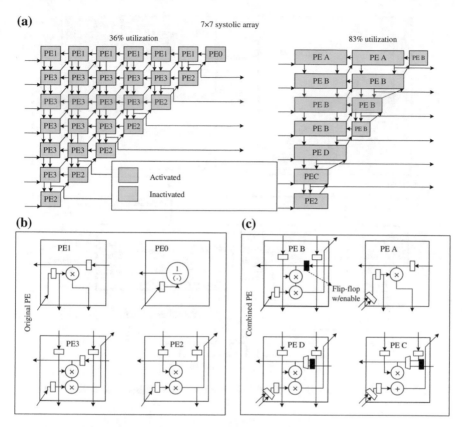

Fig. 1.21 **a** LDL systolic array, **b** original PE design, **c** combined PE design. © [2018] IEEE. Reprinted, with permission, from Ref. [87]

In the literature [88], a 1.1 mm², 128 × 8, massive MIMO baseband chip is designed, which realizes twice the spatial multiplexing gain. Figure 1.22 shows the pre-coder structure of the chip, which can be divided into three sub-modules:

① The triangle array performs operation and QR decomposition (QRD) of the Gram matrix.
② The vector projection module and backward replacement module complete the implicit inversion of the matrix.
③ Perform matched filtering, inverse fast IFFT and optional threshold clipping operation for peak-to-average ratio (PAR) pre-coding. Highly pipelined sub-carriers can be implemented between sub-modules. The pipeline registers are assigned to the vector projection module and backward replacement module, which store the Givens rotation (GR) coefficients to provide high access bandwidth. The thin pipeline between the PEs ensures 0.95 ns critical path delay within each PE.

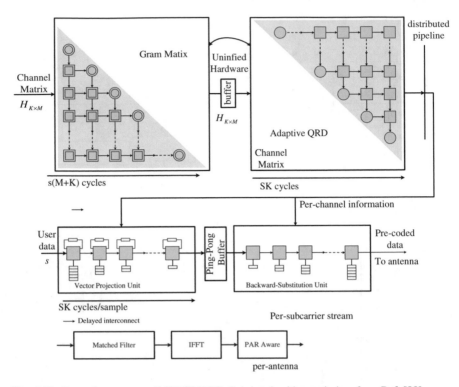

Fig. 1.22 Pre-coder structure. © [2018] IEEE. Reprinted, with permission, from Ref. [88]

During design, highly-unified basic processing units (Fig. 1.23) are used to calculate the Gram matrix and QRD. With such reuse design, the number of gates
(2700) per basic processing unit can be reduced. The unified triangular systolic array
first calculates the Gram matrix and feeds it back to the QRD module for calculation through vertical interconnection. Besides, a single general multiplier is used
to implement highly time division multiplexing of the basic processing unit. An
accurate GR requires 16 clock cycles, and an approximate GR uses a constant multiplier and requires 8 clock cycles. The two accumulator units complete the matrix
multiplication by reusing the general multiplier. With the vector projection unit, the
matrix Q will not be explicitly calculated and the data stream is processed using the
pre-calculated GR coefficients. The total storage capacity required for accurate calculation is 1.7 KB; when approximate rotation is used, half of the storage is gated.
A 0.4 KB ping-pong buffer is used for pipelining and reordering the user vector
stream of the backward replacement unit. The backward replacement unit uses Newton iteration blocks and reuses the multiplier during initialization to improve the area
efficiency.

Usually, the massive MIMO channel matrix approximates independent uniform
and identical distribution. However, under highly correlated channel conditions, massive MIMO detection requires a non-linear solution, such as dense user deployment

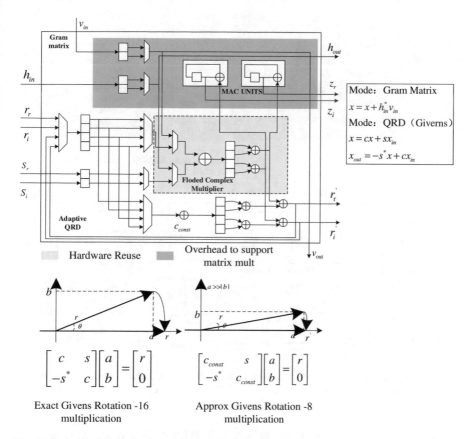

Fig. 1.23 Architecture of basic processing unit and accurate/approximate QRD method. © [2018] IEEE. Reprinted, with permission, from Ref. [88]

in the system. QRD following a tree search is almost the optimal method for the small-scale MIMO system. In the massive MIMO, it is of great importance to conduct MF to reduce the dimensionality of the detection matrix. However, the MF will produce noise. In the article, a flexible framework is designed to support linear and nonlinear detection, as shown in Fig. 1.24. The Cholesky decomposition unit facilitates the operation of linear equations to perform the MMSE detection operation, and provides a data basis for subsequent tree search algorithms. Since the division unit decides the accuracy and time constraint, the author designs the bit-level division unit of the pipeline. It provides a highly-accurate decomposition, and can implement 51 dB SNR under 12-bit internal word length. In the 325 cycles, The Cholesky decomposition of the 8×8 g matrix is first calculated, and then the result is used for calculation in the forward replacement unit and backward replacement unit for linear detection.

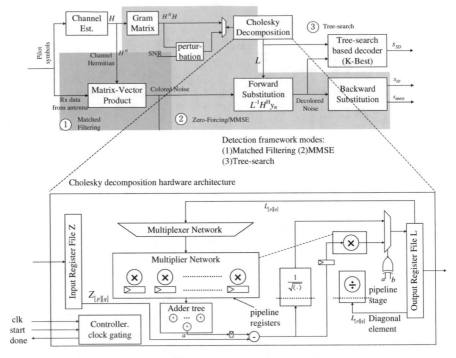

Fig. 1.24 Linear/nonlinear detection architecture. © [2018] IEEE. Reprinted, with permission, from Ref. [88]

In the 28 nm process, the chip works at 300 Mbit/s on 300 MHz, with the power consumption for uplink detection and downlink pre-coding 18 mW and 31 mW, respectively. The QRD unit is used for downlink pre-coding, and its performance and energy efficiency are 34.1MQRD/(s.kGE) and 6.56 nJ/QRD, respectively. Meanwhile, compared with other design, the area overhead is reduced by 17–53%.

In the literature [89], an architecture integrating the message passing detector (MPD) and polarization decoder is proposed. First, soft output MPD detector is proposed. Compared with other design, the proposed MPD detector's data throughput rate is improved by 6.9 times, power consumption is reduced by 49%, and soft output result can be obtained. The proposed polarization decoder's data throughput rate is improved by 1.35 times under the same power consumption. The proposed chip provides 7.61 Gbit/s data throughput rate for the massive MIMO system having 128 antennas and 32 users. Figure 1.25 shows the architecture of the proposed iterative detection and decoding receiver. The architecture includes soft output MPD detector and bidirectional polarization decoder. A high-throughput polarization decoder is used to support K users. The MPD completes detecting the symbol between the interference cancellation and the messaging state. The high computing complexity of the MPD comes from a series of MAC operations (mean and square error) required for calculating the symbol. Finally, the adaptive square error and reliable symbol

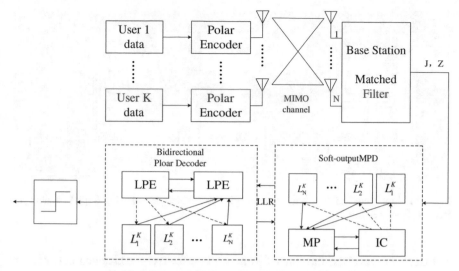

Z: Match filter output, L_i: Internal log-likelihood ratio（LLR）for K users

Fig. 1.25 Architecture of iterative detection and decoding receiver. © [2018] IEEE. Reprinted, with permission, from Ref. [89]

detection technology is proposed. Due to channel hardening, the non-diagonal elements of the Gram matrix are much less than the diagonal elements as the number of iterations increases. Therefore, the symbol square error can be approximated by the largest symbol square error with a scale factor. Compared with the real square error calculation, this approximation saves 73.3% of the multiplication operation. The mean value of the symbol can be evaluated effectively by increasing/decreasing the reliable hard symbols through the given fixed threshold, thus reducing 93.8% of the multiplication operation. Besides, 50% memory overhead can be reduced by using the symmetric characteristics of the Gram matrix.

Figure 1.26 shows the architecture of the proposed bidirectional polarization decoder. The polarization code with a length of 1024 bits and a code rate of 1/2 is considered, and the variable length critical path can be reduced by using the dual-column bidirectional propagation architecture. Where, L PEs sequentially generate L messages from phase 0 to phase $m-1$; and R messages from phase $m-1$ to phase 0, L messages in phase-1 and R messages in phase $m-1-i$ are updated and propagated at the same time. On the same technical node, the critical path delay is reduced by 27.8%. According to the memory access mode, the L messages generated in phase $m-1$ are used for hard decisions only, and the R messages generated in phase 0 are fixed. Then, these operations are removed from the iterative decoding process, and the decoding cycle of each iteration is shortened from 10 to 9. Therefore, the data throughput rate is improved by 11.1%.

The architecture of the MPD detector and polarization decoder proposed in the literature [89] also supports iterative detection and decoding. In the iterative detection

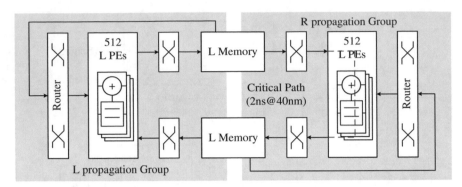

Fig. 1.26 Architecture of bidirectional polarization decoder. © [2018] IEEE. Reprinted, with permission, from Ref. [89]

and decoding receiver, the soft information is effectively exchanged between the MPD and polarization decoder. Through the iterative detection and decoding that is based on the joint factor, the soft information of the K users is effectively exchanged to reduce multi-user interference. In the 40 nm CMOS process, the chip proposed in the literature [89] occupies 1.34 mm² and uses 1,167 K logic gates (including external memory). It consumes 501 mW when working on 500 MHz. The MPD detector provides a maximum of 8 Gbit/s data throughput rate for the massive MIMO (32×8 to 128×8) QPSK system. When the SNR is 4.0 dB, the polarization decoder provides a peak data throughput rate as high as 7.61 Gbit/s by terminating the average 7.48 iteration times in advance. Despite the soft output, the normalized data throughput rate of the proposed MPD is 6.9 times higher than those of other architectures and power consumption is reduced by 49%. After the normalization, the data throughput rate of the proposed polarization decoder is improved by 1.35 times, and the area and power consumption overhead are comparable.

In the literature [90], a 0.58 mm² MPD is designed for the 256-QAM massive MIMO system that supports 32 concurrent mobile users on each clock frequency. Based on the channel hardening technology in the massive MIMO, a symbol hardening technology is proposed and it can reduce the MPD complexity by more than 60% while reducing SNR loss. As the MPD uses the four-layer bidirectional interweaving system architecture, the MPD area is 76% smaller than that of a complete parallel system. The data throughput rate of the designed architecture is 2.76 Gbit/s (4.9 iterations in average, SNR 27 dB). By optimizing the architecture using dynamic precision control and gating clock technology, the chip's energy efficiency can reach 79.8 pJ/bit (or 2.49 pJ/bit/number of receiving antennas). Besides, an MPD detector is designed for the 128×32 256-QAM system in the literature. With the channel hardening in the massive MIMO system, the square error convergence of the symbol evaluation is quickened. Therefore, small fixed square error can be used to replace complex square error calculation, thus saving 4 K MACs of the 32 interference cancellation units and 1 K MACs of the 32 constellation matching units. Reduce

CPE processing by using small square error, therefore make a hard symbol decision based on its distribution. With the symbol hardening technology, the 1 K MACs of the 32 constellation matching units and the 1 K Gaussian estimation operations are removed. With the proposed method, 0.25 dB SNR is sacrificed when BER is 10^{-4}, but the optimized MPD still boasts 1 dB higher detection accuracy than the MMSE detector.

The MPD detection algorithms can completely parallelize the 32 interference cancellation units and 32 constellation matching units (Fig. 1.27a). Up to 4 K MACs and 10 K interconnections are needed. Though the data throughput rate is high, the complete parallel architecture is mainly controlled based on global wiring, therefore, the chip area is large, and problems of low clock frequency and high power consumption exist. In the literature [90], a compact design (Fig. 1.27b) is selected, and the 32 users are divided into four layers with each layer 8 users. The number of MACs used by each interference cancellation unit is reduced to 1/4 of the original number of MACs. On each layer, the 32 interference cancellation units calculate the total interference caused by 8 users and update the symbol estimated value. Then, the updated estimated value is forwarded to the next layer. Compared with the complete parallel architecture, the method proposed in the literature [90] doubles the convergence speed. Based on the experimental results, the area and power consumption of the four-layer architecture are reduced by 66 and 61%, respectively. Meanwhile, the convergence speed is increased, and data throughput rate is reduced by 28%. Since the layered architecture increases the data dependency between layers, the number of interference cancellation units is reduced to 16 in order to reduce the data dependency and decrease the area, and the interference cancellation units are multiplexed in the level-2 pipeline (Fig. 1.27c). In each cycle, each pipeline level calculates the symbol estimated value for group 1 or group 2 users. The estimated value calculation process is interlaced to avoid pipeline delay. Based on the experimental results, the chip area and power consumption of the four-layer two-way interlaced architecture (Fig. 1.28) are reduced by 76 and 65%, respectively.

The power consumption of the data path in the architecture is controlled by 512 MACs. To save dynamic power consumption, the multiplier precision is adjusted based on the convergence of the MPD. In the earlier iteration, the MPD uses low-precision (6 bit × 2 bit) multiplication for rough symbolic estimation. In the later iteration, the MPD uses full-precision (12 bit × 4 bit) multiplication for symbolic

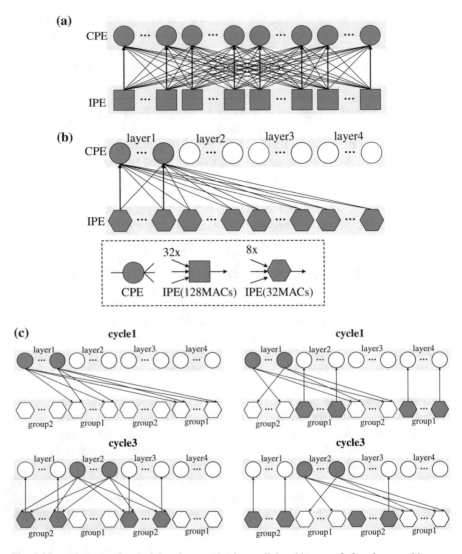

Fig. 1.27 a Methods of optimizing the completely parallel architecture, **b** four-layer architecture, **c** four-layer two-way architecture. © [2018] IEEE. Reprinted, with permission, from Ref. [90]

estimation tuning (Fig. 1.29a). This design helps save 75% switch activities and relevant dynamic power consumption. Besides, the register in the article is used as data memory to support the data access needed by the architecture. The memory access is regular (Fig. 1.29b), for example, 3 KB interfering memory is updated every 8 cycles. When the memory is not updated, the clock gating technology is used to close the clock input and save dynamic power. In the TSMC 40 nm CMOS process, the manufactured massive MIMO MPD chip includes a 0.58 mm^2 MPD core,

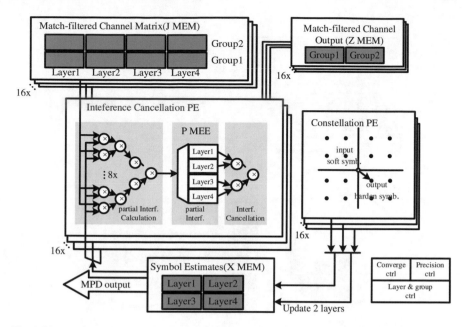

Fig. 1.28 Four-layer two-way interlaced architecture. © [2018] IEEE. Reprinted, with permission, from Ref. [90]

a PLL for generating clock, a test memory for storing test vectors, and I/O ports. Under 0.9 V voltage, the chip works on 425 MHz and consumes 221 mW. With the architecture technology and dynamic precision control and gating clock technologies, the MPD power consumption is reduced by 70%, and energy consumption per bit is decreased by 52%. By enabling the advanced termination technology on the chip, the detection is conducted at an average of 5.7, 5.2, and 4.9 iterations to achieve different performance (23, 25, and 27 dB SNR), resulting in a data throughput rate of 2.76 Gbit/s per mobile user. By deploying multiple MPD modules and applying the interlace technology, the data throughput rate of massive MIMO can be further improved.

1.3.3   Limitation of Traditional MIMO Detection Chips

Boasting robust instruction sets and flexible architecture, the ISAP-based MIMO detection chip supports different algorithms. However, the instruction set architecture does not match the MIMO detection chip highly, so the MIMO detection chip cannot be customized. Therefore, the ISAP-based MIMO detection chip has a low processing rate, a low data throughput rate, and high delay. What's more, it is based on the traditional MIMO system entirely whose scale is far smaller than that of a massive MIMO system in future wireless communication. Since the data that needs

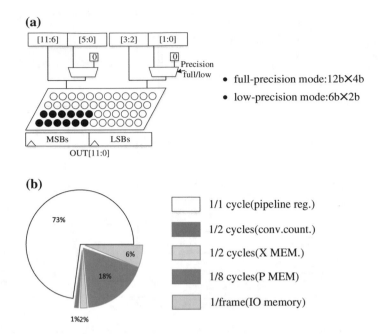

Fig. 1.29 Low-power consumption design technology. **a** 12 bit × 4 bit full-precision multiplier, **b** register update frequency statistics. © [2018] IEEE. Reprinted, with permission, from Ref. [90]

to be processed by the existing ISAP-based MIMO detection chip increases exponentially, the system cannot process the data in real time, and the data throughput rate is reduced and power consumption per area is increased, which cannot be resolved in a short time. This will limit the application of the existing MIMO detector in future wireless communication system. The ASIC-based MIMO detection chip is designed as a custom hardware circuit based on the different MIMO detection algorithms. During the circuit customization, the characteristics of the different algorithms are considered to optimize the circuit. Therefore, the ASIC boasts high data throughput rate, low delay, small power consumption per area, and high energy efficiency. However, with the continuous evolution of the MIMO detection algorithms, the communication algorithm standards and protocols are updated continuously, which requires the hardware to be able to adapt to the requirements of these changes. Since the functions of the ASIC-based MIMO detection chip cannot be changed in form after it is manufactured, you need to redesign and produce the chip in order that it supports different algorithms. This will result in a very large consumption of manpower, material and financial resources. Meanwhile, as traditional MIMO system is evolved to massive MIMO system, the scale of data processed by the MIMO detection algorithms changes; therefore, the ASIC-based MIMO detector needs to adapt to new systems. That is, the hardening hardware of the ASIC-based MIMO detector cannot meet the requirements of flexibility and scalability. In some MIMO detection chips, ASIC and ISAP are combined to implement complex systems and algorithms. In such form, however, the inherent defects of the two methods coexist.

1.4 Dynamic Reconfigurable Chip Technologies of MIMO Detection

1.4.1 Overview of Reconfigurable Computing

1.4.1.1 Development and Definition of Reconfigurable Computing

The dynamic reconfigurable chip of MIMO detection is an application instance of reconfigurable computing in wireless baseband processing. In wireless communication, the computing characteristics of MIMO detection algorithms decide their applicability of reconfigurable computing technologies. To enable readers to better understand the dynamic reconfigurable chip of MIMO detection, the following content gives a comprehensive introduction of reconfigurable computing [91, 92].

Before reconfigurable computing emerged, common computing architectures include GPPs and ASICs. The GPP is based on the von Neumann architecture and composed mainly of arithmetic logic units (ALU), memory, control units, and input/output (I/O) interfaces. During computation, tasks are scheduled and processed based on instruction set compilation through software. Therefore, the GPP boasts high flexibility. However, the performance and energy efficiency of the GPPs are usually low, mainly because:

① The von Neumann architecture is based on TDM, and the spatial parallelism is poor.
② A lot of time and energy are spent on instruction fetching and decoding, and register access, execution and data write-back in the von Neumann architecture. When the transistor size is gradually reduced, the leakage current and power consumption problems begin to become serious. Using new process nodes in the internal architecture of the traditional processor to improve the performance of the chip and therefore improve processor performance is far from meeting requirements. To implement performance improvement without increasing the working frequency, the only method is to increase the core number. In recent years, multi-core and many-core architecture becomes a research hotspot. Single-chip multiprocessors are beginning to replace the more complex single-threaded processors in the server and PC field step by step. Performance improvement is at the expense of area and power consumption, and the energy efficiency problem is always the shortcoming of GPPs [93–95].

The ASIC is a hardware circuit designed for specific application. The characteristics of the ASIC circuit computing mode is that hardware is used to implement operations defined in applications. Since the ASIC circuit is designed for specific applications, it executes at a high speed, efficiency and accuracy. The ASIC is driven by the data stream during execution without the process of instruction translation. Besides, redundancy of the PE can be reduced by special customization, which greatly reduces execution time and area and power consumption. But, the ASIC requires a long development period, which indicates a high cost. What's more, a hardware circuit cannot be changed at random once it is designed. This means that if functional requirements change, even the repair of a very small part of the chip requires redesign and reprocessing of a new ASIC. If dedicated circuit chips need to be designed for each application, costs of development will be high. Only batch production of the ASICs can result in low costs.

Abstractly, a GPP is a solution in which time is used as an extension method and its computational power is reduced in order to increase its versatility. An ASIC is a solution in which space is used as an extension method. It provides maximum computing power at a reduced resource at the expense of computational flexibility. Space extension should be considered during parts manufacturing, and time extension is decided by users after that. Reconfigurable computing enables free customization in time and space extension. Through software programming, the reconfigurable technology can change reconfiguration information to alter hardware functions, and therefore it boasts both space and time extension [96, 97]. Though the generality of the reconfigurable technology is reduced, this technology can still meet the requirements of specific fields and provide operation efficiency as approximate as the ASIC. Besides, the reconfigurable technology helps significantly shorten the product listing time. After configured based on specific application, the prototype system of the reconfigurable processing architecture becomes an asset and can be directly put into the market. This eliminates the time overhead resulted from separate design for specific application.

In view of the current multi-mode and multi-standard coexistence in the communication field, and the emergence of new standards, the reconfigurable processing architecture enables switch between various protocols and algorithms in real time, maintains status quo, and adjusts system functions in time using the existing resources to meet market demand according to actual conditions. Specifically, the reconfigurable processing architecture has the following three levels of flexibility and adaptability:

(1) Adaptability of the protocol layer, that is, it enables flexible switch between different protocols.
(2) Adaptability of the algorithm selection layer, that is, it can flexibly select an algorithm to implement the function.
(3) Adaptability of the algorithm parameter layer, that is, it can flexibly control the parameters of a specific algorithm.

Therefore, the dynamic reconfigurable processing architecture applied to the baseband signal processing in the communication field has become a hot research direction and market trend, and will promote the development of software radio, cognitive radio and future communication technologies.

The concept of reconfigurable computing was first proposed and implemented by the University of California, Los Angeles in the 1960s [98]. This architecture was composed of a fixed, unchangeable main processor and a variable digital logic structure. In this architecture, the main processor unit was responsible for task loading and scheduling, and the variable digital logic structure was responsible for accelerating and optimizing the key algorithms. In this architecture, the concept of reconfigurable computing was first proposed. This concept was the system prototype of the reconfigurable processor currently. Due to the technical limits at that time, reconfigurable computing was re-emphasized until the mid-1990s. At the International Conference on Design Automation in 1999, the University of California at Berkeley proposed a general definition of reconfigurable computing, and regarded it as a type of computer organization structure, with the ability to customize the chip after manufacturing (different from ASIC) and the ability to implement spatial mapping of algorithm to computational engine (different from GPP and DSP).

Most reconfigurable structures have the following two features:

(1) The control is separated from the data. In the process of operation, the reconfigurable PE is used to perform data operation, and the processor is used to control the data flow and complete the reconstruction work of the reconfigurable processing unit operation.
(2) In the reconfigurable architecture, the array structures consisting of basic reconfigurable processing units are used mostly.

In recent years, more and more research institutions and companies have chosen reconfigurable computing to carry out the comprehensive processor architecture innovation, because reconfigurable computing integrates the advantages of both GPP architecture and ASIC architecture. Reconfigurable computing uses GPP architecture and ASIC architecture to complement respective defects, and makes a proper compromise in flexibility, performance, power consumption, cost, and programmability. Figure 1.30 shows the distributions of different computing forms in the five indexes.

In the 1990s, the most outstanding reconfigurable computing architecture was FPGA, which was one of the main computation forms for reconfigurable computing. In 1986, Xilnix developed the first FPGA chip in the world which showed a good application effect. The potential value of reconfigurable chips in both technology and business has attracted some scholars and companies to conduct research on reconfigurable computing. Since then, FPGA has been valued and studied as one of the representatives of reconfigurable computing. FPGA is a semi-customized form of ASIC that performs hardware function configuration using a programmable logic unit which is based on a lookup table structure. Compared with ASICs, FPGAs are highly flexible and user-developable, so they have gained a lot of room for development as logic devices in the programmable field. From a fast-scaling point of view, when system functionality needs to be upgraded, device performance requirements are increased, and configurable features allow the system to be quickly upgraded without hardware changes, thus meeting performance requirements. Current mainstream products include Altera's Stratix series, Xilinx's Virtex and Spartan series.

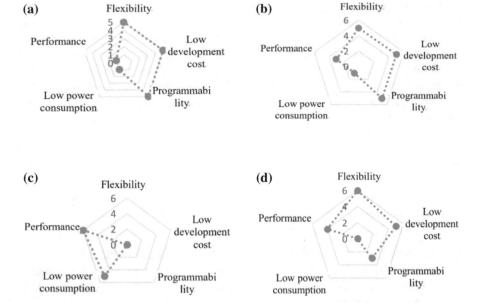

Fig. 1.30 Compromise of performance indexes for various solutions in digital integrated circuits. **a** GPU, **b** DSP, **c** ASIC, **d** FPGA

Meanwhile, there are also reconfigurable chips that can be reconfigured in other different ways, such as Altera's EEPROM-based complex programmable logic device (CPLD) and Actel's anti-fuse FPGA. They have been also well applied and promoted in other fields. The research on reconfigurable computing in the academia is very comprehensive and extensive, focusing on fine-grained reconfigurable architectures such as Ramming machine, PAM machine, and GARP in the earlier phases. Subsequent research focuses on coarse-grained reconfigurable architectures such as MATRIX, MorphoSys, ADRES, DySER, Remus, and others. In the meantime, the fine- and coarse-grained hybrid reconfigurable architecture is also an important research direction, such as TRIPS, TIA, GReP, and so on.

1.4.1.2   General Architecture Model of Reconfigurable Computing

The reconfigurable computing architecture integrates GPP and ASIC advantages, and can be seen as a combination of the two or a combination of the advantages of the two. Figure 1.31 shows the comparison results of the GPP architecture and ASIC architecture. The following content describes the general architecture model of reconfigurable computing based on GPP and ASIC. From the perspective of computational mode, GPP's computational modes are characterized in that they all have their own instruction sets. By executing the relevant instructions in the instruction set to complete the calculation, the software rewriting instructions can change the

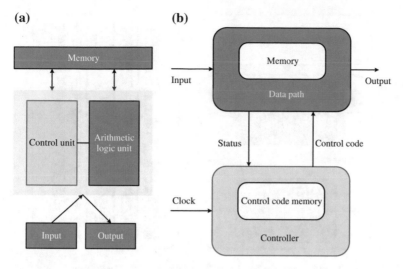

Fig. 1.31 Fundamental principle and architecture comparison of GPP and ASIC **a** GPP architecture, **b** ASIC architecture

functions implemented by the system without changing the underlying hardware environment. Figure 1.31a is a GPP architecture that includes an arithmetic logic unit, control unit, memory, input and output, and so on. The processor is much slower than the ASIC because the processor must read each instruction from memory, decode it, and execute it, so each individual operation has a higher execution overhead. The characteristics of the ASIC circuit computing mode is that hardware is used to implement operations in applications. Figure 1.31b is an ASIC architecture. The main differences of the ASIC from the GPP are:

(1) data path enhancement
(2) control unit weakening.

Regarding the first point, a large number of data paths of the ASIC are hardware mapped spatially compared with those of the GPP. Thus, a large number of fixed arithmetic logic resources, memory units and interconnect resources are available to implement operations and processing. Regarding the second point, the control unit of the ASIC is usually an FSM, which outputs a control code according to the status signal fed back by the data path, and only controls the critical system state of the data path. The control unit of the GPP needs to first obtain an instruction from the memory, and then decode, fetch, and execute it. Due to the two customization features, the ASIC boasts higher rate, energy efficiency, and precision during application execution than the GPP. The reconfigurable processor is a compromise of the GPP and ASIC, which is enhanced in terms of the computing capability of the data path while maintaining some flexibility compared with the GPP; and simplified on the controller while maintaining the capability of controlling the data path.

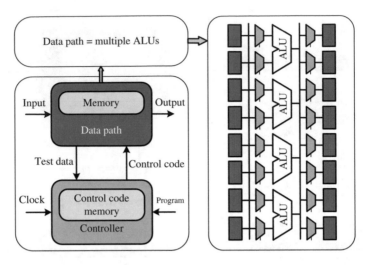

Fig. 1.32 General architecture model of reconfigurable computing processor

The general architecture model of reconfigurable computing generally consists of two major parts: the reconfigurable data path (RCD) and the reconfigurable controller (RCC). Figure 1.32 shows a general architecture model of a reconfigurable computing processor. In the entire model, the RCD is responsible for the parallel computation and processing of the data stream, and the reconfigurable controller is responsible for scheduling and allocating tasks. Through the joint operation of the two, the computing capability and flexibility are improved.

(1) Reconfigurable data path

The RCD is composed of the processing element array (PEA), memory, data interface, and configuration interface, as shown in Fig. 1.33. The control signal, configuration word, and state quantity generated by the reconfigurable controller are transmitted to the reconfigurable PE through the configuration interface. On the configuration interface, the configuration word is parsed, the PEA function is configured, and the sequence of executing tasks is scheduled on an array. The PEA functions are defined using the configuration word. After the configuration, the PEA starts to be driven by the data streams like an ASIC within the set time. During execution, the PE obtains data through the data interface, and the memory caches the intermediate data. In addition to completing external data access and write-back, the data interface can receive signals from the configuration interface, transform the data streams (transposition, concatenation, or other operations), and coordinate with the PEA.

The PEA consists of a large number of PEs and configurable interconnection structure, and completes parallel computing, as shown in Fig. 1.34. The PE is composed of an ALU and a register group generally. The ALU performs the basic operation, and the registers cache the internal data of the PE. During the parallel computing, the external storage interface usually becomes a bottleneck to system performance

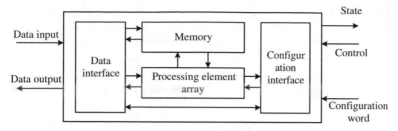

Fig. 1.33 Hardware architecture of RCD

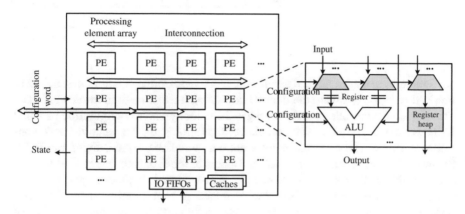

Fig. 1.34 PEA and PE architectures

on the precondition that computing resources are sufficient. The system performance is subject to the efficient data cache and read method as well as external memory access interval and duration. In the PE, a layered distributed storage structure is used to resolve the system cache problem. Interconnection is an important characteristics of the RCD. Flexible and configurable interconnection structure can help implement space mapping of algorithms. The data streams of different PEs can be quickly completed through the interconnection. Compared with the register read/write method of the superscalar, VLIW processor, the interconnection structure implements more efficient hardware wiring. The specific organization mode of the interconnection is not fixed. The more flexible the interconnection is, the higher the hardware cost. Usually, the interconnection organization mode is customized based on the algorithm characteristics of a field to implement high efficiency and flexibility.

(2) Reconfigurable controller

The reconfigurable controller is composed of the memory, configuration management and control unit, and configuration interface, as shown in Fig. 1.35. The internal configuration information is stored in the storage module, and accessed and transmitted to the RCD through the configuration interface if needed. The configuration interface is used to send the configuration word and control signal to the RCD. The

Fig. 1.35 Hardware
architecture of
reconfigurable controller

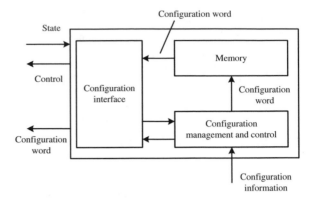

configuration management and control unit receives external configuration information and parses it to obtain the internal control signal and configuration word. The
reconfigurable controller mainly manages the reconfigurable configuration path and
RCD. By managing the reconfigurable configuration path, the reconfigurable controller schedules and coordinates different configurations. By managing the RCD,
the reconfigurable controller controls the status of the data path and critical system,
and coordinates PE required resources. In the traditional single-core processor, the
controller focuses on time scheduling of nodes. A large number of parallel optimization technologies such as pipeline are used due to the repeated execution of
instruction flow on a single node, the time requirements of the controller are high.
Usually, the reconfigurable computing processor employs array form and schedules
computing resources of multiple nodes. The reconfigurable PE is not as complex as
a single-core processor, and the controller implements simple time control of nodes.
The entire spatial and timing utilization is more important than node scheduling,
which raises new design requirements for the controller. If the reconfigurable computing unit array is maximally configured, a custom acceleration unit even control
unit array can be added to the reconfigurable controller.

1.4.1.3 Classification of Reconfigurable Computing

The general forms of reconfigurable computing have been discussed in the above. As
for different architectures, their reconfigurable computing categories differ greatly.
Based on the reconfiguration granularity, reconfigurable computing falls into fine-
grained, coarse-grained, and medium-grained reconfigurable computing. Based on
time, reconfigurable computing falls into static and dynamic reconfigurable computing. Based on space, reconfigurable computing falls into partial and entire reconfigurable computing. It is noticeable that reconfigurable computing categories are
not fully independent, so you can say reconfigurable computing is categorized based
on both granularity and time. Fine-grained reconfigurable computing processors are
difficult to implement dynamic reconfiguration due to long reconfiguration time.

Usually, the reconfigurable processors we talk about are fine-grained static reconfigurable processors (such as FPGA) and coarse-grained dynamic reconfigurable processors.

(1) Categorization based on reconfiguration granularity

The data bit width of the PE in the data path of the reconfigurable computing processor is referred to as granularity. Based on the processor granularity, the reconfigurable computing processors fall into fine-grained, coarse-grained, medium-grained, and mix-grained processors. In general cases, the smaller the granularity is, the more the configuration information required for reconfigurable computing processors, the slower the processor reconfiguration speed, and the higher the functional flexibility. Otherwise, the situation is reversed. Fine granularity is less than or equal to 4 bits, and coarse granularity is more than or equal to 8 bits. The traditional FPGA is a common fine-grained reconfigurable computing processor. The PE of the traditional FPGA is 1 bit. Since the PE is a single-bit component (fine-grained reconfigurable computing processor), the FPGA boasts high flexibility. Without considering capacity, digital logic can be implemented in any form. This is one of the reasons for the commercial success of the FPGA. The data bit width of the PE in the coarse-grained reconfigurable computing processor is 8 or 16 bits. The 4-bit width is referred to as medium-grained granularity, which is less common. If the reconfigurable computing processor contains more than 1 type of granularity PEs, it is referred to as a mix-grained reconfigurable computing processor. It is noticeable that the definitions of mix-grained and coarse-grained reconfigurable computing processors are confused sometimes. For example, a PEA containing both 8-bit PEs and 16-bit PEs can be referred to as a mix-grained or coarse-grained PEA.

(2) Categorization based on reconfiguration time

Based on the reconfiguration time, the reconfigurable computing processors fall into static and dynamic processors.

In static reconfiguration, the data path of the reconfigurable computing processor can be reconfigured for its functions before computing. During the computing, the functions of the data path cannot be reconfigured due to the price cost. In the common work mode, the FPGA loads the configuration bit streams from the off-chip memory during power-on to conduct function reconfiguration, so the FPGA is a typical static reconfigurable processor. After function reconfiguration, the FPGA can conduct corresponding computing. During the computing, the FPGA functions cannot be reconfigured. If the FPGA needs to be reconfigured, interrupt the current computing task of the FPGA. The fine-grained FPGA brings a massive amount of configuration, which makes the reconfiguration time and power consumption extremely large. For example, the FPGA reconfiguration time is tens to hundreds of milliseconds, even seconds. Typical dynamic reconfigurable processor requires several to tens of nanoseconds of reconfiguration time. Dynamic reconfiguration and static reconfiguration are relative. Since the reconfiguration time is less, the data path of the reconfigurable computing processor can be reconfigured for its functions during computing. This is referred to as dynamic reconfiguration. The typical reconfigurable computing processor boasting dynamic reconfiguration is the coarse-grained reconfigurable array (CGRA). In the common work mode, the CGRA loads the new configuration bit streams immediately after completing a specific computing task to conduct function reconfiguration. Since the configuration bit streams of the CGRA are small, the reconfiguration process usually lasts several to hundreds of clock cycles. After function reconfiguration, the CGRA can conduct new computing tasks. From the aspect of the application layer, switch time of two computing tasks is very short (function reconfiguration time), and the two computing tasks are executed one after another, so the reconfiguration can be regarded as real time. In some literature, dynamic reconfiguration is referred to as real-time reconfiguration as well.

(3) Categorization based on reconfiguration space

Based on the reconfiguration space, the reconfigurable computing processors fall into partial and entire processors.

The data path of the reconfigurable computing processor can be divided into multiple areas, and each area can be reconfigured into a specific function engine to execute a specific computing task, without affecting the current status of other areas. This is referred to as partial reconfiguration. Based on time, the partial reconfiguration can be further divided into static and dynamic partial reconfiguration. When one or multiple areas in the RCD are executing computing tasks, other areas support function reconfiguration on the precondition that the computing tasks are not interrupted. This is referred to as dynamic partial reconfiguration or real-time partial reconfiguration. When one or multiple areas in the RCD undergo function reconfiguration, other areas cannot process computing tasks and must be in the sleep or inactive status. This is referred to as static partial reconfiguration. The typical reconfigurable computing processor boasting dynamic partial reconfiguration is the CGRA. Usually, the CGRA is divided into multiple different areas, and each area can be reconfigured as different function engines to execute different computing tasks. The reconfiguration and computing of each area are separate. With the dynamic partial reconfiguration, the hardware utilization of the data path of reconfigurable computing processor is improved, thus the energy efficiency of the entire processor is improved. It is declared that some commercial FPGAs support static partial reconfiguration, which is meant to shorten the FPGA reconfiguration time by reducing the configuration bit streams. A reconfigurable computing processor which supports dynamic partial reconfiguration must support static partial reconfiguration. The reverse saying, however, is not necessarily true. Unless otherwise specified, partial reconfiguration mentioned in the following content is referred to as dynamic partial reconfiguration. At present, it is declared that some commercial FPGAs support dynamic reconfiguration or dynamic partial reconfiguration. Due to the scale of the configuration bit streams, the FPGA cannot support dynamic and static reconfiguration actually.

1.4.2 Status Quo of Dynamic Reconfiguration Chip of MIMO Detection

Most MIMO detection algorithms are compute-intensive and data-intensive algorithms, which apply to the implementation of the reconfigurable processors and boast high efficiency, flexibility and scalability. Therefore, dynamic reconfigurable processors of MIMO detection attract more and more attention.

In the literature [99], a heterogeneous reconfigurable array processor is proposed to implement signal processing of the MIMO system. To achieve high performance and high energy efficiency, and keep the high flexibility of reconfigurable processors, a heterogeneous and layered resource design is used in the literature. This architecture employs optimized vector computing and flexible memory access mechanism to support MIMO signal processing. With heterogeneous resource deployment and layered network topology, an efficient mixed data computing is implemented and communication cost is reduced significantly. The flexible memory access mechanism helps reduce the register access times of the non-core computing part. What's more, the coordinate optimization of the algorithms and architecture help further improve the hardware efficiency. On the basis of the processing unit array framework, this architecture is composed of four heterogeneous parts and they are divided into scalar processor and vector processing domain, as shown in Fig. 1.36. Data trans-

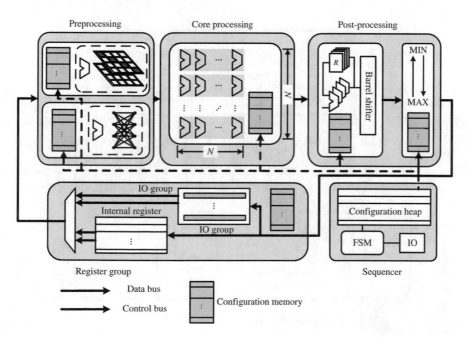

Fig. 1.36 Scalar- and vector-computing heterogeneous architecture. © [2018] IEEE. Reprinted, with permission, from Ref. [99]

mission between two domains is bridged via a storage unit which provides finer data access than physical memory. This function can effectively support mixed data transmission without additional control of the processor. Figure 1.36 shows the scalar- and vector-computing heterogeneous architecture which consists of three processing units (preprocessing unit, core processing unit, and post-processing unit), one register file, and one sequencer. In the upper part in Fig. 1.36, the three processing units are used for vector computing, and the register file provides data access to the internal register and other modules through the mapped I/O ports of the register. The sequencer controls other units' operations through the control bus, as shown in Fig. 1.36 (dotted line). In wireless baseband processing, the SIMD is used usually as a baseline architecture which employs fixed data levels for paralleling. Similarly, the core processing units employ SIMD-based architecture which is composed of N × N complex MAC units. Figure 1.37 shows the basic architecture of the complex MAC units. By analysis, it can be found that tight coupling operations processed by the vectors are existent in the algorithms. The long mapping on a single SIMD core needs to be completed by multiple operations, which increases not only the execution time but also the redundant accesses of the middle result register. In the literature, a multi-level computing chain in the form of very long instruction set is used to expand the SIMD core and complete several continuous operations with a single instruction. The preprocessing unit and post-processing unit are arranged and centered around the SIMD core, as shown in Fig. 1.36. Such arrangement reduces the register accesses by

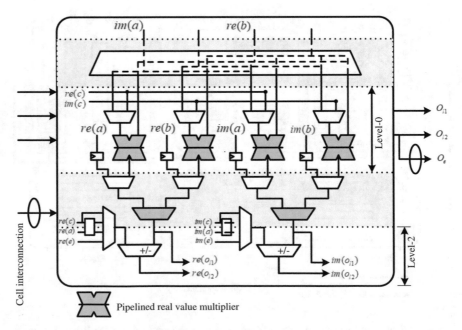

Fig. 1.37 Complex MAC unit structure. © [2018] IEEE. Reprinted, with permission, from Ref. [99]

more than 60%. In the CMOS 65 nm process, this architecture occupies 8.88 mm^2, and works at 367.88 Mbit/s on the 500 MHz frequency. Working on the 500 MHz frequency and under 1.2 V power supply voltage, the architecture consumes 548.78 mW in average to process a signal. Where, 306.84 mW is consumed by the logic module, and 241.94 mW is consumed by the data cache module. Therefore, a bit corresponds to 0.83 and 1.49 nJ/bit power consumption, respectively.

In addition to vector computing, the efficiency of the vector processor is subject to the access bandwidth and the memory access flexibility. In this architecture, it is required that the SIMD core can access multiple matrices and vectors in each operation to avoid a low resource utilization rate and a low data throughput rate. To meet the requirements, mixed memory and flexible matrix access mechanisms are used in the vector data memory block in the literature, as shown in Fig. 1.38. To meet the high memory access bandwidth requirement, the vector access is separate from the matrix access so that they can be conducted at the same time, as shown in Fig. 1.38a. The memory operation and access mode of each unit and page are managed by the local controller, and configured and stored in the embedded register, as shown in Fig. 1.38b. To further improve the flexibility of matrix access and implement the data circuit in each memory page shown in Fig. 1.38c, this architecture loads data from each memory page, and caches it in the local register file. Based on the access index related to the matrix memory, the data can be rearranged vertically. Therefore, this architecture supports free access to a whole line or whole column of the matrices in any order, without physical exchange of data.

In the literature [100], a baseband processing accelerator is introduced, and it contains C-based programmable coarse grained array-SIMD (CGA-SIMD). This accelerator employs the high-instruction parallelism in the SDR core and the simple and efficient high data parallelism. Its programming flow is fully integrated in the main CPU. Figure 1.39 shows the top-layer architecture of the accelerator, and it is composed of 16 interconnected 64-bit FUs. Some FUs have a local distributed register, and a larger global register.

The 16 64-bit core processing units perform the computing of loop bodies. These FUs can execute all conventional arithmetic and logic operations as well as common instructions. They can execute special instructions as well to implement four-bit 16-bit SIMD operations such as parallel shift, parallel subtraction, and parallel addition. All these basic operations have the delay of one cycle. All the FUs can perform signed multiplication and unsigned multiplication of 16-bit integers, all of which have the delay of three cycles. In addition, a FU can perform 24-bit division, which will have the delay of eight cycles. All operations are done in complete pipeline mode. Of the 16 FUs, three FUs are connected to global registers, and each FU performs data interaction through two read ports and one write port. Each FU has a 64-bit local register with two read ports and one write port. Because the local register has small volume and less ports, its power consumption is far less than the global register.

In the literature [101], an ADRES-based MIMO detection chip is proposed. This MIMO detector can provide a data throughput rate as high as that of an ASIC. It

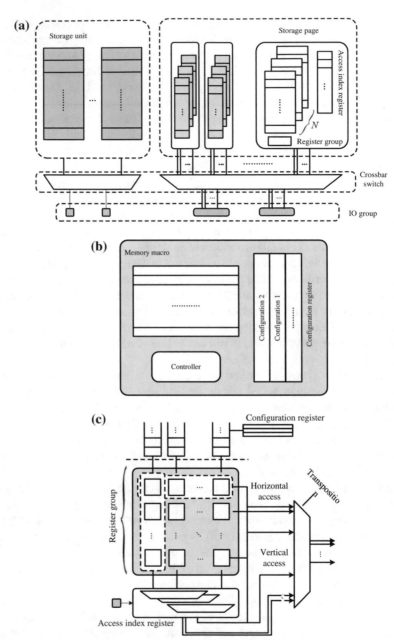

Fig. 1.38 Access mechanism of mixed memory and flexible matrix. **a** Concurrent access mechanism of vector and matrix, **b** memory unit structure, **c** data circuit structure of memory page. ©
[2018] IEEE. Reprinted, with permission, from Ref. [99]

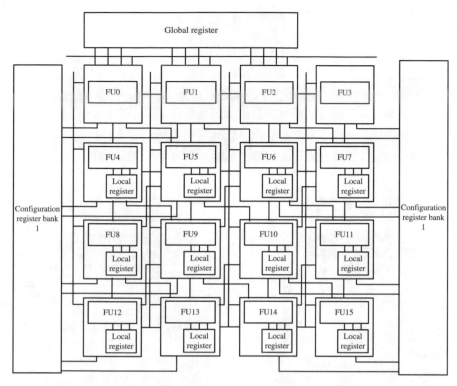

Fig. 1.39 Top-layer architecture of the accelerator. © [2018] IEEE. Reprinted, with permission, from Ref. [100]

contains 2 similar cores (Fig. 1.40) and each core has three scalar FUs. The FUs are combined to form a VLIW which shares a register file. The CGA consists of three vector unit FUs (FU0, FU1, FU2) and two vector loading/storage (LD/ST) units. The vector units are interconnected to form a CGA which contains shared vector registers as well. The data between the VLIW and CGA is connected via the packing/unpacking unit. Each core has two scalar memories and two vector memories.

The highly-optimized special instruction set is used for the CGA to execute complex operation. These special instructions support each vector unit of the CGA, therefore it can implement a high data throughput rate. With the design of the special instruction set, the hardware overhead of arithmetic operators, such as bit-level shifting and addition, can be reduced. In MIMO detection, the size specification is one of the common algorithms, which needs to complete more complex multiplication operations. However, these multiplication operations can be decomposed into low-cost arithmetic shift operations and addition operations, further reducing

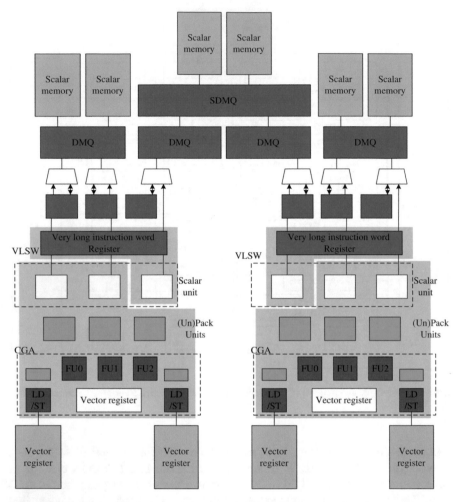

Fig. 1.40 ADRES-based MIMO detection chip. © [2018] IEEE. Reprinted, with permission, from Ref. [101]

computing complexity. The size specification computing can be implemented using the architecture shown in Fig. 1.41. A single execution cycle is shown in the figure.

The literature [102] introduces a MIMO detector which is designed using the reconfigurable ASIP (rASIP) and the detector supports multiple types of MIMO detection algorithms under different antenna configuration and modulation situations. The rASIP is mainly composed of a CGRA and a processor. The MIMO detection implements some important computing steps (for example, preprocessing) using the matrix operation, and even includes the entire detection algorithms. For the rASIP, an MIMO detector is designed in the CGRA and can be used in different matrix operations. The Markov chain Monte Carlo (MCMC) based MIMO detection can

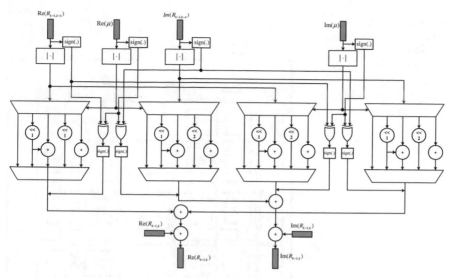

Fig. 1.41 Architecture for size specification computing. © [2018] IEEE. Reprinted, with permission, from Ref. [101]

be efficiently implemented through mapping in the proposed architecture to evaluate the flexibility of the method. In the literature, the multimode-based MIMO detection rASIP architecture helps improve the energy Efficiency by 1.6–5.4 times and provides performance as high as that of an ASIC.

Figure 1.42 shows a CGRA which is built using 20 PEs and one Center Alpha unit. Of the 20 PEs, 16 PEs constitute a 4 × 4 PE array and are aggregated as four PE 2 × 2 clusters. The PE array size is determined based on the maximum supported antenna configuration (4 × 4 or 4 × 8). With the 4 × 4 configuration, the PE array can be used maximally to implement the maximum data throughput rate. Since the matrix size for the two types of antenna configuration is 4 × 4 after the post-processing phase, the PE array can be used to conduct complete storage. The left 4 PEs are arranged as a line and inserted between the PE 2 × 2 clusters. The four PEs are aggregated as two PE clusters. After the PE line is added, it can be used to execute the final accumulation and allows two multipliers to be available on the accumulation path when the matrix vector multiplication is mapped to the 4 × 4 PE array. Furthermore, the PE line allows matrix operation and vector operation to be conducted and separate from each other on the 4 × 4 PE array. By the concurrent matrix operation and vector operation, performance can be improved. The PE line also provides separate storage space for the vector results, without occupying the matrix storage resources on the 4 × 4 PE array. The CGRA has four input ports and five outputs, and each port can input/output a complex value. The output information of the five output ports comes from the Center Alpha unit and the four PEs in the PE clusters. The global interconnection propagates the four data elements d1-d4 to the PEs of the 4 × 4 PE array. These four data elements can come from either

Fig. 1.42 CGRA architecture. © [2018] IEEE. Reprinted, with permission, from Ref. [102]

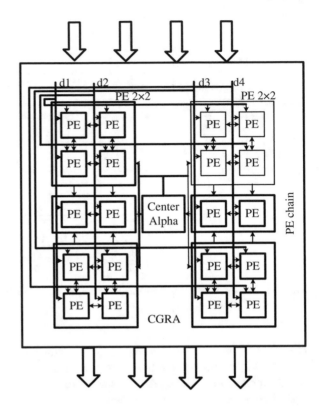

the four input ports of the CGRA or the four PEs in the PE chain. The PE chain can acquire data from its upper and lower PE 2 × 2 clusters. The results of the Center Alpha unit can be passed to the PE 2 × 2 cluster and used by the internal 16 PEs. A PE consists of four basic functional units, namely a complex multiplier, a complex ALU, a barrel shifter, and a local register file, as shown in Fig. 1.43. Flexible interconnections allow communication between different functional units. Through the functional units and interconnections, PEs can be configured differently to perform more complex functions, such as multiply-accumulate operations. The local registration file in the PE is used to store the intermediate results generated by the PE. The PE can output the result of any functional unit. The Center Alpha unit is another basic module. Center Alpha has a similar structure to the PE, including all the basic functional units of the PE. Center Alpha calculates the zoom bit width based on the given fixed point number by removing the sign bits. This zoom width can be performed using the barrel shifter in the PE and Central Alpha to perform dynamic zooming of the data stored in the PE and Center Alpha.

The matrix operations used in the algorithms can be mapped using the CGRA. In order to effectively apply the CGRA to different algorithms, integrate several additional components and the CGRA into a multimode detection architecture (MDA), as shown in Fig. 1.44, including a data register file, CGRA configuration memory,

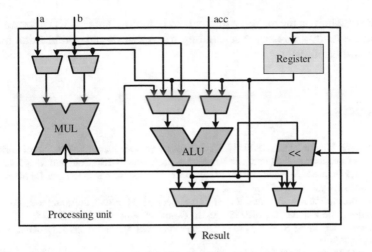

Fig. 1.43 PE architecture. © [2018] IEEE. Reprinted, with permission, from Ref. [102]

Fig. 1.44 Multimode detection architecture. © [2018] IEEE. Reprinted, with permission, from Ref. [102]

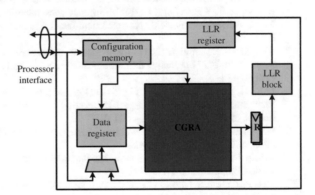

an LLR block which is used to calculate soft information and the LLR register file used to store the soft information.

It can be inferred from the above that reconfigurable MIMO detection chip architectures have been proposed inside and outside China, but there is still a certain distance for the architectures to be applied to processors. Besides, there are still many key scientific issues that need to be solved:

(1) No mathematical model is available for the design method of the reconfigurable MIMO detection chip.
(2) The existing reconfigurable MIMO processor architecture is only designed for the traditional MIMO systems, and no consideration has been given to the processor design for massive MIMO systems.
(3) The range of MIMO detection algorithms supported by a single processor is small, and the processing data size is relatively fixed, so the flexibility and scalability need to be improved.

(4) The research on the mapping methods of reconfigurable MIMO detection chips
     and the hardware resource scheduling management and optimization methods
     are still insufficient.

References

1. Niyato D, Maso M, Dong IK et al (2017) Practical perspectives on IoT in 5G networks: from
    theory to industrial challenges and business opportunities. IEEE Commun Mag 55(2):68–69
2. Le NT, Hossain MA, Islam A et al (2016) Survey of promising technologies for 5G networks.
    Mobile Inf Syst 2016(2676589):1–25
3. Osseiran A, Boccardi F, Braun V et al (2014) Scenarios for 5G mobile and wireless commu-
    nications: the vision of the METIS project. Commun Mag IEEE 52(5):26–35
4. Andrews JG, Buzzi S, Wan C et al (2014) What will 5G be? IEEE J Sel Areas Commun
    32(6):1065–1082
5. Peltier WR (2004) Geoide height Time dependence and global glacial isostasy: the ICE-
    5G(VM2) model and GRACE. In: AGU Spring meeting, 2004
6. Roh W, Seol JY, Park J et al (2014) Millimeter-wave beamforming as an enabling technology
    for 5G cellular communications: theoretical feasibility and prototype results. Commun Mag
    IEEE 52(2):106–113
7. Alberio M, Parladori G (2017) Innovation in automotive: a challenge for 5G and beyond
    network. In: 2017 International Conference of Electrical and electronic technologies for auto-
    motive, pp 1–6
8. Jiang H, Liu H, Guzzino K et al (2012) Digitizing the Yuan Tseh Lee array for microwave
    background anisotropy by 5 Gsps ADC boards. In: IEEE international conference on elec-
    tronics, circuits and systems, pp 304–307
9. Manyika J, Chui M, Brown B et al (2011) Big data: the next frontier for innovation, compe-
    tition, and productivity. Analytics
10. Walker SJ (2013) Big data: a revolution that will transform how we live, work, and think.
    Math. Comput. Educ. 47(17):181–183
11. Mell PM, Grance T (2011) SP 800-145. The NIST definition of cloud computing. National
    Institute of Standards & Technology, p 50
12. Buyya R, Yeo CS, Venugopal S et al (2009) Cloud computing and emerging IT platforms:
    vision, hype, and reality for delivering computing as the 5th utility. Future Gener Comput
    Syst 25(6):599–616
13. Lewenberg Y, Sompolinsky Y, Zohar A (2015) Inclusive block chain protocols[C]. Interna-
    tional conference on financial cryptography and data security. Springer, Berlin, Heidelberg,
    2015:528–547
14. Li X, Baki F, Tian P et al (2014) A robust block-chain based tabu search algorithm for the
    dynamic lot sizing problem with product returns and remanufacturing. Omega 42(1):75–87
15. Hussein A, Elhajj IH, Chehab A et al (2017) SDN VANETs in 5G: an architecture for resilient
    security services. In: International conference on software defined systems, pp 67–74
16. Pan F, Wen H, Song H et al (2017) 5G security architecture and light weight security authenti-
    cation. In: IEEE/CIC International conference on communications in china—workshops, pp
    94–98
17. Bastug E, Bennis M, Medard M et al (2017) Toward interconnected virtual reality: opportu-
    nities, challenges, and enablers. IEEE Commun Mag 55(6):110–117
18. Parsons TD, Courtney CG (2018) Interactions between threat and executive control in a virtual
    reality stroop task. IEEE Trans Affect Comput 9(1): 66–75
19. Al-Shuwaili A, Simeone O (2017) Energy-efficient resource allocation for mobile edge
    computing-based augmented reality applications. IEEE Wireless Commun Lett PP(99):1

20. Chatzopoulos D, Bermejo C, Huang Z et al (2017) Mobile augmented reality survey: from where we are to where we go. IEEE Access 5(99):6917–6950

21. Azuma R, Baillot Y, Behringer R et al (2001) Recent advances in augmented reality. IEEE Comput Graphics Appl 21(6):34–47

22. Lin C, Dong F, Hirota K (2015) A cooperative driving control protocol for cooperation intelligent autonomous vehicle using VANET technology. In: Int Symp Soft Comput Intell Syst, 275–280

23. Guan Y, Wang Y, Bian Q et al (2017) High efficiency self-driven circuit with parallel branch for high frequency converters. IEEE Trans Power Electron PP(99):1

24. Scanlon JM, Sherony R, Gabler HC (2017) Models of driver acceleration behavior prior to real-world intersection crashes. IEEE Trans Intell Transp Syst PP(99):1–13

25. Marques M, Agostinho C, Zacharewicz G et al (2017) Decentralized decision support for intelligent manufacturing in Industry 4.0. J Ambient Intell. Smart Environ 9(3):299–313

26. Huang J, Xing CC, Wang C (2017) Simultaneous wireless information and power transfer: technologies, applications, and research challenges. IEEE Commun Mag 55(11):26–32

27. Shafi M, Molisch AF, Smith PJ et al (2017) 5G: a tutorial overview of standards, trials, challenges, deployment and practice. IEEE J Sel Areas Commun PP(99): 1

28. Tran TX, Hajisami A, Pandey P et al (2017) Collaborative mobile edge computing in 5G networks: new paradigms, scenarios, and challenges. IEEE Commun Mag 55(4):54–61

29. Benmimoune A, Kadoch M (2017) Relay technology for 5G networks and IoT applications. Springer International Publishing

30. Schulz P, Matthe M, Klessig H et al (2017) Latency critical IoT applications in 5G: perspective on the design of radio interface and network architecture. IEEE Commun Mag 55(2):70–78

31. Mehmood Y, Haider N, Imran M et al (2017) M2M communications in 5G: state-of-the-art architecture, recent advances, and research challenges. IEEE Commun Mag 55(9):194–201

32. Zhang X, Liang YC, Fang J (2017) Novel Bayesian inference algorithms for multiuser detection in M2M communications. IEEE Trans Veh Technol PP(99):1

33. Akpakwu GA, Silva BJ, Hancke GP et al (2017) A survey on 5G networks for the internet of things: communication technologies and challenges. IEEE Access PP(99):1

34. Wang CX, Haider F, Gao X et al (2014) Cellular architecture and key technologies for 5G wireless communication networks. Commun Mag IEEE 52(2):122–130

35. Islam SMR, Avazov N, Dobre OA et al (2016) Power-domain non-orthogonal multiple access (NOMA) in 5G systems: potentials and challenges. IEEE Commun Surveys Tutorials, PP(99):1

36. Pham AV, Nguyen DP, Darwish M (2017) High efficiency power amplifiers for 5G wireless communications. In: 2017 Global symposium on millimeter-waves, pp 103–107

37. Pedersen K, Pocovi G, Steiner J et al (2018) Agile 5G scheduler for improved E2E performance and flexibility for different network implementations. IEEE Commun. Mag. PP(99):2–9

38. Simsek M, Zhang D, Öhmann D et al (2017) On the flexibility and autonomy of 5G wireless networks. IEEE Access PP(99):1

39. Chaudhary R, Kumar N, Zeadally S (2017) Network service chaining in fog and cloud computing for the 5G environment: data management and security challenges. IEEE Commun Mag 55(11):114–122

40. Pan F, Jiang Y, Wen H et al (2017) Physical layer security assisted 5G network security. IEEE Veh Technol Conf, 1–5

41. Monserrat JF, Mange G, Braun V et al (2015) METIS research advances towards the 5G mobile and wireless system definition. Eurasip J Wireless Commun Networking 2015(1):53

42. Yuan Y, Zhao X (2015) 5G: vision, scenarios and enabling technologies. ZTE Commun (English edition) 1:3–10

43. Yonggang Ren, Liang Zhang (2014) Prospect of the fifth generation mobile communication system. Inf Commun 8:255–256

44. Xiaohu You, Zhiwen Pan, Xiqi Gao et al (2014) Development trend and some key technologies of 5G mobile communications. Sci China Inf Sci 44(5):551–563

45. Rappaport TS, Sun S, Mayzus R et al (2013) Millimeter wave mobile communications for 5G cellular: it will work! IEEE Access 1(1):335–349
46. Jungnickel V, Manolakis K, Zirwas W et al (2014) The role of small cells, coordinated multipoint, and massive MIMO in 5G. IEEE Commun Mag 52(5):44–51
47. Swindlehurst AL, Ayanoglu E, Heydari P et al (2014) Millimeter-wave massive MIMO: the next wireless revolution? IEEE Commun Mag 52(9):56–62
48. Björnson E, Sanguinetti L, Hoydis J et al (2014) Optimal design of energy-efficient multi-user MIMO systems: is massive MIMO the answer? IEEE Trans Wireless Commun 14(6):3059–3075
49. Gao X, Edfors O, Rusek F et al (2015) Massive MIMO performance evaluation based on measured propagation data. IEEE Trans Wireless Commun 14(7):3899–3911
50. Ngo HQ, Ashikhmin A, Yang H et al (2015) Cell-Free Massive MIMO: Uniformly great service for everyone. In: IEEE International Workshop on Signal Processing Advances in Wireless Communications, pp 201–205
51. Ngo H, Ashikhmin A, Yang H et al (2016) Cell-free massive MIMO versus small cells. IEEE Trans Wireless Commun PP(99):1
52. Rao X, Lau VKN (2014) Distributed compressive CSIT estimation and feedback for FDD multi-user massive MIMO systems. IEEE Press, pp 3261–3271
53. Björnson E, Larsson EG, Marzetta TL (2015) Massive MIMO: ten myths and one critical question. IEEE Commun Mag 54(2):114–123
54. Larsson EG, Edfors O, Tufvesson F et al (2014) Massive MIMO for next generation wireless systems. IEEE Commun Mag 52(2):186–195
55. Zhang K, Mao Y, Leng S et al (2017) Energy-efficient offloading for mobile edge computing in 5G heterogeneous networks. IEEE Access 4(99):5896–5907
56. Sabharwal A, Schniter P, Guo D et al (2014) In-band full-duplex wireless: challenges and opportunities. Sel Areas Commun IEEE J 32(9):1637–1652
57. Zhou M, Song L, Li Y et al (2015) Simultaneous bidirectional link selection in full duplex MIMO systems. IEEE Trans Wireless Commun 14(7):4052–4062
58. Liao Y, Wang T, Song L et al (2017) Listen-and-talk: protocol design and analysis for full-duplex cognitive radio networks. IEEE Trans Veh Technol 66(1):656–667
59. Sharma A, Ganti RK, Milleth JK. Joint backhaul-access analysis of full duplex self-backhauling heterogeneous networks. IEEE Trans Wireless Commun 16(3):1727–1740
60. Duy VH, Dao TT, Zelinka I et al (2016) AETA 2015: recent advances in electrical engineering and related sciences. Springer Publishing Company, Incorporated
61. Kieu TN, Do DT, Xuan XN et al (2016) Wireless information and power transfer for full duplex relaying Networks: performance analysis. Springer International Publishing
62. Zheng G (2014) Joint beamforming optimization and power control for full-duplex MIMO two-way relay channel. IEEE Trans Signal Process 63(3):555–566
63. Yue Yao (2015) Key technology prospect of the fifth generation mobile communication system. Telecommun Technol 1(1):18–21
64. Hosseini K, Hoydis J, Ten Brink S et al (2014) Massive MIMO and small cells: How to densify heterogeneous networks. In: IEEE International Conference on Communications, pp 5442–5447
65. Yang HH, Geraci G, Quek TQS (2016) Energy-efficient design of MIMO heterogeneous networks With wireless backhaul. IEEE Trans Wireless Commun 15(7):4914–4927
66. Osseiran A, Braun V, Hidekazu T et al (2014) The foundation of the mobile and wireless communications system for 2020 and beyond: challenges, enablers and technology solutions. In: Vehicular Technology Conference, pp 1–5
67. Webb W (2007) Wireless communications: the future. Wiley, pp 11–20
68. Hwang I, Song B, Soliman SS (2013) A holistic view on hyper-dense heterogeneous and small cell networks. Commun. Mag. IEEE 51(6):20–27
69. Baldemair R, Dahlman E, Parkvall S et al (2013) Future wireless communications. Veh Technol Conf, pp 1–5

70. Liu S, Wu J, Koh CH et al (2011) A 25 Gb/s(/km^2) urban wireless network beyond IMT-advanced. IEEE Commun Mag 49(2):122–129

71. Jo M, Maksymyuk T, Batista RL et al (2014) A survey of converging solutions for heterogeneous mobile networks. IEEE Wirel Commun 21(6):54–62

72. Aijaz A, Aghvami H, Amani M (2013) A survey on mobile data offloading: technical and business perspectives. IEEE Wirel Commun 20(2):104–112

73. Tabrizi H, Farhadi G, Cioffi J (2011) A learning-based network selection method in heterogeneous wireless systems. In: Global Telecommunications Conference, pp 1–5

74. Yoon SG, Han J, Bahk S (2012) Low-duty mode operation of femto base stations in a densely deployed network environment. In: IEEE international symposium on personal, indoor and mobile radio communications, pp 636–641

75. Chen M (2015) Research on key technologies of reconfigurable computing for communication baseband signal processing. Southeast University

76. Poston JD, Horne WD (2005) Discontiguous OFDM considerations for dynamic spectrum access in idle TV channels. In: IEEE international symposium on new frontiers in dynamic spectrum access networks, pp 607–610

77. Keller T, Hanzo L (2000) Adaptive modulation techniques for duplex OFDM transmission. IEEE Trans Veh Technol 49(5):1893–1906

78. Truong KT, Heath RW (2013) Effects of channel aging in massive MIMO systems. J Commun Networks 15(4):338–351

79. Choi J, Chance Z, Love DJ et al (2013) Noncoherent trellis coded quantization: a practical limited feedback technique for massive MIMO systems. IEEE Trans Commun 61(12):5016–5029

80. Li K, Sharan R, Chen Y et al (2017) Decentralized Baseband Processing for Massive MU-MIMO Systems. IEEE J Emerg Sel Top Circuits Syst PP(99):1

81. Roger S, Ramiro C, Gonzalez A et al (2012) Fully parallel GPU implementation of a fixed-complexity soft-output MIMO detector. IEEE Trans Veh Technol 61(8):3796–3800

82. Guenther D, Leupers R, Ascheid G (2016) Efficiency enablers of lightweight SDR for MIMO baseband processing. IEEE Trans Very Large Scale Integr Syst 24(2):567–577

83. Winter M, Kunze S, Adeva EP et al (2012) A 335 Mb/s 3.9 mm 265 nm CMOS flexible MIMO detection-decoding engine achieving 4G wireless data rates. In: Solid-state circuits conference digest of technical papers, pp 216–218

84. Noethen B, Arnold O, Perez Adeva E et al (2014) 10.7 A 105GOPS 36 mm 2 heterogeneous SDR MPSoC with energy-aware dynamic scheduling and iterative detection-decoding for 4G in 65 nm CMOS. In: Solid-state circuits conference digest of technical papers, pp 188–189

85. Chen C, Tang W, Zhang Z (2015) 18.7 A 2.4 mm 2 130 mW MMSE-nonbinary-LDPC iterative detector-decoder for 4 × 4 256-QAM MIMO in 65 nm CMOS. In: Solid-state circuits conference, pp 1–3

86. Studer C, Fateh S, Seethaler D (2011) ASIC implementation of soft-input soft-output MIMO detection using MMSE parallel interference cancellation. IEEE J Solid-State Circuits 46(7):1754–1765

87. Tang W, Prabhu H, Liu L et al (2018) A 1.8 Gb/s 70.6 pJ/b 128x16 Link-Adaptive Near-Optimal Massive MIMO Detector in 28 nm UTBB-FDSOI. In: Solid-state circuits conference digest of technical papers, pp 60–61

88. Prabhu H, Rodrigues JN, Liu L et al (2017) 3.6 A 60 pJ/b 300 Mb/s 128 × 8 massive MIMO precoder-detector in 28 nm FD-SOI. In: Solid-state circuits conference digest of technical papers, pp 60–61

89. Chen YT, Cheng CC, Tsai TL et al (2017) A 501 mW 7.61 Gb/s integrated message-passing detector and decoder for polar-coded massive MIMO systems. In: VLSI Circuits, pp C330–C331

90. Tang W, Chen CH, Zhang ZA 0.58 mm 2 2.76 Gb/s 79.8 pJ/b 256-QAM massive MIMO message-passing detector. In: VLSI Circuits, pp 1–2

91. Todman TJ, Constantinides GA, Wilton SJE et al (2005) Reconfigurable computing: Architectures and design methods. IEE Proc—Comput Digital Tech 152(2):193–207

92. Shaojun W, Leibo L, Shouyi Y (2014) Reconfigurable computing. Science Press

93. Liu L, Li Z, Chen Y et al (2017) HReA: an energy-efficient embedded dynamically reconfigurable fabric for 13-dwarfs processing. IEEE Trans. Circuits Syst II Express Briefs PP(99):1

94. Liu L, Wang J, Zhu J et al (2016) TLIA: efficient reconfigurable architecture for control-intensive kernels with triggered-long-instructions. IEEE Trans Parallel Distrib Syst 27(7):2143–2154

95. Radunovic B, Milutinovic VM (1998) A survey of reconfigurable computing architectures. In: International workshop on Field programmable logic and applications, pp 376–385

96. Atak O, Atalar A (2013) BilRC: an execution triggered coarse grained reconfigurable architecture. IEEE Trans Very Large Scale Integr Syst 21(7):1285–1298

97. Liu L, Chen Y, Yin S et al (2017) CDPM: context-directed pattern matching prefetching to improve coarse-grained reconfigurable array performance. IEEE Trans Comput-Aided Des Integr Circuits Syst PP(99):1

98. Estrin G (1960) Organization of computer systems-the fixed plus variable structure computer. In: AFIPS, pp 3–40

99. Zhang C, Liu L, Marković D et al (2015) A heterogeneous reconfigurable cell array for MIMO signal processing. IEEE Trans Circuits Syst I Regul Pap 62(3):733–742

100. Bougard B, Sutter BD, Verkest D et al (2008) A coarse-grained array accelerator for software-defined radio baseband processing. Micro IEEE 28(4):41–50

101. Ahmad U, Li M, Appeltans R et al (2013) Exploration of lattice reduction aided soft-output MIMO detection on a DLP/ILP baseband processor. IEEE Trans Signal Process 61(23):5878–5892

102. Chen X, Minwegen A, Hassan Y et al (2015) FLEXDET: flexible, efficient multi-mode MIMO detection using reconfigurable ASIP. IEEE Trans Very Large Scale Integr Syst 23(10):2173–2186

Chapter 2
Linear Massive MIMO Detection Algorithm

Massive MIMO signal detection is the key technology of next generation wireless communication (such as 5G) [1], and how to detect the transmitted signal from the mass MIMO system efficiently and accurately is of vital importance. As for massive MIMO signal detection, there are many algorithms to implement the signal detection. Generally, these algorithms can be divided into the linear detection algorithm and the nonlinear detection algorithm according to different calculation methods [2]. Although the linear detection algorithm is less accurate than the nonlinear detection algorithm, it is still a practical signal detection method for massive MIMO system in some cases due to its low complexity. In the linear detection algorithm, the difficulty people often encounter is the calculation to find the inverse matrix of a large-scale matrix, especially when the scale of a massive MIMO system is very large, the algorithm complexity is very high and the corresponding hardware is difficult to implement. Therefore, this chapter introduces several typical linear iterative algorithms for massive MIMO signal detection. Using these algorithms, the iterations between vectors or matrices can be effectively used to avoid direct inversion of large-scale matrices and reduce complexity of the linear detection algorithm. In the following sections, we will introduce Neumann series approximation (NSA) algorithm, the Chebyshev iteration algorithm, the Jacobi iteration algorithm and the Conjugate gradient (CG) algorithm respectively. And the optimization methods of the Chebyshev iteration algorithm, the Jacobi iteration algorithm and the CG algorithm are also introduced to get better linear detection algorithms. In addition, this chapter also compares the complexity and accuracy of these algorithms with those of other massive MIMO signal detection algorithms.

2.1 Analysis of Linear Detection Algorithm

Massive MIMO signal detection algorithms are usually divided into the linear detection algorithm and the nonlinear detection algorithm. The nonlinear detection algorithm, as the name implies, refers to the algorithm that adopts the nonlinear algorithm

© Springer Nature Singapore Pte Ltd. and Science Press, Beijing, China 2019   71
L. Liu et al., *Massive MIMO Detection Algorithm and VLSI Architecture*,
https://doi.org/10.1007/978-981-13-6362-7_2

to recover the transmitted signal s from the received signal y. Such algorithms usually have high accuracy, but the computation complexity is also high. For example, the maximum likelihood (ML) detection is a typical nonlinear detection algorithm [3]. In theory, the ML detection is very ideal for massive MIMO signal detection with high accuracy. However, in the ML detection, the number of cycles required for computation depends largely on the modulation order q and the number of the user antennas (N_t). The total number of cycles is denoted as q^{N_t}. Undoubtedly, this result is disastrous undoubtedly because the total number of cycles will still increase considerably even if the modulation order or the number of user antennas increases very little. Therefore, ML detection is not applicable in practice although it is a very ideal detection method in theory, especially in massive MIMO signal detection. In general, the nonlinear detection algorithm is more accurate than the linear detection algorithm in massive MIMO signal detection, but with higher complexity. We will introduce the nonlinear detection algorithm in detail Chap. 4.

Corresponding to the nonlinear detection algorithm, the linear detection algorithm usually estimates the signal s by operating a matrix. The common linear detection algorithms include the zero-forcing (ZF) detection algorithm and the minimum mean square error (MMSE) detection method [4], both of which transform massive MIMO signal detection into linear matrix equation solution by deforming the channel matrix H, that is $Hs = y$. According to the massive MIMO channel model in Sect. 1.2.1, Eq. (2.1) for both the received signal and the transmitted signal is

$$y = Hs + n \tag{2.1}$$

where y is the received signal, H is the channel matrix, s is the transmitted signal, and n is the additive noise. The linear detection algorithm of massive MIMO is to simultaneously left multiply the both sides of Eq. (2.1) by conjugate transpose H^H of the channel matrix, neglecting the additive noise n, so as to obtain Eq. (2.2)

$$H^H y = H^H H s \tag{2.2}$$

If Eq. (2.3) is true, then we obtain Eq. (2.4).

$$y^{MF} = H^H y \tag{2.3}$$

$$s = (H^H H)^{-1} H^H y = (H^H H)^{-1} y^{MF} \tag{2.4}$$

The detection for the transmitted signal s is implemented. However, there is an error in Eq. (2.4) due to the existence of the additive noise n. Based on the above idea, the transmitted signal s can be estimated by a matrix W which makes Eq. (2.5) hold

$$\hat{s} = Wy \tag{2.5}$$

where the $\hat{s}$ denotes the estimated transmitted signal. In this way, the massive MIMO linear detection is transformed into the estimation of matrix W.

The ZF is a common linear detection algorithm. Its main idea is to neglect the additive noise n in the massive MIMO channel model in the analysis, which will make the massive MIMO detection algorithm much simpler and easier to implement. However, considering that the noise is usually not negligible in actual situation, the result obtained by using this algorithm may not be the optimal solution. For a massive MIMO system with a scale of $N_r \times N_t$, the signal received by a receiving antenna at the base station can be expressed as

$$y = \sum_{i=1}^{N_t} h_i s_i + n, i = 1, 2, \ldots, N_t \tag{2.6}$$

where $H = (h_1, h_2, \ldots, h_{N_t})$, s_i is the ith element of the transmitted signal s, $i = 1, 2, \ldots, N_t$. Now we define a vector $w_{i,1 \times N_r}$ that follows Eq. (2.7)

$$w_i h_j = \begin{cases} 1, i = j \\ 0, i \neq j \end{cases} \tag{2.7}$$

where $i = 1, 2, \ldots, N_t$, w_i will be acted as rows to form a matrix $W_{N_t \times N_r}$, There is obviously $WH = I$ from Eq. (2.7). And combining with Eq. (2.4), we have

$$W = (H^H H)^{-1} H^H \tag{2.8}$$

In this way, the transmitted signal s can be estimated as

$$\hat{s} = W(Hs + n) = s + Wn \tag{2.9}$$

Obviously, when the additive noise is $n = 0$, $\hat{s} = s$ is strictly satisfied. Because the ZF detection algorithm meets the conditions in Eq. (2.7), it can eliminate the interference between the data sent by different transmitting antennas, and can get relatively accurate detection results when the signal-to-noise ratio is relatively high.

Although there are some deficiencies in the accuracy of the ZF detection algorithm, its derivation process also provides some ideas whether the influence of noise n can be added to the W matrix to simplify the solution of the transmitted signal s using the same method of solving linear matrix. On this basis, a MMSE detection algorithm is proposed.

The MMSE detection algorithm is another typical linear detection algorithm. Its basic idea is to make the estimated signal $\hat{s} = Wy$ as close to the real value as possible. In the MMSE detection algorithm, the adopted objective function is

$$\hat{s} = W_{MMSE} = \arg \min_W E \|s - Wy\|^2 \tag{2.10}$$

By solving the matrix W based on Eq. (2.10), the following equation can be obtained:

$$
\begin{aligned}
W_{\mathrm{MMSE}} &= \arg \min_{W} E \|s - Wy\|^2 \\
&= \arg \min_{W} E\{(s - Wy)^{\mathrm{H}}(s - Wy)\} \\
&= \arg \min_{W} E\{\mathrm{tr}[(s - Wy)(s - Wy)^{\mathrm{H}}]\} \\
&= \arg \min_{W} E\{\mathrm{tr}[ss^{\mathrm{H}} - sy^{\mathrm{H}}W^{\mathrm{H}} - Wys^{\mathrm{H}} + Wyy^{\mathrm{H}}W^{\mathrm{H}}]\} \qquad (2.11)
\end{aligned}
$$

Find the partial derivatives for Eq. (2.11) and set it equal to zero, and get

$$
\frac{\partial \mathrm{tr}[ss^{\mathrm{H}} - sy^{\mathrm{H}}W^{\mathrm{H}} - Wys^{\mathrm{H}} + Wyy^{\mathrm{H}}W^{\mathrm{H}}]}{\partial W} = 0 \qquad (2.12)
$$

By solving Eq. (2.12), we can get

$$
W = \left(H^{\mathrm{H}}H + \frac{N_0}{E_{\mathrm{s}}}I_{N_{\mathrm{t}}} \right)^{-1} H^{\mathrm{H}} \qquad (2.13)
$$

where N_0 is the spectral density of noise, E_{s} is the spectral density of signal, and the Gram matrix is defined. Similar to the ZF detection algorithm, we can deduce the MMSE detection algorithm. It also makes the estimated signal $\hat{s} = Wy$, which is different from the ZF detection algorithm. $W_{\mathrm{ZF}} = G^{-1}H^{\mathrm{H}}$ in the ZF detection algorithm and $W_{\mathrm{MMSE}} = \left(G + \frac{N_0}{E_{\mathrm{s}}}I_{N_{\mathrm{t}}} \right)^{-1} H^{\mathrm{H}}$ in the MMSE detection algorithm. Both algorithms can estimate the transmitted signal s in massive MIMO detection.

Whether in the ZF detection algorithm or in the MMSE detection algorithm, matrix inverse is involved. The scale of the channel matrix H is usually very large in massive MIMO system. But the matrix inversion is more complex, which is usually difficult to achieve in the actual signal detection circuit. In order to avoid the huge complexity of matrix inversion, many algorithms have been put forward to reduce the algorithm complexity so as to reduce the perplexities caused by large-scale matrix inversion. Among these algorithms, the typical ones are the NSA algorithm, the Chebyshev iteration algorithm, the Jacobi iteration algorithm, the CG algorithm and so on. The above four common algorithms will be introduced below in detail.

2.2 Neumann Series Approximation Algorithm

In the massive MIMO system, the number of receiving antennas is usually much greater than that of the user antennas [5], that is, N_{r} is much larger than N_{t}. Since the elements in channel matrix H are independent identically distributed, and the real and imaginary parts are subject to the Gaussian distribution with parameter N(0, 1), both Gram matrix G and MMSE matrix $A = G + N_0/E_sI_{N_t}$ are diagonally dominant matrices. Gram matrix G tends to be a scalar matrix when Nr tends to infinity, i.e., $G \rightarrow N_rI_{N_t}$ [5]. Using this property, we can simplify the inverse process of matrix in the linear massive MIMO detection algorithm.

Because of the diagonal dominance of the MMSE matrix A, if taking the diagonal elements of matrix A out and denoting it as matrix D, it is very easy to seek the inverse of matrix D. As mentioned above, the condition that A tends to D is that the number of user antennas N_t tends to infinity. However that is not actually possible. Even it is often far from that condition. Thus, we use $A \approx D$ to approximate the matrix A and finding the inverse of the matrix A will lead to relatively large errors.

2.2.1 Algorithm Design

Using Neumann series [6] can get the exact matrix inverse. The Neumann series expands the inverse matrix A^{-1} of the MMSE matrix A into

$$A^{-1} = \sum_{n=0}^{\infty} (X^{-1}(X - A))^n X^{-1} \qquad (2.14)$$

In the above Eq. (2.14) the matrix X need to satisfy

$$\lim_{n \to \infty} (I - X^{-1}A)^n = \mathbf{0}_{N_t \times N_t} \qquad (2.15)$$

The matrix A is decomposed into the sum of diagonal elements and non-diagonal elements $A = D + E$. Bring the sum into Eq. (2.14), we can get

$$A^{-1} = \sum_{n=0}^{\infty} (D^{-1}E)^n D^{-1} \qquad (2.16)$$

According to Eq. (2.15), if the condition $\lim_{n \to \infty} \left(-D^{-1}E\right)^n = \mathbf{0}_{N_t \times N_t}$ is satisfied, the expansion of Eq. (2.16) will definitely converge.

Expanding the matrix A^{-1} into a sum of numerous terms is not practical. In order to apply Neumann series to the linear massive MIMO detection algorithm, it is necessary to use NSA to estimate Eq. (2.16) [7]. The main idea of solving inverse matrix with NSA is to take out the first K items of Neumann series in Eq. (2.16), then the expression of calculating the first K terms Neumann series is

$$\tilde{A}_K^{-1} = \sum_{n=0}^{K-1} (D^{-1}E)^n D^{-1} \qquad (2.17)$$

By using Eq. (2.17), Neumann series with a certain number of terms can be calculated to change the infinite number of terms into the finite number of terms, so as to reduce the computational complexity and estimate the inverse of MMSE matrix A. By approximating A^{-1}, an approximate MMSE equilibrium matrix $\tilde{W}_K^{-1} = \tilde{A}_K^{-1} H^H$ can be obtained. A^{-1} can be expressed in different expressions according to the number of selected items K. When $K = 1, \tilde{A}_1^{-1} = D^{-1}$, now $\tilde{W}_1^{-1} = D^{-1} H^H$. When $K = 2, \tilde{A}_2^{-1} = D^{-1} + D^{-1} E D^{-1}$, the operation complexity is $O(N_t^2)$. When $K = 3$, $\tilde{A}_3^{-1}$ is

$$\tilde{A}_3^{-1} = D^{-1} + D^{-1} E D^{-1} + D^{-1} E D^{-1} E D^{-1} \tag{2.18}$$

The computational complexity of Eq. (2.18) is $O(N_t^3)$, which is comparable to the actual computational complexity of the inverse matrix, but the approximate operation of Eq. (2.18) is less. When $K > 4$, the complexity of the inverse computation of the actual matrix may be lower than that of the approximate algorithm.

2.2.2 Error Analysis

Obviously, the sum of the preceding K items in Eq. 2.17 can cause errors, which can be expressed as

$$\begin{aligned}
\Delta_K &= \tilde{A}^{-1} - \tilde{A}_K^{-1} \\
&= \sum_{n=K}^{\infty} (-D^{-1} E)^n D^{-1} \\
&= (-D^{-1} E)^K \sum_{n=0}^{\infty} (-D^{-1} E)^n D^{-1} \\
&= (-D^{-1} E)^K A^{-1}
\end{aligned} \tag{2.19}$$

By using Eq. (2.17) to estimate the transmitted signal and substituting Eq. (2.17) into Eq. (2.19), the expression about the error of the transmitted signal of Eq. (2.20) can be obtained

$$\hat{s}_K = \tilde{A}_K^{-1} H^H y = A^{-1} y^{MF} - \Delta_K y^{MF} \tag{2.20}$$

By taking the second norm of the second term of Eq. (2.20), we have

$$\begin{aligned}
\left\| \Delta_K y^{MF} \right\|_2 &= \left\| (-D^{-1} E)^K A^{-1} y^{MF} \right\|_2 \\
&\leq \left\| (-D^{-1} E)^K \right\|_F \left\| A^{-1} y^{MF} \right\|_2
\end{aligned}$$

$$\leq \left\| -\boldsymbol{D}^{-1}\boldsymbol{E} \right\|_F^K \left\| \boldsymbol{A}^{-1}\boldsymbol{y}^{\mathrm{MF}} \right\|_2 \tag{2.21}$$

We can see from Eq. (2.21) that if the condition of inequality (2.22) is satisfied in Eq. (2.21), the approximate error exponent approaches zero as the number of terms K increases, and it can be proved that inequality (2.22) is a sufficient condition for the convergence of formula (2.16).

$$\left\| -\boldsymbol{D}^{-1}\boldsymbol{E} \right\|_F < 1 \tag{2.22}$$

Now it is necessary to prove that Eq. (2.16) converges when the scale of the massive MIMO system satisfies the condition that N_r is much greater than N_t, the elements in the channel matrix $\boldsymbol{H}$ are independent and identically distributed, and all of them obey the complex Gaussian distribution with parameter N (0, 1). More specifically, it is necessary to prove that the condition of convergence of Neumann series and the condition of minimal error in Eq. (2.21) are only related to N_t and N_r. Here is a theorem and corresponding proof below.

Theorem 2.2.1 *When $N_r > 4$, the elements in the channel matrix $\boldsymbol{H}$ are independent of each other and satisfy the complex Gaussian distribution with its variance being 1, the following expression is obtained*

$$P\left\{ \left\| -\boldsymbol{D}^{-1}\boldsymbol{E} \right\|_F^K < \alpha \right\} \geq 1 - \frac{(N_t^2 - N_t)}{\alpha^{\frac{2}{K}}} \sqrt{\frac{2N_r(N_r + 1)}{(N_r - 1)(N_r - 2)(N_r - 3)(N_r - 4)}} \tag{2.23}$$

Proof To prove Theorem 2.2.1, we need to give three other lemmas and their proofs.

Lemma 2.2.1 *Let $x^{(k)}$, $y^{(k)}(k = 1, 2, \cdots, N_r)$ independent and identically distributed, and make them satisfy the complex Gauss distribution with its variance being 1.*

$$E\left[\left| \sum_{k=1}^{N_r} x^{(k)}y^{(k)} \right|^4 \right] = 2N_r(N_r + 1) \tag{2.24}$$

Proof First we can get

$$E\left[\left| \sum_{k=1}^{N_r} x^{(k)}y^{(k)} \right|^4 \right] = E\left[\left(\sum_{k=1}^{N_r} x^{(k)}y^{(k)} \sum_{k=1}^{N_r} \left(x^{(k)}y^{(k)} \right)^* \right)^2 \right]$$

$$= \binom{N_r}{2} E\left[\left| x^{(k)} \right|^2 \left| y^{(k)} \right|^2 \right] + N_r E\left[\left| x^{(k)} \right|^4 \left| y^{(k)} \right|^4 \right]$$

$$= 2N_r(N_r - 1) + 4N_r$$

$$= 2N_r^2 + 2N_r \tag{2.25}$$

The operation process in Eq. (2.25) can be described as the following steps. The nonzero term can be expressed as $\left|x^{(k)}\right|^4\left|y^{(k)}\right|^4$ and $\left|x^{(k)}\right|^2\left|y^{(k)}\right|^2$ after the Quadratic term is expanded, in which there are a total of N_r items $\left|x^{(k)}\right|^4\left|y^{(k)}\right|^4$ and $\binom{N_r}{2}$ items $\left|x^{(k)}\right|^2\left|y^{(k)}\right|^2$. According to $\mathrm{E}\left[\left|x^{(k)}\right|^4\right] = \mathrm{E}\left[\left|y^{(k)}\right|^4\right] = 2, \mathrm{E}\left[\left|x^{(k)}\right|^2\right] = \mathrm{E}\left[\left|y^{(k)}\right|^2\right] = 1$, we can get the conclusion of Lemma 2.2.1.

Lemma 2.2.2 Let $N_r > 4$, and $x^{(k)} (k = 1, 2, \cdots, N_r)$ *is independent and identically distributed, obey the complex Gauss distribution with the variance being 1, and* $g = \sum_{k=1}^{N_r} \left|x^{(k)}\right|^2$, *then*

$$E\left[\left|g^{-1}\right|^4\right] = ((N_r - 1)(N_r - 2)(N_r - 3)(N_r - 4))^{-1} \tag{2.26}$$

Proof First, g is rewritten as

$$g = \frac{1}{2}\sum_{k=1}^{2N_r} \left|s^{(k)}\right|^2 \tag{2.27}$$

Among them, $s^{(k)}$ is independent and identically distributed, also follows the real Gauss distribution with the mean being 0 and the variance being 1. Therefore, $2g^{-1}$ obeys the inverse χ^2 distribution with $2N_t$ free degrees. The inverse χ^2 distribution with $2N_r$ free degrees corresponds to the inverse Gaussian distribution with $2N_t$ free degrees. The fourth inverse χ^2 distribution can be obtained by Eq. (2.28):

$$E\left(\left|2g^{-1}\right|^4\right) = \frac{16}{(N_r - 1)(N_r - 2)(N_r - 3)(N_r - 4)} \tag{2.28}$$

Then we have the conclusion from Eq. (2.26).

Lemma 2.2.3 Let $N_t > 4$, *and the elements in the channel matrix H satisfy the complex Gaussian distribution with the independent identically distributed zero mean and unit variance.*

$$E\left[\left\|\boldsymbol{D}^{-1}\boldsymbol{E}\right\|_F^2\right] \leq \left(N_t^2 - N_t\right)\sqrt{\frac{2N_r(N_r + 1)}{(N_r - 1)(N_r - 2)(N_r - 3)(N_r - 4)}} \tag{2.29}$$

Proof The normalized Gram matrix $\boldsymbol{G}$ corresponds to matrix $\boldsymbol{A}$, $\boldsymbol{A} = \boldsymbol{D} + \boldsymbol{E} = \boldsymbol{G} + \frac{N_0}{E_s}\boldsymbol{I}_{N_t}$. Therefore, the element of ith row and jth column in matrix $\boldsymbol{A}$ can be expressed as

$$a^{(i,i)} = \begin{cases} g^{(i,i)} = \sum_{k=1}^{N_r} (h^{(k,i)})^* h^{(k,j)}, & i \neq j \\ g^{(i,i)} + \frac{N_0}{E_s} = \sum_{k=1}^{N_r} \left|h^{(k,i)}\right|^2 + \frac{N_0}{E_s}, & i = j \end{cases} \tag{2.30}$$

where $g^{(i,j)}$ is the element of ith row and jth column of the matrix $\boldsymbol{G}$. Therefore, we can find the inequality in Eq. (2.31):

$$E\left[\left\|\boldsymbol{D}^{-1}\boldsymbol{E}\right\|_F^2\right] = E\left[\sum_{i=1}^{i=N_t}\sum_{j=1,i\neq j}^{j=N_t}\left|\frac{g^{(i,j)}}{a^{(i,i)}}\right|^2\right] \leq \sum_{i=1}^{i=N_t}\sum_{j=1,i\neq j}^{j=N_t}E\left|\frac{g^{(i,j)}}{g^{(i,i)}}\right|^2 \qquad (2.31)$$

Then use the Cauchy–Schwartz inequality for Eq. (2.31), there is

$$E\left[\left\|\boldsymbol{D}^{-1}\boldsymbol{E}\right\|_F^2\right] \leq \sum_{i=1}^{i=N_t}\sum_{j=1,i\neq j}^{j=N_t}\sqrt{E\left[\left|g^{(i,j)}\right|^4\right]E\left[\left|\left(g^{(i,i)}\right)^{-1}\right|^4\right]} \qquad (2.32)$$

Use Lemmas 2.2.2 and 2.2.3 in the calculation of first and second moments respectively, get the following expression:

$$E\left[\left\|\boldsymbol{D}^{-1}\boldsymbol{E}\right\|_F^2\right] \leq \sum_{i=1}^{i=N_t}\sum_{j=1,i\neq j}^{j=N_t}\sqrt{\frac{2N_r(N_r+1)}{(N_r-1)(N_r-2)(N_r-3)(N_r-4)}}$$

$$= \left(N_t^2 - N_t\right)\sqrt{\frac{2N_r(N_r+1)}{(N_r-1)(N_r-2)(N_r-3)(N_r-4)}} \qquad (2.33)$$

Now let uss prove Theorem 2.2.1. Using Markov's inequality, we obtain

$$P\left\{\left[\left\|\boldsymbol{D}^{-1}\boldsymbol{E}\right\|_F^K \geq \alpha\right]\right\} = P\left\{\left[\left\|\boldsymbol{D}^{-1}\boldsymbol{E}\right\|_F^K \geq \alpha^{\frac{2}{K}}\right]\right\} \leq \alpha^{-\frac{2}{K}}E\left[\left\|\boldsymbol{D}^{-1}\boldsymbol{E}\right\|_F^2\right] \qquad (2.34)$$

Combining the upper bound of $P\left\{\left[\left\|\boldsymbol{D}^{-1}\boldsymbol{E}\right\|_F^K < \alpha\right]\right\} = 1 - P\left\{\left[\left\|\boldsymbol{D}^{-1}\boldsymbol{E}\right\|_F^K \geq \alpha\right]\right\}$ and $E\left[\left\|\boldsymbol{D}^{-1}\boldsymbol{E}\right\|_F^2\right]$ in Lemma 2.2.3, we can get the conclusion of Theorem 2.2.1.

As we can see from Eq. (2.23), when $N_r \gg N_t$, the probability of the condition (2.22) satisfied is greater. This theorem shows that Neumann series converges with a certain probability, the greater the ratio of N_r/N_t, the greater the probability of convergence. In addition, the theorem also provides α condition for minimizing the error of residual estimation and when alpha is less than 1, the greater the number of terms K selected, the greater the probability of convergence.

2.2.3 Complexity and Block Error Rate

The NSA reduces the computational complexity of inverse matrix. Now we discuss the advantages and limitations of the NSA from two aspects of computational complexity and block error rates. Here, we only consider the case that N_r is 64, 128, and 256 respectively.

In the exact algorithm to solve the inverse matrix, the Cholesky decomposition (CHD) algorithm [8] has lower complexity than other exact solutions, such as direct matrix inversion, QR decomposition, LU decomposition [3, 9], etc. Therefore, the CHD algorithm can be selected as the object to be compared with the NSA. The complexity of the CHD algorithm for inverse matrix solution is in the in the $O(N_t^3)$ range, while the complexity of the NSA for inverse matrix is in the $O(N_t)$ range and the $O(N_t^2)$ range respectively when $K = 1$ and $K = 2$. When $K > 3$, the complexity of the NSA mainly comes from the multiplication between large matrices, and the complexity increases linearly with the value of K. For example, when $K = 3$, there is one multiplication between large matrices once in the algorithm, and when $K = 4$, there are two multiplications between large matrices, that is, when $K > 3$, you need to calculate $K - 2$ multiplications between large matrices in the NSA. So when $K > 3$, the complexity of the NSA algorithm is $O\big((K - 2)N_t^3\big)$. It can be seen that when $K \geq 3$, the NSA has no advantage in complexity compared with the CHD algorithm.

The complexity of an algorithm mainly depends on the number of real number multiplications in the algorithm. Figure 2.1 depicts the variation curve of the number of real number multiplications with the number of antennas N_t in the CHD algorithm and the NSA algorithm with different K values. Figure 2.1 shows that the complexity of the NSA algorithm is lower than that of the CHD algorithm at $K \leq 3$, while the complexity of the Newman series approximation algorithm is higher when $K > 3$.

NSA is the solution of approximating matrix inversion by taking the first K terms of the Neumann series. Obviously, the more the number of terms is taken, the closer it is to the exact result, but the cost is the increase in complexity. Thus, the precision and complexity are a pair of contradictions. In order to compare the block error rate between Neuman series approximation and Holesky decomposition algorithm, the uplink of massive MIMO system is selected here. At the base station, MMSE detection using the above NSA and CHD algorithm is adopted, and SNR $= N_r \frac{E_s}{N_0}$ is defined. Figure 2.2 shows at a different N_r, the block error rate of the NSA and the CHD algorithm when N_t is equal to 4, 8, and 12 respectively.

As we can see from Fig. 2.2, when $K = 1$ or $K = 2$, the NSA algorithm has a large block error rate, when the number of antennas at the base station is large, it can make up for part of block error rate. Considering the requirement of 10% block error rate [10] in LTE, it is not suitable for practical applications when $K = 1$ and $K = 2$ with modulation order of 64-QAM. The simulation result shows that the block error rate is less than 10^{-2} when $K = 1$, $N_r = 512$, $N_t = 4$, and the number of terms required by the NSA algorithm is fewer when the modulation order is 16-QAM. When the modulation order is 64-QAM and $K = 3$, the result of the NSA algorithm is close to that of the CHD algorithm. For example, when $K = 3$, $N_t = 4$ and $K = 3$, $N_t = 8$,

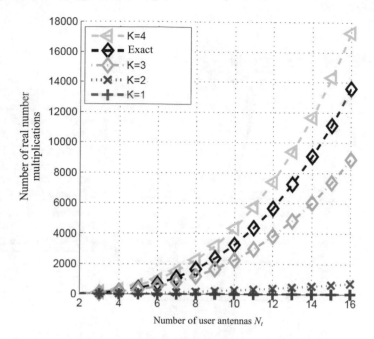

Fig. 2.1 The Relationship between the number of the user antennas N_t and the number of the real number multiplications in the algorithm. © [2018] IEEE. Reprinted, with permission, from Ref. [7]

$N_r = 256$, and the block error rate is 10^{-2}, the SNR loss of NSA algorithm is less than 0.25 dB. Therefore, when the N_r/N_t ratio of the massive MIMO system is large, the NSA algorithm is used to find the inverse of the matrix and the term K is 3, so that the lower block error rate can be reduced under the condition of low computational complexity.

In summary, in the massive MIMO system, when the N_r/N_t ratio is small, the CHD algorithm and other exact arithmetic are required to seek the inverse of the matrix. When N_r/N_t ratio is large, the NSA algorithm can be used to approximate the inversion. By using the NSA algorithm, we can find the relatively accurate inverse matrix results with low computational complexity. This makes the NSA an efficient and accurate method for massive MIMO detection in some specific cases.

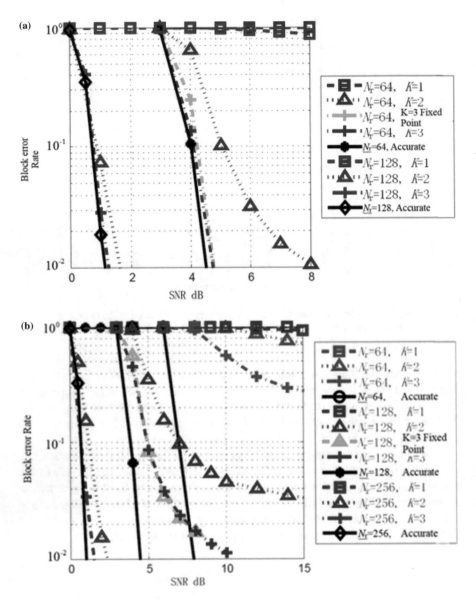

Fig. 2.2 Block error Rate Curve at **a** $N_t = 4$, **b** $N_t = 8$, **c** $N_t = 12$. © [2018] IEEE. Reprinted, with permission, from Ref. [7]

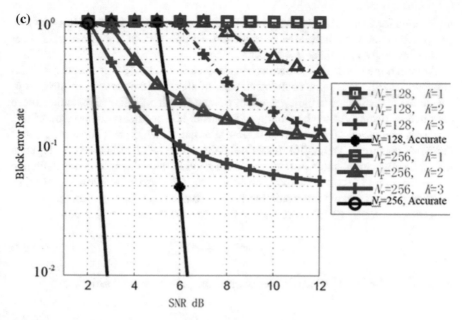

Fig. 2.2 (continued)

2.3 Chebyshev Iteration Algorithm

2.3.1 Algorithm Design

The Chebyshev iteration algorithm [11] is an algorithm for solving matrix equation $Ax = b$ by using iteration computation to avoid large matrix inversion. Its basic iteration form is

$$x^{(K)} = x^{(K-1)} + \sigma^{(K)} \tag{2.35}$$

where σ is the correction matrix and K is the number of iterations. The σ can be expressed as

$$\sigma^{(0)} = \frac{1}{\beta} r^{(0)} \tag{2.36}$$

$$\sigma^{(K)} = \rho^{(K)} r^{(K)} + \varphi^{(K)} \sigma^{(K-1)} \tag{2.37}$$

where $r^{(K)} = b - Ax^{(K)}$ denotes the residual vector, $\rho^{(K)}$ and $\varphi^{(K)}$ are the Chebyshev polynomial parameters for two iterations, and β is an iterative parameter related to the eigenvalue of matrix A. Therefore, the Chebyshev iteration can be used to solve the linear equation in the MMSI of massive MIMO detection, so as to avoid the operation complexity caused by large matrix inversion. In this section, for convenience, set $A = H^H H + \frac{N_0}{E_s} I_{N_t}$, so there is $A\hat{s} = y^{\text{MF}}$.

Although the Chebyshev iteration can be used in the MMSE of massive MIMO detection, it still faces some challenges. First of all, since the parameters such as β, $\rho^{(K)}$ and $\varphi^{(K)}$ are related to the eigenvalues of matrix A, it is difficult to calculate these parameters. Second, at the beginning of iteration, it is necessary to solve matrix A. Matrix A involves multiplication of large-scale matrix, which will consume a lot of hardware resources. Third, different initial values of iteration affect the convergence rate of the algorithm, and how to determine a good initial value is also the challenge of the algorithm. To solve the above problems, the Chebyshev iteration is optimized in this section to make it more suitable for massive MIMO signal detection.

According to Eq. (2.35), the iteration form in MMSE can be written as

$$\hat{s}^{(K)} = \hat{s}^{(K-1)} + \sigma^{(K)} \tag{2.38}$$

where $\hat{s}^{(0)}$ is the initial value of iteration, and the parameter $\sigma^{(K)}$ satisfies:

$$\sigma^{(K)} = \frac{1}{\beta}\left(y^{\text{MF}} - A\hat{s}^{(K)}\right) \tag{2.39}$$

In order to reduce computational complexity, matrix A and $\hat{s}^{(K)}$ in Eq. (2.39) can be split into

$$A\hat{s}^{(K)} = \frac{N_0}{E_s}\hat{s}^{(K)} + H^H\left(H\hat{s}^{(K)}\right) \tag{2.40}$$

In Eqs. (2.36) and (2.37), the parameters $\rho^{(K)}$ and $\varphi^{(K)}$ can be expressed as

$$\rho^{(K)} = \frac{2\alpha}{\beta}\frac{T_K(\alpha)}{T_{K+1}(\alpha)} \tag{2.41}$$

$$\varphi^{(K)} = \frac{T_{K-1}(\alpha)}{T_{K+1}(\alpha)} \tag{2.42}$$

where T is Chebyshev polynomial, and α is the parameter related to the eigenvalues of matrix A. According to Chebyshev polynomial [11], the expression of T is

$$T_0(\alpha) = 1 \tag{2.43}$$

$$T_1(\alpha) = \alpha \tag{2.44}$$

$$T_K(\alpha) = 2\alpha T_{K-1}(\alpha) - T_{K-2}(\alpha), \quad K \geq 2 \tag{2.45}$$

Combining formula (2.41) and (2.42), we can get Eqs. (2.46)–(2.49):

$$\rho^{(1)} = \frac{2\alpha^2}{(2\alpha^2 - 1)\beta} \tag{2.46}$$

$$\rho^{(K)} = \frac{4\alpha^2}{4\alpha^2\beta - \beta^2\rho^{(K-1)}} \tag{2.47}$$

$$\varphi^{(1)} = \frac{1}{2\alpha^2 - 1} \tag{2.48}$$

$$\varphi^{(K)} = \frac{\beta^2\rho^{(K-1)}}{4\alpha^2\beta - \beta^2\rho^{(K-1)}} \tag{2.49}$$

where α and β satisfy:

$$\alpha = \frac{\lambda_{\max} + \lambda_{\min}}{\lambda_{\max} - \lambda_{\min}} \tag{2.50}$$

$$\beta = \frac{\lambda_{\max} + \lambda_{\min}}{2} \tag{2.51}$$

where $\lambda_{\max}$ and $\lambda_{\min}$ are the maximum and minimum eigenvalues of matrix A respectively. Since computing the eigenvalues of matrix A is complicated, an approximation is adopted here. As N_r and N_t increase, $\lambda_{\max}$ and $\lambda_{\min}$ can be approximately expressed

$$\lambda_{\max} \approx N_r\left(1 + \sqrt{\frac{N_t}{N_r}}\right)^2 \tag{2.52}$$

$$\lambda_{\min} \approx N_r\left(1 - \sqrt{\frac{N_t}{N_r}}\right)^2 \tag{2.53}$$

So far, all the parameters used in the Chebyshev iteration can be expressed in the scale parameters of the channel matrix H in the massive MIMO system. According to Eq. (2.38), we still need an iterative initial value if we want to use the Chebyshev iterative algorithm to estimate the signal s. Theoretically, although using any initial values we can get the final estimation, the convergence rates of the algorithm corresponding to different initial values are not the same. A good initial value can make the algorithm converge faster and generate the desired results, achieving twice the result with half the effort.

As described in Sect. 2.2, in the massive MIMO system, the number of receiving antennas is often much larger than the number of the user antennas, i.e., $N_r \gg N_t$, the elements in the channel matrix $\boldsymbol{H}$ are subject to Gaussian distribution that is independent and identically distributed and with parameters N(0, 1), so that matrix $\boldsymbol{A}$ is a diagonally dominant matrix and satisfies

$$A_{i,j} = \begin{cases} \frac{\lambda_{max} + \lambda_{min}}{2} = \beta, & i = j \\ 0, & i \neq j \end{cases} \tag{2.54}$$

So the initial value $\hat{s}^{(0)}$ can be approximated as

$$\hat{s}^{(0)} \approx \frac{1}{\beta} \boldsymbol{H}^H \boldsymbol{y} = \frac{2}{\lambda_{max} + \lambda_{min}} \boldsymbol{H}^H \boldsymbol{y} \tag{2.55}$$

This initial value enables the Chebyshev iteration to achieve a faster convergence rate, and the computational complexity of the initial value is very low. In addition, the computation of the initial value can be executed in parallel.

In massive MIMO signal detection, it is often necessary to output the log-likelihood ratio for the use of the next stage circuit, so it is necessary to discuss how to use Chebyshev iteration to find the approximate log-likelihood ratio. The estimated transmitted signal can be expressed as

$$\hat{s} = \boldsymbol{A}^{-1} \boldsymbol{H}^H \boldsymbol{y} = \boldsymbol{A}^{-1} \boldsymbol{H}^H \boldsymbol{H} \boldsymbol{s} + \boldsymbol{A}^{-1} \boldsymbol{H}^H \boldsymbol{n} \tag{2.56}$$

Set $\boldsymbol{X} = \boldsymbol{A}^{-1} \boldsymbol{H}^H \boldsymbol{H}$ and $\boldsymbol{Y} = \boldsymbol{X} \boldsymbol{A}^{-1}$, they can be used to solve equivalent channel gain and NPI respectively. In combination with Eqs. (2.37) and (2.38), the estimated received $\hat{s}$ can be expressed as.

$$\hat{s} \approx \hat{s}^{(K)} = \hat{s}^{(K-1)} + \rho^{(K-1)} \boldsymbol{r}^{(K-1)} + \varphi^{(K-1)} \sigma^{(K-2)}$$
$$= \left[(1 + \varphi^{(K-1)}) \boldsymbol{I}_{N_t} - \rho^{(K-1)} \boldsymbol{A} \right] \hat{s}^{(K-1)} - \varphi^{(K-1)} \hat{s}^{(K-2)} + \rho^{(K-1)} \boldsymbol{y}^{MF} \tag{2.57}$$

Set $\boldsymbol{y}^{MF} = \boldsymbol{e}_{(N_t, 1)}$, the iteration in Eq. (2.57) can be approximated as the inverse of matrix $\boldsymbol{A}$, $\boldsymbol{X}$, and $\boldsymbol{Y}$. For example

$$\tilde{\boldsymbol{A}}^{-1} \approx \left(\tilde{\boldsymbol{A}}^{-1} \right)^{(K)}$$
$$= \left[(1 + \varphi^{(K-1)}) \boldsymbol{I}_{N_t} - \rho^{(K-1)} \boldsymbol{A} \right] \left(\tilde{\boldsymbol{A}}^{-1} \right)^{(K-1)} - \varphi^{(K-1)} \left(\tilde{\boldsymbol{A}}^{-1} \right)^{(K-2)} + \rho^{(K-1)} \boldsymbol{I} \tag{2.58}$$

where $(\tilde{\boldsymbol{A}}^{-1})^{(0)} = \frac{1}{\beta} \boldsymbol{I}_{N_t}$, and all $(\tilde{\boldsymbol{A}}^{-1})^{(K)}$ are diagonal matrices. Similarly, matrices $\boldsymbol{X}$ and $\boldsymbol{Y}$ are shown as

$$\hat{X} \approx \hat{X}^{(K)}$$

$$= \left[\left(1 + \varphi^{(K-1)} \right) I_{N_t} - \rho^{(K-1)} A \right] \hat{X}^{(K-1)} - \varphi^{(K-1)} \hat{X}^{(K-2)} + \rho^{(K-1)} H^{\mathrm{H}} H \quad (2.59)$$

$$\hat{Y} \approx \hat{Y}^{(K)} = \hat{X} (\hat{W}^{-1})^{(K)} \quad (2.60)$$

The equivalent channel gain μ_i and NPI can be approximately expressed as

$$\mu_i = \hat{X}_{i,i} \approx \hat{X}_{i,i}^{(K)} \quad (2.61)$$

$$v_i^2 = \sum_{j \neq i}^{N_t} |X_{i,j}| E_s + Y_{i,i} N_0 \approx N_0 \hat{X}_{i,i}^{(K)} \tilde{A}_{i,i}^{(K)} \quad (2.62)$$

The signal-to-interference-plus-noise ratio (SINR) can be calculated by combining Eqs. (2.58)–(2.62). However, this algorithm has a high computational complexity. An algorithm based on initial eigenvalue solution is presented here to solve LLR, which can reduce the computational complexity of LLR [12]. The LLR is shown as below.

$$L_{i,b}(\hat{s}_i) = \gamma_i \left(\min_{s \in S_b^0} \left| \frac{\hat{s}_i}{\mu_i} - s \right|^2 - \min_{s \in S_b^1} \left| \frac{\hat{s}_i}{\mu_i} - s \right|^2 \right) \quad (2.63)$$

where the parameter γ_i meets

$$\gamma_i = \frac{\mu_i^2}{v_i^2} = \frac{(\hat{X}_{i,i}^{(K)})^2}{N_0 \hat{X}_{i,i}^{(K)} \tilde{A}_{i,i}^{(K)}} = \frac{\hat{X}_{i,i}^{(K)}}{N_0 \tilde{A}_{i,i}^{(K)}} \approx \frac{1}{\beta} \frac{1}{N_0} \quad (2.64)$$

where γ_i indicates the SINR of the ith user, S_b^0 and S_b^1 are the set of modulation points $Q(|Q| = 2^\vartheta)$ in the constellation diagram, and the bth bits of S_b^0 and S_b^1 are 0 and 1 respectively. For the sake of convenience, write $L_{i,b}$ as $L_{i,b}(\hat{s}_i) = \gamma_i \xi_b(\hat{s}_i)$ expressed as a form of linear equation. Here, the approximate SINR no longer depends on the result of $\hat{X}_{i,i}^{(K)}$ and $\tilde{A}_{i,i}^{(K)}$.

Based on the above analysis, we can obtain the optimized Chebyshev iterative algorithm that approximates an MMSE detector in a massive MIMO system and name it as parallelizable Chebyshev iteration (PCI). The algorithm [13] is shown in Algorithm 2.1.

Algorithm 2.1 The parallelizable Chebyshev iteration (PCI) algorithm for soft-output MMSE detection

---

Input :

1 : The Tayleigh flat-fading channel matrix, H of $N_r \times N_t$;

2 : The received signal vector at the BS, y of $N_r \times 1$;

3 : Number of iterations T;

Output:

    log-likelihood ratio(LLR) for each digit;

Initialization:

1 : $\lambda_{\max} = N_r \left(1 + \sqrt{\dfrac{N_t}{N_r}}\right)^2$, $\lambda_{\min} = N_r \left(1 - \sqrt{\dfrac{N_t}{N_r}}\right)^2$;

2 : $\alpha = \dfrac{\lambda_{\max} + \lambda_{\min}}{\lambda_{\max} - \lambda_{\min}}$, $\beta = \dfrac{\lambda_{\max} + \lambda_{\min}}{2}$;

3 : $\mathbf{y}^{MF} = \mathbf{H}^H \mathbf{y}$, $\rho^{(1)} = \dfrac{2\alpha^2}{(2\alpha^2 - 1)\beta}$, $\varphi^{(1)} = \dfrac{1}{2\alpha^2 - 1}$;

4 : $\hat{\mathbf{s}}^{(0)} = \dfrac{1}{\beta} \mathbf{y}^{MF}$;

5 : $\dot{\mathbf{h}}^{(0)} = \mathbf{H}\hat{\mathbf{s}}^{(0)}$;

6 : $\mathbf{r}^{(0)} = \mathbf{y}^{MF} - \dfrac{N_0}{E_s}\hat{\mathbf{s}}^{(0)} - \mathbf{H}^H \dot{\mathbf{h}}^{(0)}$;

7 : $\sigma^{(0)} = \dfrac{1}{\beta}\mathbf{r}^{(0)}$;

Iteration algorithm:

8 : **for** $K=1,2,\cdots,T\text{-}1$ **do**

9 : $\hat{\mathbf{s}}^{(K)} = \hat{\mathbf{s}}^{(K-1)} + \sigma^{(K)}$;

10: $\dot{\mathbf{h}}^{(K)} = \mathbf{H}\hat{\mathbf{s}}^{(K)}$;

11: $\mathbf{r}^{(K)} = \mathbf{y}^{MF} - \dfrac{N_0}{E_s}\hat{\mathbf{s}}^{(K)} - \mathbf{H}^H \dot{\mathbf{h}}^{(K)}$;

12: $\sigma^{(K)} = \rho^{(K)}\mathbf{r}^{(K)} + \varphi^{(K)}\sigma^{(K-1)}$;

13: $\rho^{(K)} = \dfrac{4\alpha^2}{4\alpha^2\beta - \beta^2\rho^{(K-1)}}$, $\varphi^{(K)} = \dfrac{\beta^2\rho^{(K-1)}}{4\alpha^2\beta - \beta^2\rho^{(K-1)}}$;

14: **end**

Approximate LLR computation method:

15: **for** $i=1:N_r$ **do**

16:    **for** $b=1:\vartheta$ **do**

17:       $\gamma_i = \dfrac{1}{\beta N_0}$, $\xi_b(\hat{s}_i) = \min\limits_{s \in S_b^0}\left|\dfrac{\hat{s}_i}{\mu_i} - s\right|^2 - \min\limits_{s \in S_b^1}\left|\dfrac{\hat{s}_i}{\mu_i} - s\right|^2$;

18:       $L_{i,b}(\hat{s}_i) = \gamma_i \xi_b(\hat{s}_i)$;

19:    **end**

20: **end**

2.3.2 Convergence

Now let us discuss the problem of the convergence rate of the Chebyshev iteration. After K steps, the approximation error can be expressed as [14]

$$e^{(K)} = s - \hat{s}^{(K)} = P^{(K)}(A)e^{(0)} \tag{2.65}$$

In the above equation, $P^{(K)}$ satisfies

$$P^{(K)}(\lambda_i) = \frac{T^{(K)}\left(\alpha - \frac{\alpha}{\beta}\lambda_i\right)}{T^{(K)}(\alpha)} \tag{2.66}$$

So the error can be expressed as

$$\left\|e^{(K)}\right\| \le \left\|P^{(K)}(A)\right\| \left\|e^{(0)}\right\| \tag{2.67}$$

From the previous description, we know that matrix A is a diagonally dominant matrix, so there is

$$P^{(K)}(A) = P^{(K)}(SJS^{-1}) = SP^{(K)}(J)S^{-1}$$

$$= S \begin{bmatrix} P^{(K)}(\lambda_1) & & \\ & \ddots & \\ & & P^{(K)}(\lambda_{N_t}) \end{bmatrix} S^{-1} \tag{2.68}$$

where S is a complex matrix of $N_t \times N_t$ and meets the condition $S^{-1} = S^H$. J is an upper triangular matrix. Now two lemmas are proposed and their corresponding proofs are given, followed by corresponding conclusions.

Lemma 2.3.1 *In the massive MIMO system, there is*

$$\left|P^{(K)}(\lambda_i)\right| \approx \left|\frac{N_r + N_t - \lambda_i + \sqrt{(N_r + N_t - \lambda_i)^2 - 1}}{2N_r}\right|^K \tag{2.69}$$

where $P^{(K)}(\lambda_i)$ is *the Kth normalized Chebyshev polynomial.*

Proof The Chebyshev polynomials in Eq. (2.45) can be rewritten as [15]

$$T_K(\alpha) = \cosh(K\operatorname{arcosh}(\alpha)) = \frac{e^{K\operatorname{arcosh}(\alpha)} + e^{-K\operatorname{arcosh}(\alpha)}}{2} \tag{2.70}$$

Combining Eqs. (2.66) and (2.70), the Chebyshev polynomial is converted to

$$P^{(K)}(\lambda_i) = \frac{e^{K\operatorname{arcosh}\left(\alpha - \left(\frac{\alpha}{\beta}\right)\cdot\lambda_i\right)} + e^{-K\operatorname{arcosh}\left(\alpha - \left(\frac{\alpha}{\beta}\right)\cdot\lambda_i\right)}}{e^{K\operatorname{arcosh}(\alpha)} + e^{-K\operatorname{arcosh}(\alpha)}}$$

$$= \left(\frac{e^{\operatorname{arcosh}\left(\alpha - \left(\frac{\alpha}{\beta}\right)\cdot\lambda_i\right)}}{e^{\operatorname{arcosh}(\alpha)}}\right)^K \cdot \left(\frac{1 + e^{-2K\operatorname{arcosh}\left(\alpha - \left(\frac{\alpha}{\beta}\right)\cdot\lambda_i\right)}}{1 + e^{-2K\operatorname{arcosh}(\alpha)}}\right) \tag{2.71}$$

On this basis, using the identity $\operatorname{arcosh}(x) = \ln\left(x + \sqrt{x^2 - 1}\right)$ to transform Eq. (2.71) into

$$P^{(K)}(\lambda_i) = (V(\lambda_i))^k \cdot Q^{(K)}(\lambda_i) \tag{2.72}$$

where $V(\lambda_i)$ and $Q^{(K)}(\lambda_i)$ meet:

$$V(\lambda_i) = \frac{e^{\operatorname{arcosh}\left(\alpha - \left(\frac{\alpha}{\beta}\right)\cdot\lambda_i\right)}}{e^{\operatorname{arcosh}(\alpha)}} = \frac{\alpha - \frac{\alpha}{\beta}\lambda_i + \sqrt{\left(\alpha - \frac{\alpha}{\beta}\lambda_i\right)^2 - 1}}{\alpha + \sqrt{\alpha^2 - 1}} \tag{2.73}$$

$$Q^{(K)}(\lambda_i) = \frac{e^{\operatorname{arcosh}\left(\alpha - \left(\frac{\alpha}{\beta}\right)\cdot\lambda_i\right)}}{e^{\operatorname{arcosh}(\alpha)}} = \frac{1 + e^{-2K\operatorname{arcosh}\left(\alpha - \left(\frac{\alpha}{\beta}\right)\cdot\lambda_i\right)}}{1 + e^{-2K\operatorname{arcosh}(\alpha)}} \tag{2.74}$$

From Eqs. (2.50)–(2.53), the parameters satisfy $\alpha \notin (-\infty, 1]$ and $\left(\alpha - \frac{\alpha}{\beta}\lambda_i\right) \notin [-1, 1]$. As a result. Now $Q^{(K)}(\lambda_i)$ satisfies

$$0 \leq \left|Q^{(K)}(\lambda_i)\right| \leq \frac{2}{1 - \tau^K} \tag{2.75}$$

When the iteration number K increases, the value $\left|Q^{(K)}(\lambda_i)\right|$ is limited, so there is

$$P^{(K)}(\lambda_i) \approx (V(\lambda_i))^{(K)} \tag{2.76}$$

Considering that $V(\lambda_i)$ is an operator similar to that in Eq. (2.72), set $V = V(A)$. When K is very large, we can get

$$P^{(K)}(A) \approx V^{(K)} \tag{2.77}$$

If the eigenvalues of A satisfy $\lambda_{\min} < \lambda_i < \lambda_{\max}$, then the value of $|V(\lambda_i)|$ keeps a constant, so the eigenvalues of V and ψ_i satisfy

$$|\psi_i| = \left|\frac{\alpha - \frac{\alpha}{\beta}\lambda_i + \sqrt{\left(\alpha - \frac{\alpha}{\beta}\lambda_i\right)^2 - 1}}{\alpha + \sqrt{\alpha^2 - 1}}\right| \tag{2.78}$$

From Eqs. (2.50)–(2.53), Eqs. (2.76)–(2.78). The following equation is obtained:

$$\left|P^{(K)}(\lambda_i)\right| \approx \left|\frac{\frac{N_r+N_t-\lambda_i}{2\sqrt{N_r N_t}} + \sqrt{\left(\frac{N_r+N_t-\lambda_i}{2\sqrt{N_r N_t}}\right)^2 - 1}}{\frac{N_r+N_t}{2\sqrt{N_r N_t}} + \sqrt{\left(\frac{N_r+N_t}{2\sqrt{N_r N_t}}\right)^2 - 1}}\right|^K \tag{2.79}$$

Therefore, the conclusion of Lemma 2.3.1 is obtained.

According to Eqs. (2.52) and (2.53), when N_t remains constant but N_r increases, the maximum eigenvalue and the minimum eigenvalue of matrix A will approach to N_r. It is proved in Lemma 2.3.1 that the value $\left|P^{(K)}(\lambda_i)\right|$ is less than 1 and will decrease as the ratio of N_r to N_t increases. According to Eqs. (2.67) and (2.68), the estimation error is very small when the number of iterations K is limited. And from Lemma 2.3.1, we know that the estimation error will decrease with the increase of the number of iterations K. Therefore, we can get

$$\lim_{K\to\infty} \left|P^{(K)}(\lambda_i)\right| = 0 \tag{2.80}$$

$$\lim_{K\to\infty} \left|P^{(K)}(A)\right| = 0 \tag{2.81}$$

In other words, in massive MIMO signal detection, using Chebyshev iterative algorithm to estimate the transmitted signal s, the calculation error is very small, even close to zero.

Lemma 2.3.2 *In the massive MIMO system, $|V_{ch}(\lambda_i)| \leq |V_{cg}(\lambda_i)|$ and $|V_{ch}(N_r, N_t)| \leq |V_{ne}(N_r, N_t)|$ are satisfied, where V_{ch}, V_{cg} and V_{ne} are respectively the normalized Chebyshev polynomials corresponding to the Chebyshev iteration algorithm, the CG algorithm and the NSA algorithm.*

Proof The convergence rate of the Chebyshev iteration algorithm is

$$R(A) = -\log_2 \left(\lim_{K\to\infty} \left(\left\|P^{(K)}(A)\right\|^{\frac{1}{k}} \right) \right) \tag{2.82}$$

In order to make the convergence faster, $\lim_{K\to\infty} \left(\left\|P^{(K)}(A)\right\|^{\frac{1}{K}} \right)$ should be as great as possible in Eq. (2.82). The problem is transformed to find the minimization of the maximum value $|V(\lambda_i)|$ by Eqs. (2.76) and (2.78), i.e.,

$$\text{minmax}|V(\lambda_i)| = \text{minmax}\left|\frac{\alpha - \frac{\alpha}{\beta}\lambda_i + \sqrt{\left(\alpha - \frac{\alpha}{\beta}\lambda_i\right)^2 - 1}}{\alpha + \sqrt{\alpha^2 - 1}}\right| \tag{2.83}$$

Set $\lambda_i = \beta$, and combining with Eq. (2.51), we can find the minimization of the maximum value $|V(\lambda_i)|$. Thus $|V_{ch}(\lambda_i)|$ can be summarized as

$$|V_{ch}(\lambda_i)| = \left| \frac{1}{\alpha + (\alpha^2 - 1)^{\frac{1}{2}}} \right| = \left| \frac{1}{\frac{\lambda_{max} + \lambda_{min}}{\lambda_{max} - \lambda_{min}} + \sqrt{\frac{\lambda_{max} + \lambda_{min}}{\lambda_{max} - \lambda_{min}}^{-2} - 1}} \right|$$

$$= \left| \frac{\lambda_{max} - \lambda_{min}}{\lambda_{max} + \lambda_{min} + 2\sqrt{\lambda_{max} \cdot \lambda_{min}}} \right| \tag{2.84}$$

$\theta = \sqrt{\frac{\lambda_{min}}{\lambda_{max} - \lambda_{min}}}$, the $|V_{cg}(\lambda_i)|$ of the CG algorithm can be summarized as

$$|V_{cg}(\lambda_i)| = \left| \left[\frac{2}{\left(1 + 2\theta + \sqrt{(1 + 2\theta)^2 - 1}\right)^K + \left(1 + 2\theta + \sqrt{(1 + 2\theta)^2 - 1}\right)^{-K}} \right]^{\frac{1}{K}} \right|$$

$$\geq \left| \frac{1}{1 + 2\theta + \sqrt{(1 + 2\theta)^2 - 1}} \right| = \left| \frac{\lambda_{max} - \lambda_{min}}{\lambda_{max} + \lambda_{min} + 2\sqrt{\lambda_{max} \cdot \lambda_{min}}} \right|$$

$$= |V_{ch}(\lambda_i)| \tag{2.85}$$

We can solve $|V_{ch}(\lambda_i)| \leq |V_{cg}(\lambda_i)|$ from Eq. (2.85). This inequality indicates that compared with CG algorithm and Chebyshev iteration algorithm, $|V(\lambda_i)|$ has a smaller maximum value using Chebyshev iteration algorithm.

In combination with Eqs. (2.50)–(2.53) and (2.73), the $|V_{ch}(N_r, N_t)|$ of Chebyshev iteration algorithm can be approximated to

$$|V_{ch}(N_r, N_t)| = \left| \frac{2\sqrt{N_r N_t}}{N_r + N_t + \sqrt{(N_r + N_t)^2 - \left(2\sqrt{N_r N_t}\right)^2}} \right| = \sqrt{\frac{N_t}{N_r}} \tag{2.86}$$

According to literature [7], The $|V_{ne}(N_r, N_t)|$ of NSA algorithm is

$$|V_{ne}(N_r, N_t)| = \left\| D^{-1}\left(L + L^H\right) \right\|_F \geq \frac{1}{N_r\sqrt{N_t}} \left\| L + L^H \right\|_F = \frac{1}{N_r\sqrt{N_t}} \sqrt{N_r N_t (N_t - 1)}$$

$$\approx \sqrt{\frac{N_t}{N_r}} = |V_{ch}(N_t, N_r)| \tag{2.87}$$

Equation (2.87) indicates that compared with the Chebyshev iteration algorithm and the NSA algorithm, $|V(N_r, N_t)|$ has a smaller maximum value.

Combined with Eqs. (2.68), (2.82) and (2.83), a smaller maximum value of $|V(\lambda_i)|$ results in a faster convergence rate. Therefore, as we know from Lemma 2.3.2, the convergence rate of Chebyshev iteration algorithm is faster than CG algorithm and NSA algorithm.

2.3.3 Complexity and Parallelism

After discussing the convergence and convergence rate of the Chebyshev iteration algorithm, we will analyze the computational complexity of the Chebyshev iteration algorithm. As computational complexity mainly refers to the number of multiplications in the algorithm, we can count the number of multiplications required by the algorithm to evaluate the computational complexity. In the Chebyshev iteration algorithm, the first part of the calculation is the parameter calculation based on Eqs. (2.46)–(2.53). Because of the fixed scale of the massive MIMO system, these parameters only need to be computed once and stored in memory as constants when calculating multiple sets of data so the actual number of multiplications is negligible. The second part is to match the filter vector y^{MF} and find the approximate initial solution by using the eigenvalues. The $N_t \times N_r$ matrix H^H needs to be multiplied by $N_r \times 1$ vector y to get the $N_t \times 1$ vector y^{MF}, then calculate $\frac{1}{\beta}$ of its value. There are $4N_rN_t$ actual multiplications in this process. There are three steps in the third part of the computation. First, multiply the $N_r \times N_t$ matrix H by the $N_t \times 1$ transmitted vector $s^{(K)}$. Then solve the residual vector $r^{(K)}$ and finally find the correction vector $\sigma^{(K)}$. These three steps require $4KN_rN_t$, $4KN_rN_t + 2KN_t$ and $4KN_t$ real number multiplications respectively. The last part of the computation is the calculation of approximate log-likelihood ratio. This step requires $N_t + 1$ multiplications when the modulation order is 64-QAM. Therefore, the total number of multiplications required by the Chebyshev iteration algorithm is $(8K + 4)N_rN_t + (6K + 1)N_t$.

As mentioned before, the method to directly find the inverse matrix will result in the computational complexity increase. Such computation is time-consuming, which makes hardware implementation difficult. A large-scale matrix multiplication takes 36 clock cycles when many pieces of hardware are parallel. These defects affect the energy and area efficiency of the detector hardware. In addition, the PCI contains the initial value, whose computation process can be considered as one iteration. Therefore, in order to balance, the iteration number K of PCI contains a calculation of the initial value and K-1 iterations. The comparison of computational complexity of different methods is listed in Table 2.1. The computational complexity of accurate MMSE detection methods such as the CHD algorithm is $O(N_t^2)$. The computational complexity of NSA increases with the increase of the number of iterations K, and the specific manifestation is that when $K < 3$ the complexity is $O(N_t^2)$, when $K = 3$ the complexity increases to $O(N_t^3)$ and the computational complexity of NSA is higher than that of accurate MMSE detection when $K > 3$. Usually, to ensure the detection accuracy, the number of the iterations of NSA K should be greater than 3. In addition, the NSA in literature [7] adopts the explicit method of large-scale matrix multiplication as $N_t \times N_r$ matrix H^H multiplied by $N_r \times N_t$ matrix H, which consumes a lot of computing resources, i.e. $O(N_rN_t^2)$. Since the results of direct inversion can be used repeatedly in downlink, if the parallelizable Chebyshev algorithm is used for calculation, the complexity will be doubled, but still lower than the results in literature [7]. The following methods directly compute matrix A and indirectly compute A^{-1}, including implicit version of the Neumann series (INS) approximation and implicit

Table 2.1 Computational complexity analysis

Algorithm	$K = 2$	$K = 3$	$K = 4$	$K = 5$
NSA [7]	$2N_rN_t^2 + 6N_tN_r$ $+ 4N_t^2 + 2N_t$	$2N_rN_t^2 + 6N_tN_r$ $+ 2N_t^3 + 4N_t^2$	$2N_rN_t^2 + 6N_tN_r$ $+ 6N_t^3$	$2N_rN_t^2 + 6N_tN_r$ $+ 10N_t^3 - 6N_t^2$
INS [17]	$2N_rN_t^2 + 6N_tN_r$ $+ 10N_t^2 + 2N_t$	$2N_rN_t^2 + 6N_tN_r$ $+ 14N_t^2 + 2N_t$	$2N_rN_t^2 + 6N_tN_r$ $+ 18N_t^2 + 2N_t$	$2N_rN_t^2 + 6N_tN_r$ $+ 22N_t^2 + 2N_t$
GAS [12]	$2N_rN_t^2 + 6N_tN_r$ $+ 10N_t^2 - 2N_t$	$2N_rN_t^2 + 6N_tN_r$ $+ 14N_t^2 - 6N_t$	$2N_rN_t^2 + 6N_tN_r$ $+ 18N_t^2 - 10N_t$	$2N_rN_t^2 + 6N_tN_r$ $+ 22N_t^2 - 14N_t$
CG [16]	$2N_rN_t^2 + 6N_tN_r$ $+ 8N_t^2 + 33N_t$	$2N_rN_t^2 + 6N_tN_r$ $+ 12N_t^2 + 49N_t$	$2N_rN_t^2 + 6N_tN_r$ $+ 16N_t^2 + 65N_t$	$2N_rN_t^2 + 6N_tN_r$ $+ 20N_t^2 + 81N_t$
CGLS [18]	$16N_tN_r + 20N_t^2$ $+ 32N_t$	$20N_tN_r + 28N_t^2$ $+ 48N_t$	$24N_tN_r + 36N_t^2$ $+ 64N_t$	$28N_tN_r + 44N_t^2$ $+ 80N_t$
OCD [19]	$16N_tN_r + 4N_t$	$20N_tN_r + 6N_t$	$32N_tN_r + 8N_t$	$40N_tN_r + 10N_t$
PCI	$2N_rN_t^2 + 6N_tN_r$ $+ 8N_t^2 + 8N_t$	$2N_rN_t^2 + 6N_tN_r$ $+ 12N_t^2 + 12N_t$	$2N_rN_t^2 + 6N_tN_r$ $+ 16N_t^2 + 16N_t$	$2N_rN_t^2 + 6N_tN_r$ $+ 20N_t^2 + 20N_t$
	$12N_tN_r + 7N_t$	$20N_tN_r + 13N_t$	$28N_tN_r + 19N_t$	$36N_tN_r + 25N_t$

version of CG [16]. The proposed PCI can also be modified to compute the explicit version of the matrix A. Table 2.1 lists the associated computational complexity of PCI, which achieves lower or equal complexity compared with literatures [12], [16] and [17]. Comparing with PCI under the same conditions, the least square conjugate gradient least square (CGLS) algorithm [18] and the optimized coordinate descent (OCD) algorithm [19] implement large-scale matrix multiplication and inversion in a completely implicit manner. Compared with the CGLS algorithm and the OCD algorithm, the PCI has advantages in computational complexity.

Now consider the parallelism of PCI. In the Gauss–Seidel (GAS) algorithm [12], there is a strong correlation between elements, which means that the method has low parallelism. In the calculation of GAS algorithm, when calculating $s_i^{(K)}$ at the Kth iteration, it requires $s_j^{(K)}, j = 1, 2, \cdots, i - 1$ and $s_{j-1}^{(K)}, j = i, i + 1, \cdots, N_r$ of the previous K iterations. Moreover, in the CHD algorithm as well as NSA, GAS, and CG methods, there is a strong correlation between large-scale matrix inverse and multiplication, which requires the calculation of large-scale matrix multiplication first. This results in their reduced parallelism. PCI parallelism is an important problem in algorithm design and hardware implementation. According to the algorithm, when calculating $s^{(K)}$ and correcting vector $\sigma^{(K)}$ and residual vector $r^{(K)}$, each element in them can be done in parallel. Besides this, we can find that the implicit method reduces the correlation between large-scale matrix multiplication and inverse calculation, and improves the parallelism of the algorithm.

2.3.4 Bit Error Rate

To evaluate the performance of PCI, BER's simulation results are compared with NSA, CG, GAS, OCD and the Richardson (RI) algorithm below. Furthermore, the BER performance comparison also includes the MMSE algorithm based on the CHD. SNR is defined at the receiving antenna [7]. The number of iterations K of PCI is an initial value calculation of the K-1th iterations.

Figure 2.3 shows the BER performance comparisons (PCI and other methods) and the simulation results when $N_t = 16$, with SNR set to 14 dB. As we can see from the figure, the PCI only uses a small number of iterations as N_r increases, and achieves the near-optimal performance (compared with accurate MMSE). This result demonstrates the reasonability of the reduced computational complexity. Figure 2.3 also exhibits the performance of the NSA. When the number of iterations is relatively small, the detection accuracy loss cannot be ignored. However when the number of iterations is large, the system consumes a lot of hardware resources. Hence, this figure validates that PCI in massive MIMO system is superior to that in NSA.

Figure 2.4 shows the BER comparisons between the PCI using the initial value for the iteration and the traditional zero vector initial solution. In this simulation, the numbers of antennas are 16 and 128 respectively. The initial solution based on the eigenvalue approximation achieves low detection accuracy loss with the same number of iterations. The simulation result in the PCI when $K = 3$ is very close to the BER when $K = 4$, which means that the initial value for the iteration reduces the amount of computation while maintaining the similar detection accuracy.

Figure 2.5 exhibits the performance of PCI, CG, OCD, GAS, RI, and NSA. It also provides the MMSE similar to the CHD algorithm for reference. The comparison of the three simulation results shows that PCI achieves the near-optimal performance

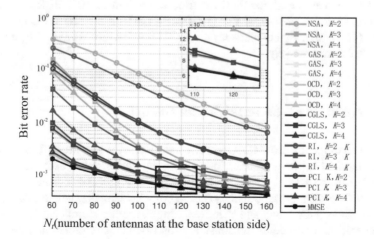

Fig. 2.3 BER simulation results of various algorithms at $N = 0$, SNR = 14 dB. © [2018] IEEE. Reprinted, with permission, from Ref. [13]

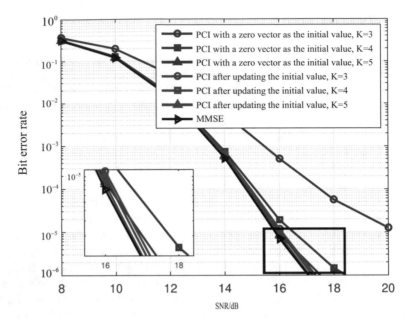

Fig. 2.4 BER performance comparisons between PCI after updating initial value and PCI for conventional zero vector value. © [2018] IEEE. Reprinted, with permission, from Ref. [13]

under different antenna configurations. To implement the same BER in PCI, SNR is required to be almost identical to GAS and OCD methods, but smaller than RI and NSA. When the ratio of N_r to N_t is small, CG is slightly better than PCI. When the ratio of N_r to N_t is larger, PCI has better detection performance. For example, in Fig. 2.5 (c), when $K = 3$, the SNR required to implement 10^{-6} BER is 17.25 dB, which is close to the precise MMSE (16.93 dB), GAS (17.57 dB), OCD (17.67 dB), and CG (17.71 dB). By comparison, the SNRs required for RI and NSA are 18.45 and 19.26 dB, respectively.

2.3.5 Analysis on Channel Model Impact

The channel of the massive MIMO system also affects the algorithm. The Kronecker channel model [20] is often used to evaluate the performance because it is more practical than the Rayleigh fading channel. In the Kronecker channel model, the elements of the channel matrix satisfy $N\left(0, d(z)\mathbf{I}_{N_r}\right)$, where $d(z)$ indicates the channel fading (such as path fading and shadow fading). Another significant feature of the Kronecker channel model is the consideration of channel correlation. Specifically, $\mathbf{R}_r$ and $\mathbf{R}_t$ denote the channel correlation of the receiving antenna and the transmitted

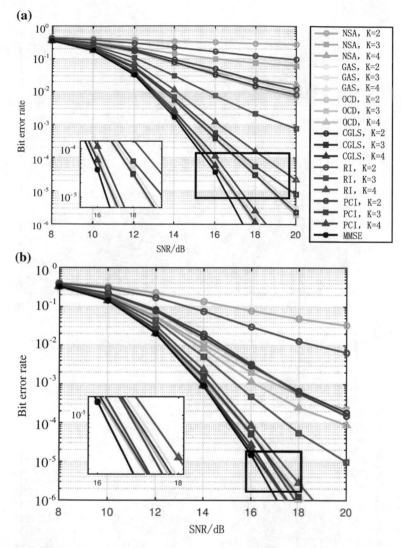

Fig. 2.5 BER comparisons between PCI and other algorithms. **a** $N_r = 64$, $N_t = 16$, **b** $N_r = 90$, $N_t = 16$, **c** $N_r = 128$, $N_t = 16$, **d** $N_r = 162$, $N_t = 16$ © [2018] IEEE. Reprinted, with permission, from Ref. [13]

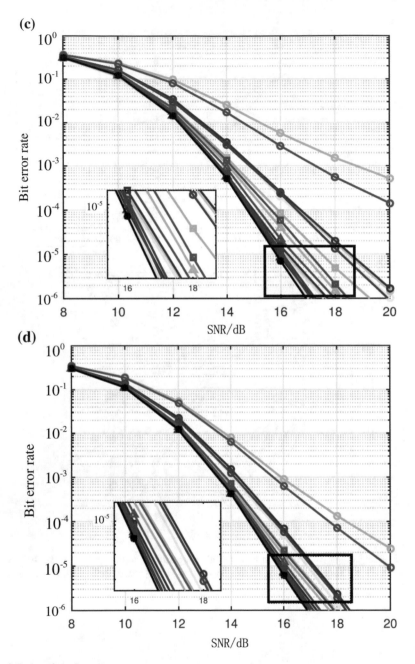

Fig. 2.5 (continued)

antenna respectively. This part is also based on the Kronecker channel model, and the channel H can be expressed as

$$H = R_r^{\frac{1}{2}} H_{\text{i.i.d.}} \sqrt{d(z)} R_t^{\frac{1}{2}} \tag{2.88}$$

where $H_{\text{i.i.d.}}$ is a random matrix, the elements of which are independent and identically distributed and are subject to the complex Gaussian distribution with a zero mean and a unit variance.

In the simulation, the radius of each hexagonal region is $r = 500$ m, and the users' locations are independent and random. The independent shadow fading C satisfies

$$10 \lg C \sim N\left(0, \sigma_{\text{sf}}^2\right) \tag{2.89}$$

Considering the channel fading variance $d(z) = \frac{C}{\|z-b\|^{\kappa}}$, where $b \in R^2$, κ and $\|\cdot\|$ are the base station location, path loss index and Euclidean norm respectively. The simulation adopts the following assumption: $\kappa = 3.7$, $\sigma_{\text{sf}}^2 = 5$ and the transmitted power is $\rho = \frac{\gamma^{\kappa}}{2}$.

Now we will discuss the influence of the Kronecker channel model on eigenvalue approximation. When the channel is independent and identically distributed Rayleigh fading, the value of the diagonal element of matrix A approximates to β, ξ is the channel correlation coefficient. When the correlation coefficient increases (the Kronecker channel model), the approximate error also increases slightly. Therefore, the eigenvalue approximation is still applicable to more practical channel models, such as the Kronecker channel model. The influence of different channel models on computational complexity is also considered. In order to satisfy the requirement of low approximation error, the number of iterations of the proposed PCI should be slightly increased for enhanced channel correlation and increased eigenvalue approximation error. Therefore, there is a slight increase in the computational complexity of PCI, which is a limitation of this method. As discussed earlier, there are three ways to compute large-scale matrix multiplication and inverse. When channel frequency is flat and changes slowly [21], as channel hardening becomes obvious, the result of matrix multiplication and inverse of explicit and partially implicit methods can be partially reused, such as NSA, INS, GAS, and CG. However, these methods also have some limitations. For example, the high computational complexity at the beginning of the detector can affect subsequent operations (such as matrix decomposition and reciprocal of diagonal elements), which need to start only after matrix multiplication is completed. In addition, the hardware utilization of large-scale multiplication is not high, which reduces the energy and area efficiency of detector. These methods (i.e., CGLS, OCD, and PCI) use implicit methods to solve large-scale matrix multiplication and inversion. The computation of these methods is much lower and the parallelism is higher than the explicit method. However, the results cannot be reused in the case of low frequency selectivity due to small-scale fading [22] average and channel hardening, which is another limitation of PCI. Finally, Fig. 2.6 shows the

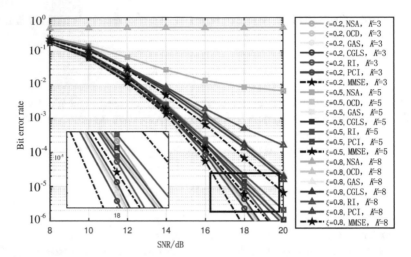

Fig. 2.6 BER comparisons of various algorithms under the Kronecker channel model. © [2018] IEEE. Reprinted, with permission, from Ref. [13]

impact of large-scale fading and spatial correlation on the MIMO channel (Kronecker channel model), which is an important problem in the actual MIMO system. Simulation results show that compared with MMSE, the accuracy loss of PCI is less. As the channel correlation increases (the channel correlation coefficient increases), the number of iterations of all methods increases to reduce the approximation error. With the same number of iterations, PCI achieves lower error compared to the NSA and RI methods. Although PCI has similar detection performance compared with CG, OCD, and GAS, its main advantages are higher parallelism and lower computational complexity than the other three methods. In a word, PCI is superior to other methods under more practical channel model conditions.

2.4 Jacobi Iteration Algorithm

2.4.1 Weighted Jacobi Iteration and Convergence

This section introduces an optimized Jacobi iteration algorithm named the weighted Jacobi iteration (WeJi) algorithm. As discussed earlier, matrix G and matrix W are diagonally dominant matrices. Here, we can decompose matrix W into $W = P + Q$, where P is a diagonal matrix, and Q is a matrix with the diagonal elements as 0. Using the WeJi to solve the linear equation, the transmitted signal can be estimated as

$$\hat{s}^{(K)} = B\hat{s}^{(K-1)} + F = ((1 - \omega)I - \omega P^{-1}Q)\hat{s}^{(K-1)} + \omega P^{-1}y^{\text{MF}} \qquad (2.90)$$

where $B = ((1 - \omega)I - \omega P^{-1}Q)$, $F = \omega P^{-1}y^{\text{MF}}$ is an iteration matrix, K is the number of iterations, and $\hat{s}^{(0)}$ is the initial solution. In addition, $0 < \omega < 1$, it plays a crucial role in the convergence and convergence rate of WeJi. In WeJi, the range of the parameter ω is set as $0 < \omega < \frac{2}{\rho}(P^{-1}W)$ [9]. Because P is a diagonal matrix, its inverse matrix is very easy to find, the computational complexity of WeJi is greatly reduced.

The initial value of the iteration will affect the convergence and convergence rate of the iteration, so we need to find a good initial value when we use the WeJi to solve the massive MIMO signal detection problem. Here, the initial value with low computational complexity can be obtained by using NSA. Therefore, we can set the initial value of iteration as Eq. (2.91), which can make WeJi converge at a faster rate.

$$\hat{s}^{(0)} = (I - P^{-1}Q)P^{-1}y^{\text{MF}} = (I - R)T \qquad (2.91)$$

In addition, it is because of the increase of algorithm convergence rate that this initial value also reduces hardware resource consumption and increases data throughput rate.

According to Eq. (2.90), when the iteration number K tends to be infinite, the error of the transmitted signal estimated by using the WeJi is [7]

$$\Delta = s - \hat{s}^{(K)} \approx \hat{s}^{(\infty)} - \hat{s}^{(K)} = B^K \left(s - \hat{s}^{(0)} \right) \qquad (2.92)$$

Here, obviously $s = \hat{s}^{(\infty)}$. So the convergence rate of WeJi is

$$R(B) = -\ln\left(\lim_{K \to \infty} \|B^K\|^{\frac{1}{K}} \right) = -\ln(\rho(B)) \qquad (2.93)$$

where $\rho(B)$ is the spectral radius of the matrix B. We can see that when $\rho(B)$ is very small the convergence rate of the algorithm is higher. As for the convergence rate of the WeJi, two lemmas and their corresponding proofs are given here.

Lemma 2.4.1 *In massive MIMO systems, $\rho(B_{\text{W}}) \leq \omega\rho(B_{\text{N}})$, $\rho(B_{\text{W}})$ and $\rho(B_{\text{N}})$ are the iterative matrices of the WeJi and NSA respectively.*

Proof The spectral radius of B_{W} is defined as

$$\rho(B_{\text{W}}) = \rho((1 - \omega)I - \omega P^{-1}Q) \qquad (2.94)$$

In the WeJi, $0 < \omega < 1$ and ω are close to 1. $0 < \omega < 1$ can be converted to $0 < 1 - \omega < 1$, so Eq. 2.94 can also be written as

$$\rho(\boldsymbol{B}_{\mathrm{W}}) = \omega\rho\boldsymbol{P}^{-1}\boldsymbol{Q} - (1 - \omega)\boldsymbol{I} \leq \omega\rho\boldsymbol{P}^{-1}\boldsymbol{Q} \tag{2.95}$$

In the Newman series approximation, there are

$$\rho(\boldsymbol{B}_{\mathrm{N}}) = \rho\big(\boldsymbol{P}^{-1}\boldsymbol{Q}\big) \tag{2.96}$$

Combined with Eq. (2.95), we can get

$$\rho(\boldsymbol{B}_{\mathrm{W}}) \leq \omega\rho(\boldsymbol{B}_{\mathrm{N}}) \tag{2.97}$$

Lemma 2.4.1 shows that the WeJi converges faster than that of the Newman series approximation. Without loss of generality, l_2 norm is used to estimate the error of iteration, for example

$$\|\Delta\|_2 \leq \big\|\boldsymbol{B}_{\mathrm{W}}^K\big\|_{\mathrm{F}}\big\|\boldsymbol{s} - \hat{\boldsymbol{s}}^{(0)}\big\|_2 \leq \|\boldsymbol{B}_{\mathrm{W}}\|_{\mathrm{F}}^K\big\|\boldsymbol{s} - \hat{\boldsymbol{s}}^{(0)}\big\|_2 \tag{2.98}$$

According to the above expression (2.98), if $\|\boldsymbol{B}_{\mathrm{W}}\|_{\mathrm{F}} < 1$ is satisfied, the approximate error of WeJi will exponentially approach 0 with the increase of the number of iterations K.

Lemma 2.4.2 *In the massive MIMO system, the probability of* $\|\boldsymbol{B}_{\mathrm{W}}\|_{\mathrm{F}} < 1$ *satisfies*:

$$P\{\|\boldsymbol{B}_{\mathrm{W}}\|_{\mathrm{F}} < 1\} \geq 1 - \omega\sqrt[4]{\frac{(N_{\mathrm{r}} + 17)(N_{\mathrm{t}} - 1)N_{\mathrm{r}}^2}{2N_{\mathrm{r}}^3}} \tag{2.99}$$

Proof We can get the following formula according to the Markov inequality:

$$P\{\|\boldsymbol{B}_{\mathrm{W}}\|_{\mathrm{F}} < 1\} \geq 1 - P\{\|\boldsymbol{B}_{\mathrm{W}}\|_{\mathrm{F}} \geq 1\} \geq 1 - E(\|\boldsymbol{B}_{\mathrm{W}}\|_{\mathrm{F}}) \tag{2.100}$$

Note that if the parameter ω satisfies and approaches 1 in WeJi, then $1 - \omega$ is close to 0, and satisfies $0 < 1 - \omega < 1$, so the effect of $(1 - \omega)\boldsymbol{I}$ can be ignored. The formula (2.99) satisfies

$$P\{\|\boldsymbol{B}_{\mathrm{W}}\|_{\mathrm{F}} < 1\} \geq 1 - E\big(\|\omega\boldsymbol{P}^{-1}\boldsymbol{Q}\|_{\mathrm{F}}\big) \tag{2.101}$$

So the probability of $\|\boldsymbol{B}_{\mathrm{W}}\|_{\mathrm{F}} < 1$ is related to $E\big(\|\omega\boldsymbol{P}^{-1}\boldsymbol{Q}\|_{\mathrm{F}}\big)$. Consider $\|\boldsymbol{P}^{-1}\boldsymbol{Q}\|_{\mathrm{F}}$, the element of the ith row and jth column in the matrix A satisfies

$$a_{ij} \rightarrow \begin{cases} \displaystyle\sum_{t=1}^{N_{\mathrm{r}}} h_{ti}^* h_{tj}, & i \neq j \\ \displaystyle\sum_{m=1}^{N_{\mathrm{r}}} |h_{mi}|^2 + \frac{N_0}{E_{\mathrm{s}}}, & i = j \end{cases} \tag{2.102}$$

As a result, $E\left(\left\|\omega P^{-1}Q\right\|_{\mathrm{F}}\right)$ can be expressed as

$$E\left(\left\|\omega P^{-1}Q\right\|_{\mathrm{F}}\right) = E\left(\sqrt{\sum_{i=1}^{N_t}\sum_{j=1,j\neq i}^{N_t}\left|\omega\frac{a_{ij}}{a_{ii}}\right|^2}\right) = E\left(\sqrt[4]{\left(\sum_{i=1}^{N_t}\sum_{j=1,i\neq j}^{N_t}\left(\omega^2\frac{|a_{ij}|^2}{|a_{ii}|^2}\right)\right)^2}\right) \quad (2.103)$$

Then using the Cauchy–Schwartz inequality for Eq. (2.103), and we have

$$E\left(\left\|\omega P^{-1}Q\right\|_{\mathrm{F}}\right) \leq \omega\sqrt[4]{\sum_{i=1}^{N_t}\sum_{j=1,i\neq j}^{N_t}E\left(|a_{ij}|^4\right)\cdot\sum_{i=1}^{N_t}E\left(\frac{1}{|a_{ii}|^4}\right)} \quad (2.104)$$

Obviously, there are two key items $E\left(|a_{ij}|^4\right)$ and $\sum_{i=1}^{N_t}E\left(\frac{1}{|a_{ii}|^4}\right)$. In massive MIMO system, the diagonal elements of the matrix A are close to N_t, so $E\left(\frac{1}{|a_{ii}|^4}\right)$ term can be approximated as

$$E\left(\frac{1}{|a_{ii}|^4}\right) = \frac{1}{N_r^4} \quad (2.105)$$

Now consider $E\left(|a_{ij}|^4\right)$. We can express it according to formula (2.102) as follows:

$$E\left(|a_{ij}|^4\right) = E\left(\left|\sum_{m=1}^{N_r}h_{mi}^*h_{mj}\right|^4\right)$$

$$= \sum_{\substack{q_1+q_2\\+\cdots+q_{N_r}=N_r}}\binom{N_r}{q_1,q_2,\cdots,q_{N_r}}\times E\left(\prod_{1\leq m\leq N_r}\left(h_{mi}^*h_{mj}\right)^{q_m}\right) \quad (2.106)$$

Let X and μ satisfy Eqs. (2.107) and (2.108), where μ_m is the average of $h_{mi}^*h_{mj}$.

$$X = \left[h_{1i}^*h_{1j}, h_{2i}^*h_{2j}, \cdots, h_{N_ri}^*h_{N_rj}\right]^{\mathrm{T}} \quad (2.107)$$

$$\mu = \left[\mu_1, \mu_2, \cdots, \mu_{N_r}\right]^{\mathrm{T}} \quad (2.108)$$

Therefore, there is

$$\varphi\left(h_{1i}^*h_{1j}, h_{2i}^*h_{2j}, \cdots, h_{N_ri}^*h_{N_rj}\right) = \frac{e^{-\frac{(X-\mu)^{\mathrm{T}}C^{-1}(X-\mu)}{2}}}{(2\pi)^{\frac{N_r}{2}}\sqrt{\det C}} \quad (2.109)$$

The matrix C is the covariance matrix. Note that the elements in the channel matrix H are independent and identically distributed and follow $N(0, 1)$, so when $m \neq p$, $h_{mi}^* h_{mj}$ and $h_{pi}^* h_{pj}$ are also independent and identically distributed and obey $N(0, 1)$. Therefore, expression (2.109) can be written as follows:

$$\varphi\left(h_{1i}^* h_{1j}, h_{2i}^* h_{2j}, \cdots, h_{N_r i}^* h_{N_r j}\right) = \frac{1}{(2\pi)^{\frac{N_r}{2}}} e^{-\frac{x^T x}{2}} \tag{2.110}$$

The expression above shows that when $m \neq p$, $h_{mi}^* h_{mj} h_{pi}^* h_{pj}$ obeys $N(0, 1)$ and $(h_{ii})^2$ obeys $\chi^2(1)$. Therefore, the probability density function of the random variable $(h_{ii})^2$ is

$$f(h_{mi}; 1) = \begin{cases} \frac{1}{2\Gamma(\frac{1}{2})} \sqrt{\frac{2}{h_{mi}}} e^{-\frac{h_{mi}}{2}} & h_{mi} > 0 \\ 0 & h_{mi} < 0 \end{cases} \tag{2.111}$$

where Γ is the gamma function [23]. We can get $E(|h_{ii}|^2) = 1$, $D(|h_{ii}|^2) = 2$. When $i \neq j$, since h_{mi}^* and h_{mj} are independent, and $E(|h_{mi}|^2) = D(|h_{mi}|^2) + [E(|h_{mi}|^2)] = 3$, $E(|h_{mi}^* h_{mj}|^2) = E(|h_{mi}^*|^2) E(|h_{mj}|^2) = 1$, $E(|h_{mi}^* h_{mj}|^4) = E(|h_{mi}^*|^4) E(|h_{mj}|^4) = 9$. After neglecting the zero items, $E(|a_{ii}|^4)$ can be expressed as

$$E\left(|a_{ij}|^4\right) = N_r E\left(\left(h_{mi}^* h_{mj}\right)^4\right) + \binom{N_r}{2}\left(E\left(h_{mi}^* h_{mj}\right)^2\right)^2 = \frac{1}{2}\left(N_r^2 + 17N_r\right) \tag{2.112}$$

By substituting Eqs. (2.105) and (2.112) into Eqs. (2.101) and (2.104), we can obtain the conclusion in Lemma 2.4.2.

Lemma 2.4.2 demonstrates that when N_t is fixed, the probability of $\|B_W\|_F < 1$ increases with the increase of N_r. Because of $N_r \gg N_t$ in the massive MIMO system, the probability of $\|B_W\|_F < 1$ is close to 1.

2.4.2 Complexity and Frame Error Rate

Some analysis needs to be done for the WeJi. The WeJi first performs the calculation of the matrices $R(R = P^{-1}Q)$ and $T(T = P^{-1}y^{MF})$. In order to facilitate the calculation for the WeJi, the initial solution should be gotten as soon as possible. So the vector T and the matrix R should be ready in the allocated time. The initial value $\hat{s}^{(0)}$ of the WeJi needs to be calculated based on Eq. (2.91). It is worth noting that the architecture of the initial solution can be reused in the next iteration block when considering the hardware design. Finally, the iteration of the final value $\hat{s}^{(K)}$ is executed in Eq. (2.90) of the algorithm. In the iteration section, the matrix multiplication is implemented by vector multiplication. All elements of the vector can be executed in parallel. The computational complexity of the complex matrix inversion can reduce the number of iterations. In addition, the weighted parameters reduce the number of iterations and achieve similar performance, which also reduces the computational complexity of the detector. Now we are going to compare the WeJi with the recently developed algorithms in terms of computational complexity, parallelism, and hardware design realizability. Because the MMSE and the WeJi need to compute matrices G and y^{MF}, the work mainly focuses on the computational complexity [7, 12, 24, 25] of the matrix inversion and the LLR calculation. The computational complexity is estimated based on the number of real number multiplications required, and each complex multiplication requires four real number multiplications. The computational complexity of the first calculation comes from the product of the diagonal $N_t \times N_t$ matrix P^{-1} and the $N_t \times N_t$ matrix Q and the $N_t \times 1$ vector y^{MF}, the results are $2N_t(N_t - 1)$ and $2N_t$ respectively. The computational complexity of the second calculation comes from the multiplication of the iterative matrices B and F, which involves $4N_t$ real number multiplications. The computation complexity of the third part of comes from the calculation of initial solution. The computation complexity of the last part comes from the computation of channel gain, NPI variance and LLR. Therefore, the total number of multiplications required by the WeJi is $(4K + 4)2N_t^2 - (4K - 4)2N_t$. Figure 2.7 shows the comparison of the numbers of the real number multiplications between the WeJi and other methods. The WeJi has lower computational complexity compared with that of GAS, SOR, and SSOR methods. When $K = 2$, the computational complexity of NSA is relatively low. In general, K should not be less than 3 in NSA to ensure the accuracy of detection. When $K = 3$, NSA shows higher computation. As a result, the reduction in computation of NSA is negligible.

On the other hand, we need to consider the hardware implementation of the WeJi so that it can be executed as parallel as possible. The solution of the WeJi in Eq. (2.90) can be written as

$$\hat{s}_i^{(K)} = \frac{\omega}{A_{i,i}} y_i^{MF} + \frac{\omega}{A_{i,i}} \sum_{j \neq i} \left[A_{i,j} \hat{s}_j^{(K-1)} + (1 - \omega) \hat{s}_j^{(K-1)} \right] \qquad (2.113)$$

The computation of $\hat{s}_i^{(K)}$ only requires the elements of the previous iterations, so all the elements of $\hat{s}^{(K)}$ can be computed in parallel. However, in the GAS, the successive over-relaxation (SOR) method [26] and the symmetric successive over-relaxation (SSOR) method [27], each transmitted signal has a strong correlation in the iterative steps. When calculating $\hat{s}_i^{(K)}$, we need $\hat{s}_j^{(K)}(j = 1, 2, \cdots, i - 1)$ of the Kth iteration and $\hat{s}_j^{(K-1)}(j = i, i + 1, \cdots, N_t)$ of the (K-1)th iteration. This means that the computation for each element cannot be executed in parallel. Therefore, neither the GAS [25] nor the SOR [26] can achieve a high data throughput rate and their throughput rates are far lower than that of the WeJi.

It was noted that the detection method is also proposed based on the Jacobi iteration in the literature [28]. Compared with this method, the WeJi described in this section achieves better performance in the following three aspects. First, the WeJi is a method based on hardware architecture design consideration, that is, the hardware implementation is fully considered in the process of algorithm optimization and

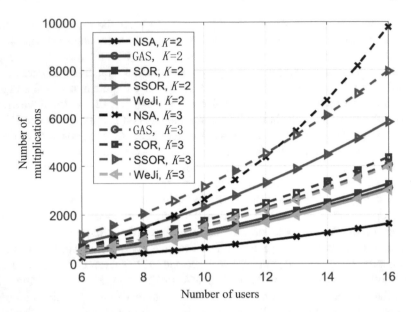

Fig. 2.7 Comparison of the numbers of the actual multiplications between the WeJi and other algorithms. © [2018] IEEE. Reprinted, with permission, from Ref. [29]

improvement. In the process of algorithm design for the WeJi, the detection accuracy, computational complexity, parallelism, and hardware reusability are considered. On the contrary, hardware implementation problems are not considered in the literature [28], such as parallelism and hardware reuse. Second, the initial iterative solutions in WeJi in this section are different from that in literature [28]. The initial value in literature [28] is fixed. By contrast, the method described in this section takes into account the characteristics of the massive MIMO system, including a computational method for the initial solution. According to Eq. (2.91), the initial iterative solution is close to the final result, so the number of iterations can be reduced and the hardware consumption can be minimized. Furthermore, the method of the initial solution for the WeJi is similar to the algorithm in the later iterative steps, and the hardware resources can be reused. Because the Gram matrix G computation will occupy a large number of clock cycles before the iteration, the reuse of hardware resources will not affect the system throughput rate. Third, compared with the literature [28], the WeJi introduces a weighed factor, as shown in Eq. (2.90), so that improves the accuracy of the solution and consequently reduces hardware resource consumption. In addition, the same unit can be reused to increase unit utilization during the pre-iteration and the iteration, and this reuse does not affect the data throughput of hardware.

Next, we will discuss the frame error rate (FER) of the WeJi and the latest other signal detection algorithms. The FER performance of the exact matrix inversion (the CHD) algorithm is also used as a comparison object. In comparison, we consider modulation scheme of 64-QAM. The channel is assumed to be an independent and identically distributed Rayleigh fading matrix. The output (LLR) is adopted by Viterbi decoding. In the receiver, the LLR is the soft-input of the viterbi decoding. As for 4G and 5G, we have discussed a kind of parallel cascade convolution code, the Turbo code [1]. The Turbo scheme currently used in 4G is also an important encoding scheme for 5G and is widely used. Furthermore, these emulation settings are often used in many massive MIMO detection algorithms and architectures in the 5G communications.

Figure 2.8 shows the FER performance curves for the WeJi [29], NSA [7], RI [14], intra-iterative interference cancelation (IIC) [30], CG [18], GAS [12, 25], OCD [31] and MMSE [8, 24]. In Fig. 2.8a, the algorithm in the 64 * 8 massive MIMO system with 1/2 bit rate is simulated. Figure 2.8b shows the FER performance of the 128 * 8 massive MIMO system with 1/2 bit rate. To demonstrate the advantages of the proposed method at higher bit rates, Fig. 2.8c shows the performance of a 128 *8 massive MIMO system with a 3/4 bit rate. These simulation results show that the WeJi can achieve near-optimal performance at different MIMO scales and bit rates. To achieve the same FER, the SNR required by the WeJi is almost the same as that required by the MMSE, but lower than that required by the OCD, CG, GAS, IIC, RI, and NSA. By Fig. 2.8, the proposed WeJi can achieve better FER performance than the existing technical methods in different MIMO scales.

2.4.3 Analyses on Channel Model Effects

The previous simulation results are obtained based on the Rayleigh fading channel model. In order to prove the superiority of the proposed algorithm in a more real channel model, Fig. 2.9 shows the effects of large-scale fading and spatial correlation of MIMO channels on the FER performance of different algorithms. The Kronecker channel model [20] was used to evaluate the FER performance of algorithm, because it is more practical than the independent and equally distributed Rayleigh fading channel model. The Kronecker channel model assumes that transmission and reception are separable, and the measurements show that the Kronecker model is a good approximation to the nonline-of-sight scenario. Therefore, this model is widely used in the literature. In the channel model, the elements of the channel matrix satisfy $N(0, d(z)I_B)$, where $d(z)$ is an arbitrary function that interprets channel attenuation such as shadow and path loss. Consider the channel attenuation variance $d(z) = \frac{C}{\|z-b\|^\kappa}$, where $z \in R^2$, $z \in R^2$, κ and $\|\cdot\|$ denote the user's location, base station's location, path loss index and Euclidean norm respectively. The independent shadow fading C satisfies $10 \lg N\left(0, \sigma_{sf}^2\right)$. Combined with the correlation matrix (R_r), the Kronecker channel matrix H can be written as

$$H = R_r^{\frac{1}{2}} H_{i.i.d.} \sqrt{d(z)} R_t^{\frac{1}{2}}$$

(2.114)

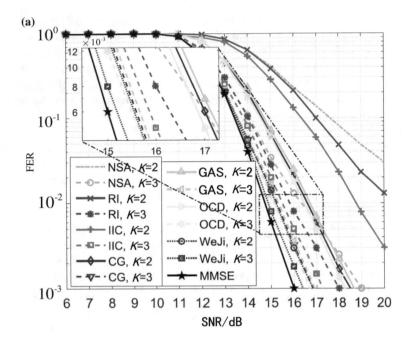

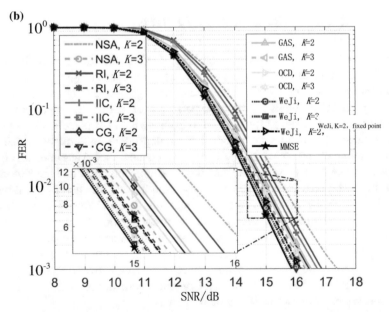

Fig. 2.8 Performance diagram of various algorithms. **a** $N_r = 64$, $N_t = 8$, 1/2 bit rate, **b** $N_r = 128$, $N_t = 8$, 1/2 bit rate, **c** $N_r = 128$, $N_t = 8$, 3/4 bit rate. © [2018] IEEE. Reprinted, with permission, from Ref. [29]

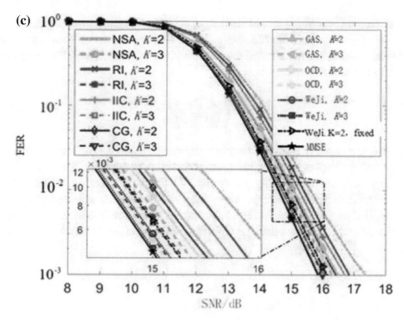

Fig. 2.8 (continued)

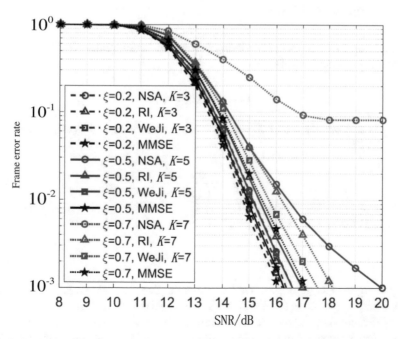

Fig. 2.9 FER performance of Kronecker channel model. © [2018] IEEE. Reprinted, with permission, from Ref. [29]

$H_{\text{i.i.d.}}$ is a random matrix whose elements are independent and identically distributed. It is a complex Gaussian distribution with the zero mean and unit variance. Exponential correlation is a model used to generate correlation matrices. The elements in correlation matrix R_r can be written as

$$r_{ij} = \begin{cases} \xi^{j-i}, & i \leq j \\ \left(\xi^{j-i}\right), & i > j \end{cases} \tag{2.115}$$

where ξ is the correlation factor between adjacent branches. The users in the same cell are evenly distributed in the hexagon with radius $r = 500$ m. Now we assume $\kappa = 3.7$, $\sigma_{\text{sf}}^2 = 5$, the transmitting power $\rho = \frac{\gamma^\kappa}{2}$ for the simulation. The correlation factors are 0.2, 0.5, and 0.7 respectively. Figure 2.9 shows that in order to achieve the same FER, the SNR required by the proposed algorithm is also smaller than CG, RI and NSA, which proves that the algorithm can maintain its advantages in a real model.

2.5 Conjugate Gradient Algorithm

2.5.1 Algorithm Design

CG is an iterative algorithm for solving linear matrix equations [9]. Its approximate solution is presented as follows:

$$x_{K+1} = x_K + \alpha_K p_K \tag{2.116}$$

where p is an auxiliary vector, K is the number of iterations, and α_K can be calculated in Eq. (2.117)

$$\alpha_K = \frac{(r_K, r_K)}{(Ap_K, r_K)} \tag{2.117}$$

Here r is the residual vector, which is expressed by Eq. (2.118)

$$r_{K+1} = r_K + \alpha_K Ap_K \tag{2.118}$$

By using the CG algorithm, we can solve the linear matrix equation. Hence this method can be applied in the massive MIMO signal detection. However, the conventional CG algorithm still has some deficiencies. First of all, in the original algorithm, although every element in the vector can be multiplied and accumulated in parallel with each row element of the matrix when matrix times vector, there is a strong data dependence between the steps, so the computation must be carried out step by step according to the order of the algorithm, and the degree of parallelism is not high.

Secondly, the conventional CG algorithm does not provide information about the initial value of iteration, but the zero vector is usually used in the iteration. Obviously, zero vector can only satisfy the feasible requirements, but it is not optimal. Therefore, finding a better initial iterative value can make the algorithm converge faster and reduce the number of iterations, thus indirectly reducing the computational complexity of the algorithm. The CG algorithm has been improved to satisfy better parallelism and convergence for the above two points. The improved CG algorithm is named as three-term-recursion conjugate gradient (TCG) algorithm [32].

The conventional CG algorithm is equivalent to Lanczos orthogonalization algorithm [9], so it is presented as

$$r_{K+1} = \rho_K(r_K - \gamma_K A r_K) + \mu_K r_{K-1} \tag{2.119}$$

Here, polynomial q_j can be used to make $r_j = q_j A r_0$. When $A = 0, r_j = b - Ax_j \equiv b, b = \rho_K b + \mu_K b$, we can get $\rho_K + \mu_K = 1$ and Eq. (2.120)

$$r_{K+1} = \rho_K(r_K - \gamma_K A r_K) + (1 - \rho_K)r_{K-1} \tag{2.120}$$

Since r_{K-1}, r_K and r_{K+1} are orthogonal, i.e., $(r_{K+1}, r_K) = (r_{K-1}, r_K) = (r_{K-1}, r_{K+1}) = 0$, We can derive from Eq. (2.120)

$$\gamma_K = \frac{(r_K, r_K)}{(A r_K, r_K)} \tag{2.121}$$

$$\rho_K = \frac{(r_{K-1}, r_{K-1})}{(r_{K-1}, r_{K-1}) + \gamma_K(A r_K, r_{K-1})} \tag{2.122}$$

And since $(A r_K, r_{K-1}) = (r_K, A r_{K-1})$, and $A r_{K-1} = -\frac{r_K}{\rho_{K-1}\gamma_{K-1}} + \frac{r_{K-1}}{\gamma_{K-1}} + \frac{(1-\rho_{K-1})r_{K-1}}{\rho_{K-1}\gamma_{K-1}}$, we can get

$$\rho_K = \frac{1}{1 - \frac{\gamma_K}{\gamma_{K-1}}\frac{(r_K, r_K)}{(r_{K-1}, r_{K-1})}\frac{1}{\rho_{K-1}}} \tag{2.123}$$

We can also derive from Eq. (2.120)

$$x_{K+1} = \rho_K(x_K + \gamma_K r_K) + (1 - \rho_K)x_{K-1} \tag{2.124}$$

By processing the massive MIMO system and utilizing the above derivation process, we can get the TCG algorithm that is applied in the massive MIMO signal detection.

Algorithm 2.2 The parallelizable Chebyshev iteration algorithm used in the massive MIMO to detect the minimum mean square error method

---

Input :

1 : $N_t \times N_r$ Rayleigh flat-fading channel matrix $\mathbf{H}$;

2 : $N_r \times 1$ matched-filter vector $\mathbf{y}^{MF}$;

2 : The number of iterations t;

Output:

    Calculated transmission vector $\tilde{\mathbf{s}}_{k+1}$;

Steps:

1 : $\rho_0 = 1$;

2 : $\mathbf{W} = \mathbf{H}^H \mathbf{H}$;

3 : $\mathbf{z}_0 = \mathbf{y}^{MF} - \mathbf{W}\tilde{\mathbf{s}}_0$;

4 : $\mathbf{z}_{-1} = \mathbf{z}_0$, $\tilde{\mathbf{s}}_{-1} = \tilde{\mathbf{s}}_0$;

5 : **for** $K = 0:t\text{-}1$ **do**

6 : $\mathbf{\eta}_K = \mathbf{W}\mathbf{z}_K$;

7 : $\phi_K = (\mathbf{\eta}_K, \mathbf{z}_K)$;

8 : $\xi_K = (\mathbf{z}_K, \mathbf{z}_K)$;

9 : $\gamma_K = \dfrac{\xi_K}{\phi_K}$;

10: **if** $K>0$

11:    $\rho_K = \dfrac{1}{1 - \dfrac{\gamma_K}{\gamma_{K-1}} \dfrac{\xi_K}{\xi_{K-1}} \dfrac{1}{\rho_{K-1}}}$;

12: **end**

13: $\tilde{\mathbf{s}}_{K+1} = \rho_K \left(\tilde{\mathbf{s}}_K + \gamma_K \mathbf{z}_K\right) + \left(1 - \rho_K\right)\tilde{\mathbf{s}}_{K-1}$;

14: $\mathbf{z}_{K+1} = \rho_K \left(\mathbf{z}_K - \gamma_K \mathbf{\eta}_K\right) + \left(1 - \rho_K\right)\mathbf{z}_{K-1}$;

15: **end**

We can see from the Algorithm 2.2 that no data dependence exists between $(\mathbf{\eta}_K, z_K)$ and (z_K, z_K) and between $\tilde{s}_{K+1} = \rho_K(\tilde{s}_K + \gamma_K z_K) + (1 - \rho_K)\tilde{s}_{K-1}$ and $z_{K+1} = \rho_K(z_K - \gamma_K \mathbf{\eta}_K) + (1 - \rho_K)z_{K-1}$. So they can perform computation at the same time, increasing the parallelism of the algorithm. Besides since there is also operation of matrix multiplied by the vector, each element in the vector can still be multiplied and accumulated with each row element of the matrix in the hardware design.

2.5.2 *Convergence*

After discussing the CG and the optimization of CG, it is necessary to study the convergence of CG. First, the polynomial is defined, shown as the following (2.125):

$$C_K(t) = \cos[K \arcos(t)], \quad -1 \le t \le 1 \tag{2.125}$$

$$C_{K+1}(t) = 2t C_K(t) - C_{K-1}(t), \quad C_0(t) = 1, \quad C_1(t) = t \tag{2.126}$$

When the above constraint conditions are extended to the case of $|t| > 1$, the polynomial (2.125) becomes

$$C_K(t) = \cosh[K \arcosh(t)], \quad |t| \ge 1 \tag{2.127}$$

The polynomial can also be expressed as

$$C_K(t) = \frac{1}{2}\left[\left(t + \sqrt{t^2 - 1}\right)^K + \left(t + \sqrt{t^2 - 1}\right)^{-K}\right] \ge \frac{1}{2}\left(t + \sqrt{t^2 - 1}\right)^K \tag{2.128}$$

If $\eta = \frac{\lambda_{\min}}{\lambda_{\max} - \lambda_{\min}}$ is defined, it can be derived from Eq. (2.127)

$$C_K(t) = \frac{1}{2}C_K(1 + 2\eta) \ge \frac{1}{2}\left[1 + 2\eta + \sqrt{(1 + 2\eta)^2 - 1}\right]^K$$

$$\ge \frac{1}{2}\left[1 + 2\eta + 2\sqrt{\eta(\eta + 1)}\right]^K \tag{2.129}$$

We can find that η satisfies from the above expression (2.129).

$$1 + 2\eta + 2\sqrt{\eta(\eta + 1)} = \left(\sqrt{\eta} + \sqrt{\eta + 1}\right)^2 = \frac{\left(\sqrt{\lambda_{\min}} + \sqrt{\lambda_{\max}}\right)^2}{\lambda_{\max} - \lambda_{\min}}$$

$$= \frac{\sqrt{\lambda_{\max}} + \sqrt{\lambda_{\min}}}{\sqrt{\lambda_{\max}} - \sqrt{\lambda_{\min}}} = \frac{\sqrt{\kappa} + 1}{\sqrt{\kappa} - 1} \tag{2.130}$$

where $\kappa = \frac{\lambda_{\max}}{\lambda_{\min}}$. If x_* is for an exact solution vector, the expression of CG convergence is obtained from (2.131):

$$\|x_* - x_K\|_A \le 2\left(\frac{\sqrt{\kappa} - 1}{\sqrt{\kappa} + 1}\right)^K \|x_* - x_0\|_A \tag{2.131}$$

The observation (2.131) shows that x_* denotes the exact solution vector, which is a definite invariant vector. So when the initial value of iteration is determined, $2\|x_* - x_0\|_A$ on the right side of (2.131) can be regarded as a constant. When the number of iterations K increases, $\|x_* - x_K\|_A$ exponentially approaches 0. This indicates that CG can converge faster when K is large, and the error of CG algorithm decreases with the increase of number of iterations K.

2.5.3 Initial Iteration Value and Search

Iteration requires corresponding initial value vectors in the TCG. As mentioned earlier, how to find a good initial value vector is crucial to the convergence rate of the algorithm. When introducing PCI in Sect. 2.3, we mentioned an eigenvalue-based initial value algorithm, which can be used in the TCG to make algorithm converge faster. Besides this, another quadrant-based initial value algorithm will be discussed in this section.

As discussed above, if the influence of noise n is ignored in the massive MIMO system, the real part and the imaginary part of the received signal y, the transmitted signal s and the channel matrix H are separated and written in the real number form, and the relationship between them can be expressed as

$$\begin{bmatrix} \text{Re}\{y\} \\ \text{Im}\{y\} \end{bmatrix}_{2N_r \times 1} = \begin{bmatrix} \text{Re}\{H\} & -\text{Im}\{H\} \\ \text{Im}\{H\} & \text{Re}\{H\} \end{bmatrix}_{2N_r \times 2N_t} \begin{bmatrix} \text{Re}\{s\} \\ \text{Im}\{s\} \end{bmatrix}_{2N_t \times 1} + \begin{bmatrix} \text{Re}\{n\} \\ \text{Im}\{n\} \end{bmatrix}_{2N_r \times 1}$$

(2.132)

$$y_R = H_R s_R \tag{2.133}$$

where the subscript R indicates the expression in the form of real numbers. By using real number form in ZF, the transmitted signal s can be estimated as

$$\hat{s}_R = \left(H_R^H H_R\right)^{-1} H_R y_R = \left(H_R^H H_R\right)^{-1} y_R^{MF} \tag{2.134}$$

In the channel matrix H_R in the real number form, each element is independent and identically distributed, and obeys the standard Gaussian distribution, so the matrix $H_R^H H_R$ can be approximated as a diagonal matrix. And the diagonal elements are all nonnegative since each diagonal element is the sum of squares. If $H_R^H H_R$ is approximated as a diagonal matrix, then $\left(H_R^H H_R\right)^{-1}$ is also a diagonal matrix, and its diagonal elements are also nonnegative. Now consider the relationship between the ith element $\hat{s}_{R,i}$ of vector $\hat{s}_R$ and the jth element $y_{R,i}^{MF}$ of vector y_R^{MF}. According to Eq. (2.134), we can get

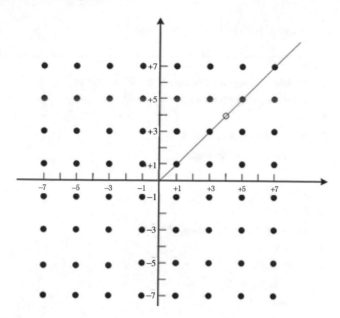

Fig. 2.10 Schematic diagram of quadrant—based initial value algorithm

$$\hat{s}_{R,i} \approx \frac{y_{R,i}^{\mathrm{MF}}}{\sum\limits_{j=1}^{N_t} \left(H_{j,i}\right)^2} \qquad (2.135)$$

where $\dfrac{1}{\sum\limits_{j=1}^{N_t} \left(H_{j,i}\right)^2}$ is nonnegative, so both $\hat{s}_{R,i}$ and $y_{R,i}^{\mathrm{MF}}$ are positive or negative at the same time. By transforming the solution from real number form to complex form, we can deduce that $\hat{s}_i$ and y_i^{MF} are in the same quadrant. Based on this conclusion, a new iteration initial value can be proposed.

Now we are considering the ith element of the transmitted vector $\hat{s}$ in the massive MIMO system with modulation order of 64-QAM. Since $\hat{s}_i$ and y_i^{MF} are in the same quadrant, assuming that y_i^{MF} is in the first quadrant, then we make $\hat{s}_i^{(0)} = 4 + 4i$. The coordinate of the hollow circle is (4,4) as in Fig. 2.10. Since $\hat{s}_i$ will eventually be located in the first quadrant, and the average distance between the point (4, 4) and all constellation points in the first quadrant is less than the average distance between the point (0, 0) and all constellation points in the first quadrant, the point (4, 4) is closer to the final solution so that algorithm 2.3 can converge as soon as possible. The quadrant-based iterative initial value algorithm can be obtained based on this principle.

Algorithm 2.3 Quadrant-based initial value algorithm

Input :

1 : Rayleigh fading channel matrix **H** with a scale of $N_r \times N_t$;

2 : $N_r \times 1$ receiving vector **y**;

Output:

1 : $N_t \times 1$ intitial solution vector $\tilde{\mathbf{s}}_0$;

2 : $N_t \times 1$ matched-filter vector $\mathbf{y}^{MF}$;

Steps:

1 : $\mathbf{y}^{MF} = \mathbf{H}^H \mathbf{y}$;

2 : **for** $k = 1 : N_t$ **do**

3 : re=real($\mathbf{y}^{MF}(k)$), im=imag($\mathbf{y}^{MF}(k)$);

4 : **if** re>0

5 :   **if** im>0

6 : $\tilde{\mathbf{s}}_0(k)$=4+4i;

7 :   **else**

8 : $\tilde{\mathbf{s}}_0(k)$=4-4i;

9 :   **end**

10:   **else**

11:      **if** im>0

12:          $\tilde{\mathbf{s}}_0(k)$=-4+4i;

13:      **else**

14:          $\tilde{\mathbf{s}}_0(k)$=-4-4i;

15:      **end**

16: **end**

17: **end**

Taking the quadrant-based initial iteration value algorithm as the initial value result of TCG, the TCG can converge faster. According to the error requirements of the massive MIMO signal detection, the number of iterations can be reduced, and the computational complexity of the algorithm can be reduced from another angle, which makes the TCG even better.

In the massive MIMO signal detection algorithm, no matter whether it is a linear detection algorithm or a nonlinear detection algorithm, each algorithm will eventually obtain the transmitted vector $\tilde{s}$ through its computation and search for the constellation points with the smallest Euclidean distance from them according to each element in $\tilde{s}$. These constellation points will serve as the final estimated trans-

Fig. 2.11 Schematic diagram of rounding off-based point seeking method

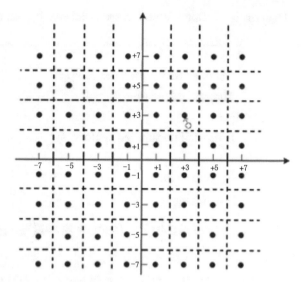

mitted vector $\tilde{s}$. It is worth noting that at present, all modulation modes of the MIMO system are two-dimensional modulation, that is, the modulated constellation points are all located in the plane rectangular coordinate system. Therefore, when calculating the minimum Euclidean distance constellation point of each element in vector $\tilde{s}$, it is actually calculating the constellation point with the minimum distance from the element plane in $\tilde{s}$. Based on this, we can simplify the traditional method, in which the Euclidean distance between the middle element and each constellation point in the constellation diagram is first obtained, and then the size is compared. As shown in Fig. 2.11, it is assumed that an element is in the position of the hollow circle in the diagram. By dividing the constellation diagram into several parts with the dotted line in the diagram, the nearest constellation point to the hollow circle can be determined according to the region in which the hollow circle is located, that is, the constellation points in the region are just the constellation points with the smallest distance from the element. By using this analysis, the operation of finding the constellation points of the minimum Euclidean distance of the elements in the transmitted vector after the iteration can be simplified correspondingly, and a point finding algorithm based on rounding is obtained. Using algorithm 2.4 to locate constellation points, we can quickly find the final result at very low computational complexity.

Algorithm 2.4 Rounding off-based point seeking method

Input :

    Obtained transmitted vector $\tilde{\mathbf{s}}_{K+1}$;

    Through calculation;

Output :

    Estimated transmitted vectors $\hat{\mathbf{s}}$;

Steps :

1 : **for** $t=1:N_t$ **do**

2 : **if** real/imag$\left(\tilde{\mathbf{s}}_{K+1}(t)\right) > 7$

3 :   real/imag$\left(\hat{\mathbf{s}}(t)\right) = 7\,\text{sgn}\left(\text{real/imag}\left(\tilde{\mathbf{s}}_{K+1}(t)\right)\right)$;

4 : **else**

5 : $re/im\left(\tilde{\mathbf{s}}_{K+1}(t)\right) = \dfrac{1}{2}\text{real}/\text{imag}\left(\tilde{\mathbf{s}}_{K+1}(t)\right) + \dfrac{m}{2}, m = 2i+1, i \in \mathrm{N}$;

6 : $re/im\left(\hat{\mathbf{s}}(t)\right) = \text{round}\left(re/im\left(\tilde{\mathbf{s}}_{K+1}(t)\right)\right)$;

7 :   real/imag$\left(\hat{\mathbf{s}}(t)\right) = 2re/im\left(\hat{\mathbf{s}}(t)\right) - m, m = 2i+1, i \in \mathrm{N}$;

8 : **end**

2.5.4 Complexity and Parallelism

In the TCG algorithm, the first part is a series of initial value calculations, including $\mathbf{y}^{\text{MF}}$, $\mathbf{W}$ and z_0. These calculations involve the $N_t \times N_r$ matrix $\mathbf{H}^H$ multiplied by the $N_r \times 1$ vector $\mathbf{y}$, the $N_t \times N_r$ matrix $\mathbf{H}^H$ multiplied by the $N_r \times N_t$ matrix $\mathbf{H}$ to get the $N_t \times N_r$ matrix $\mathbf{W}$, the $N_t \times N_t$ matrix $\mathbf{W}$ multiplied by the $N_t \times 1$ vector $\tilde{\mathbf{s}}_0$. The computational complexities of them are $4N_tN_r$, $2N_rN_t^2$ and $4N_t^2$ respectively. The complexity of matrix $\mathbf{W}$ is only half (that is $2N_rN_t^2$) because it is a symmetric matrix. Since these parameters are computed only once in the pre-iteration process, the complexity is counted only once. The calculation of the second part is the multiplication of matrices and vectors in the iteration, that is, the $N_t \times 1$ vector z is multiplied by the $N_t \times N_t$ matrix $\mathbf{W}$. The computational complexity is $4KN_t^2$. The calculation of the third part is the calculation of two inner products of $\left(\eta_K, z_K\right)$ and (z_K, z_K), and their complexity is $4KN_t$ and $4KN_t$ respectively. The last part is the update for $\tilde{s}$ and z whose complexities are $12KN_t$ and $12KN_t$ respectively. Note that when the number of iterations K is 1, the value of ρ is 1, so the step 13 and the step 14 in the algorithm do not need to be calculated. Therefore the total complexity of the algorithm is $2N_rN_t^2 + 4N_tN_r + 4N_t^2 + K\left(4N_t^2 + 32N_t\right) - 16N_t$. Table 2.2 lists the complexity of various algorithms at different iterations. We can see from the table that the complexity of

Table 2.2 Complexity comparison

	$K=2$	$K=3$	$K=4$	$K=5$
NSA [7]	$2N_rN_t^2 + 6N_tN_r + 4N_t^2 + 2N_t$	$2N_rN_t^2 + 6N_tN_r + 2N_t^3 + 4N_t^2$	$2N_rN_t^2 + 6N_tN_r + 6N_t^3$	$2N_rN_t^2 + 6N_tN_r + 10N_t^3 - 6N_t^2$
INS [17]	$2N_rN_t^2 + 6N_tN_r + 10N_t^2 + 2N_t$	$2N_rN_t^2 + 6N_tN_r + 14N_t^2 + 2N_t$	$2N_rN_t^2 + 6N_tN_r + 18N_t^2 + 2N_t$	$2N_rN_t^2 + 6N_tN_r + 22N_t^2 + 2N_t$
GAS [12]	$2N_rN_t^2 + 6N_tN_r + 10N_t^2 - 2N_t$	$2N_rN_t^2 + 6N_tN_r + 14N_t^2 - 6N_t$	$2N_rN_t^2 + 6N_tN_r + 18N_t^2 - 10N_t$	$2N_rN_t^2 + 6N_tN_r + 22N_t^2 - 14N_t$
CG [16]	$2N_rN_t^2 + 6N_tN_r + 8N_t^2 + 33N_t$	$2N_rN_t^2 + 6N_tN_r + 12N_t^2 + 49N_t$	$2N_rN_t^2 + 6N_tN_r + 12N_t^2 + 49N_t$	$2N_rN_t^2 + 6N_tN_r + 20N_t^2 + 81N_t$
CGLS [18]	$24N_tN_r + 20N_t^2 + 8N_r + 44N_t$	$32N_tN_r + 28N_t^2 + 12N_r + 66N_t$	$40N_tN_r + 36N_t^2 + 16N_r + 88N_t$	$48N_tN_r + 44N_t^2 + 20N_r + 110N_t$
OCD [19]	$16N_tN_r + 4N_t$	$24N_tN_r + 6N_t$	$32N_tN_r + 8N_t$	$40N_tN_r + 10N_t$
TCG	$2N_rN_t^2 + 4N_tN_r + 12N_t^2 + 48N_t$	$2N_rN_t^2 + 4N_tN_r + 16N_t^2 + 80N_t$	$2N_rN_t^2 + 4N_tN_r + 20N_t^2 + 112N_t$	$2N_rN_t^2 + 4N_tN_r + 24N_t^2 + 144N_t$

the TCG algorithm is low. For example, when $N_r = 128$, $N_t = 8$, $K = 2$, the OCD algorithm has the lowest complexity of 16,416 real number multiplications, and the TCG algorithm has 21,632 real number multiplications, which is higher than that of the OCD algorithm, while the complexity of other algorithms is higher than that of the TCG algorithm.

The parallelism of the TCG algorithm is also very important. According to Step 2, Step 3, Step 6, and Step 7 of the algorithm, each row element of the matrix is multiplied by the vector when the matrix is multiplied by the vector, and each multiply accumulate operation can be carried out simultaneously. There is no data dependence between the eighth and ninth steps of the algorithm and between the fourteenth and fifteenth steps of the algorithm except the parallel computation of the matrix multiplied by vectors, so the computation can be performed simultaneously. This algorithm has parallelism in two cases. Compared with other algorithms, the parallelism between steps has great advantages.

As previously analyzed, the TCG has significant advantages over other algorithms in complexity and parallelism. In addition, the TCG algorithm optimizes the initial value of iteration and the final point finding method, so that this algorithm can bring the maximum benefit. In general, complexity and accuracy are in contradiction, and lower complexity often means lower precision. Therefore, although the above performance shows the excellent performance of the TCG algorithm in this respect, it does not show that it is a relatively comprehensive algorithm, and it is also required to consider the accuracy of the algorithm.

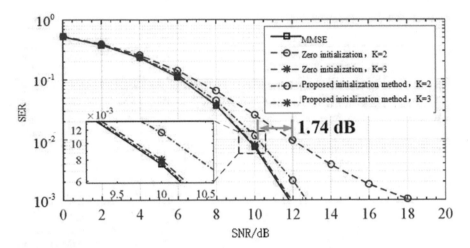

Fig. 2.12   Effect of the initial value algorithm on the SER of the CG algorithm

2.5.5   Symbol Error Rate

Figure 2.12 shows the effect of the initial value on the performance of the CG algorithm in the massive MIMO system with the size of 128×8. The initial value is calculated by the quadrant division initial value algorithm. Obviously, for CG algorithm with the same SNR, the initial value algorithm can achieve lower symbol error rate (SER) for the same number of iterations, and when the number of iterations $K = 2$, the initial value algorithm has a greater impact on the performance of the algorithm.

Figure 2.13 shows the SNR and SER curves of various algorithms for different number of iterations in the massive MIMO system with a size of 128*8 and with modulation order of 64-QAM. It can be clearly seen from the graph that, the SNR of NSA [7] is 10.17 dB, and the SNR of MMSE [8, 24], CG [12] and OCD [31] is around 9.66, 10.10, and 10.11 dB when the number of iterations $K = 2$ and SER is 10^{-2}. However the SNR of WeJi [29] is about 10.42 db, and the SNR of NSA [7] is larger than 20 dB. Figure 2.14 shows the influence of various algorithms on SER in MIMO systems of different sizes. We can see that the SER of TCG algorithm is obviously lower than that of the NSA and WeJi [29] in MIMO systems with different sizes.

Therefore, the TCG has obvious advantages compared with the three algorithms of NSA, CG and CGLS. The TCG is better than the other three methods in complexity and parallelism, while the performance of SER is still worse than the three methods under the same SNR. Compared with the PCI algorithm, the TCG algorithm has lower complexity and better parallelism than the PCI. Besides having the parallelism of the matrix multiplied by vectors in the PCI, there is also parallelism between the steps in the algorithm. Compared with the OCD algorithm, the complexity of the

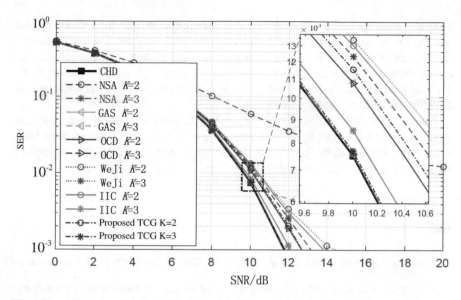

Fig. 2.13 SER curves of various algorithms for different number of iterations in massive MIMO system with 128 * 8

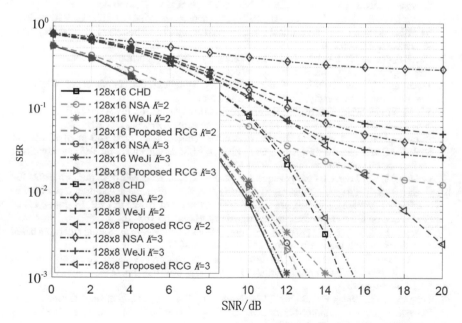

Fig. 2.14 Influence of different algorithms on SER in MIMO systems with different scales

TCG algorithm is higher than that of the OCD algorithm, and the OCD algorithm also has better performance of SER. However, the OCD method also has its inherent shortcomings: Because the OCD algorithm has too strong data dependence, the obtained data through calculation need to be stored in the register and each calculation needs to read data from the register, the parallelism is so poor that the operation can only be executed step by step sequentially.

Compared with the NSA, CG, CGLS, PCI, OCD, and other algorithms, the TCG algorithm is not the best in all performance parameters, but it overall achieves a better compromise. In the actual massive MIMO signal detection, the appropriate algorithm can be selected according to the actual application requirements.

References

1. Andrews JG, Buzzi S, Wan C et al (2014) What Will 5G Be? IEEE J Sel Areas Commun 32(6):1065–1082
2. Kim SP, Sanchez JC, Rao YN et al (2006) A comparison of optimal MIMO linear and nonlinear models for brain-machine interfaces. J Neural Eng 3(2):145–161
3. Burg A, Borgmann M, Wenk M et al (2005) VLSI implementation of MIMO detection using the sphere decoding algorithm. IEEE J Solid-State Circuits 40(7):1566–1577
4. Trimeche A, Boukid N, Sakly A et al (2012) Performance analysis of ZF and MMSE equalizers for MIMO systems. In: International conference on design & technology of integrated systems in nanoscale Era, pp 1–6
5. Rusek F, Persson D, Lau BK et al (2012) Scaling up MIMO: opportunities and challenges with very large arrays. Sig Process Mag IEEE 30(1):40–60
6. Teaching and Research Office of Computational Mathematics (2000) Fundamental of Numeric Analysis. Tongji University, Tongji University Press
7. Wu M, Yin B, Wang G et al (2014) Large-scale MIMO Detection for 3GPP LTE: algorithms and FPGA implementations. IEEE J Selected Topics Signal Process 8(5):916–929
8. Auras D, Leupers R, Ascheid GH (2014) A novel reduced-complexity soft-input soft-output MMSE MIMO detector: algorithm and efficient VLSI architecture. In: IEEE international conference on communications, pp 4722–4728
9. Golub GH, Van Loan CF (1996) Matrix computations. Mathe Gazette 47(5 Series II):392–396
10. Shahab MB, Wahla MA, Mushtaq MT (2015) Downlink resource scheduling technique for maximized throughput with improved fairness and reduced BLER in LTE. In: International conference on telecommunications and signal processing, pp 163–167
11. Zhang C, Li Z, Shen L et al (2017) A low-complexity massive mimo precoding algorithm based on chebyshev iteration. IEEE Access 5(99):22545–22551
12. Dai L, Gao X, Su X et al (2015) Low-complexity soft-output signal detection based on Gauss-Seidel method for uplink multiuser large-scale MIMO System. IEEE Trans Veh Technol 64(10):4839–4845
13. Peng G, Liu L, Zhang P et al (2017) Low-computing-load, high-parallelism detection method based on Chebyshev Iteration for massive MIMO Systems with VLSI architecture. IEEE Trans Signal Process 65(14):3775–3788
14. Gao X, Dai L, Ma Y et al (2015) Low-complexity near-optimal signal detection for uplink large-scale MIMO systems. Electron Lett 50(18):1326–1328
15. Gutknecht MH, Röllin S (2000) The Chebyshev iteration revisited. Parallel Comput 28(2):263–283
16. Yin B, Wu M, Cavallaro JR et al (2014) Conjugate gradient-based soft-output detection and precoding in massive MIMO systems. In: Global communications conference, pp 3696–3701

17. Čirkić M, Larsson EG (2014) On the complexity of very large multi-user MIMO Detection
18. Yin B, Wu M, Cavallaro JR et al (2015) VLSI design of large-scale soft-output MIMO detection using conjugate gradients. In: IEEE international symposium on circuits and systems, pp 1498–1501
19. Wu M, Dick C, Cavallaro JR et al (2016) FPGA design of a coordinate descent data detector for large-scale MU-MIMO. In: IEEE international symposium on circuits and systems, pp 1894–1897
20. Werner K, Jansson M (2009) Estimating MIMO channel covariances from training data under the Kronecker model. Sig Process 89(1):1–13
21. Sun Q, Cox DC, Huang HC et al (2002) Estimation of continuous flat fading MIMO channels. IEEE Trans Wireless Commun 1(4):549–553
22. Rappaport TS (2002) Wireless Communications—Principles and Practice. Second Edition. (The Book End). 8(1):33–38
23. Li X, Chen CP (2013) Inequalities for the gamma function. J Inequal Pure Appl Mathematics 8(1):554–563
24. Prabhu H, Rodrigues J, Liu L et al (2017) A 60 pJ/b 300 Mb/s 128×8 Massive MIMO Precoder-Detector in 28 nm FD-SOI
25. Wu Z, Zhang C, Xue Y et al (2016) Efficient architecture for soft-output massive MIMO detection with Gauss-Seidel method. In: IEEE international symposium on circuits and systems, pp 1886–1889
26. Zhang P, Liu L, Peng G et al (2016) Large-scale MIMO detection design and FPGA implementations using SOR method. In: IEEE international conference on communication software and networks, pp 206–210
27. Quan H, Ciocan S, Qian W et al (2015) Low-complexity MMSE signal detection based on WSSOR method for massive MIMO Systems. In: IEEE international symposium on broadband multimedia systems and broadcasting, pp 193–202
28. Kong BY, Park IC (2016) Low-complexity symbol detection for massive MIMO uplink based on Jacobi method. In: IEEE international symposium on personal, indoor, and mobile radio communications, pp 1–5
29. Peng G, Liu L, Zhou S, et al (2017) A 1.58 Gbps/W 0.40 Gbps/mm^2 ASIC implementation of MMSE detection for $128x8$ 64-QAM Massive MIMO in 65 nm CMOS. IEEE Trans Circuits Syst I Regular Papers PP(99):1–14
30. Chen J, Zhang Z, Lu H et al (2016) An intra-iterative interference cancellation detector for large-scale MIMO communications based on convex optimization. IEEE Trans Circuits Syst I Regul Pap 63(11):2062–2072
31. Wu M, Dick C, Cavallaro JR et al (2016) High throughput data detection for massive MU-MIMO-OFDM using coordinate descent. IEEE Trans Circuits Syst I Regul Pap 63(12):2357–2367
32. Tongxiang Gu (2015) Iterative methods and pretreatment technology. Science Press, Beijing

Chapter 3
Architecture of Linear Massive MIMO Detection

In practical massive MIMO detection, besides the influence of the algorithm's own characteristics on the detection results, the hardware circuit also affects the efficiency of signal detection. In Chap. 2, we introduce four typical iteration algorithms of massive MIMO linear detection, and illustrate their advantages by comparing them with some existing linear detection algorithms. This chapter describes how to implement the four algorithms in VLSI. First, it describes how to implement the algorithm in the hardware circuit, and the matters needing attention. Then, the optimization problems in the chip design are introduced, including how to improve the throughput rate of the chip, reduce the power consumption of the chip, and reduce the area of the chip. Finally, the parameters of the designed chip are compared with those of the existing linear detection algorithm, and the comprehensive comparison results are obtained.

3.1 NSA-Based Hardware Architecture

Based on the Neumann Series Approximation (NSA), this section details two large-scale MIMO detection VLSI architectures for 3GPP LTE-A. First, this section introduces the VLSI top-level structure, and analyzes the design method of VLSI architecture for large-scale MIMO detection as a whole. Second, this section introduces the approximate inversion and matched filtering (MF) module in detail. Then, the equalization, SINR, Inverse Fast Fourier Transform (IFFT) and LLR module are described. Finally, the design method of the exact inverse module based on Cholesky decomposition and details are introduced.

3.1.1 VLSI Top-Level Structure

Based on NSA, the proposed general architecture is shown in Fig. 3.1. The whole framework consists of the following parts: the preprocessing element, the subcarrier

© Springer Nature Singapore Pte Ltd. and Science Press, Beijing, China 2019
L. Liu et al., *Massive MIMO Detection Algorithm and VLSI Architecture*,
https://doi.org/10.1007/978-981-13-6362-7_3

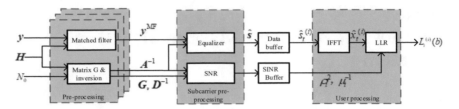

Fig. 3.1 VLSI top-level architecture for massive MIMO detection in 3GPP LTE-A. © [2018] IEEE. Reprinted, with permission, from Ref. [8]

processing element, and the approximate LLR processing unit. The preprocessing element consists of the matched filter, the Gram matrix processing element, and its inversion. This element is used to perform MF computation, i.e., to compute $y^{\mathrm{MF}} = H^H y$, and the computation of the normalized Gram matrix G and its approximate inverse matrix as well. It is worth noting that for the approximate inverse unit, we also output D^{-1} and G required by SINR computation. In order to achieve the peak throughput rate [1] required in LTE-A, multiple preprocessing elements (PEs) are used in the design. After operation of the preprocessing element, the data output by the matched filtering, the approximate inverse, and the normalized Gram matrix are transferred to the subcarrier processing element. The subcarrier processing element performs equalization processing, that is to compute $s = A^{-1} y^{\mathrm{MF}}$ and post-equalization SINR. To detect each user's data, a buffer is needed to aggregate all equalized symbols and SINR values, which are computed on each subcarrier. After operation of the subcarrier processing element, the architecture converts the equalization symbols from the subcarrier domain into the user domain (or time domain) symbols by performing IFFT. The approximate LLR -processing unit finally solves the maximum value of LLR and the NPI value at the same time. The key details on the proposed detector architecture shall be discussed later.

3.1.2 Approximate Inversion and Matched Filtering Module

1. Computing Unit for Approximate Inversion

To achieve a higher data throughput, a single systolic array is used here. Four phases are needed to calculate the normalized Gram matrix and its approximate inverse matrix. The Gram matrix calculation and the approximate inversion unit are shown in Fig. 3.2. The structure can select the number of terms of multiple Neumann series at run time. As shown in Fig. 3.2, the lower triangular systolic array consists of two different PEs: PE (PE-D) on the main diagonal and PE (PE-OD) on the non-diagonal of the systolic array, which have different modes in four computational phases.

In the first phase, the computation of the normalized Gram matrix $\frac{A}{N_r} = \frac{G + N_0 / E_s I}{N_r}$ of $N_t \times N_t$ takes N_r clock cycles. Since A is asymptotic to N_r in the diagonal direction

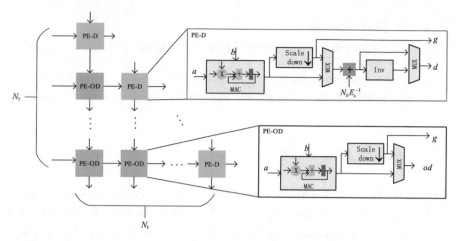

Fig. 3.2 Diagram of Gram matrix computation and approximate inversion unit. © [2018] IEEE. Reprinted, with permission, from Ref. [8]

and diagonally dominant, its dynamic range can be reduced by normalization, which is a common matrix inversion circuit and fixed-point algorithm. The systolic array also computes $D^{-1}N_r$ from the diagonal elements of $\frac{A}{N_r}$, which are computed in reciprocal units (RECUs) of the PE-D unit (expressed by "Inv" in Fig. 3.2). Then, the values of $D^{-1}N_r$ and $\frac{E}{N_r}$ will be stored in the systolic array distributed registers.

In the second phase, the systolic array computes $-D^{-1}E$ by using the solution of matrices $D^{-1}N_r$ and $\frac{E}{N_r}$ in the first phase. Since the matrix $-D^{-1}E$ is not a Hermitian matrix [2], the systolic array needs to calculate the upper triangle and the lower triangle of $-D^{-1}E$. Since D^{-1} is a diagonal matrix, the computation of $-D^{-1}E$ requires only a series of scalar multiplications (not matrix multiplication).

In the third phase, the systolic array calculates the NSA when $K = 2$, that is $\tilde{A}_2^{-1}N_r = D^{-1}N_r - D^{-1}ED^{-1}N_r$. First, it is important to understand that the matrix $D^{-1}N_r - D^{-1}ED^{-1}N_r$ is a Hermitian matrix, which means we only need to solve the lower triangle of the matrix. Moreover, the computation of $-D^{-1}ED^{-1}N_r$ only requires multiplications term by term (not matrix multiplication) because $D^{-1}N_r$ is a diagonal matrix. These scalar multiplications are performed by loading $D^{-1}N_r$ and $-ED^{-1}$ into all PEs and performing scalar multiplications to solve $D^{-1}ED^{-1}N_r$. Then $D^{-1}N_r$ is needed to add to the solution of the diagonal PE. The result of this phase, $D^{-1}N_r - D^{-1}ED^{-1}N_r$ is stored in distributed registers.

In the fourth phase, the NSA value with the number of terms equal to K is computed, and the solution is also stored in the distributed registers. In particular, the systolic array first performs matrix multiplication between $-D^{-1}E$ and $\tilde{A}_{K-1}^{-1}N_r$, and then adds $D^{-1}N_r$ to the diagonal PE. After that, the approximate value $\tilde{A}_K^{-1}N_r$ with entries K is stored in the register. In this phase, the configurable number of iterations can be repeated so that the structure can compute the NSA value with any number of terms K.

2. Matched Filtering Unit

The MF unit reads a new y input in each clock cycle and multiplies it with the H^H in each PE. In the MF unit, the input data of each PE is a row of matrix H^H, and each PE contains a multiply–accumulate (MAC) and a normalization unit for computing $\frac{y^{MF}}{N_r}$. Then it is added to the previous result and the obtained new result is normalized.

3.1.3 Equalization and SINR Module

The equalization unit consists of a linear array of MAC units and reads the normalized approximate inversion matrices $\tilde{A}_K^{-1} N_r$ and $\tilde{A}_K^{-1} N_r$ from the MF unit. The unit reads a column of $\tilde{A}_K^{-1} N_r$ each clock cycle, multiplies it by an element in $\frac{y^{MF}}{N_r}$, and then adds it to the previous solution. The unit outputs an equalization vector $\hat{s}$ every N_t clock cycles.

The SINR processing unit consists of N_t MAC units and computes the approximate effective channel gain sequentially. The unit also uses a single MAC unit to compute approximate NPI. Subsequently, the unit multiplies $\tilde{\mu}_K^{(i)}$ with the reciprocal of approximate NPI $\tilde{v}_i^2$ to obtain post-equalization SINR ρ_i^2. The same unit will also compute the reciprocal of $\tilde{\mu}_K^{(i)}$ used in the approximate LLR processing unit.

3.1.4 IFFT and LLR Module

To convert the data in each subcarrier to data in the user (or time) domain, a Xilinx IFFT IP LogiCORE unit is required to be deployed. The unit supports all FFT and IFF modes specified in 3GPP LTE, but only its IFFT functions are used in this design. The IFFT unit reads and outputs data in serial mode. For processing 1200 subcarriers of IFFT, the kernel can process a new set of data every 3779 clock cycles. This IFFT unit achieves a frequency higher than 317 MHz on Virtex-7 XC7VX980T FPGA. Therefore, a 64-QAM MIMO system with eight users can achieve a bandwidth of 20 MHz and a data throughput of over 600 Mbit/s.

The LLR computing unit (LCU) outputs maximum LLR soft-output value and effective channel gain $\mu^{(i)}$. Since LTE specifies Gray mapping of all modulation schemes (BPSK, QPSK, 16-QAM, and 64-QAM), and $\lambda_b(\cdot)$ is a piecewise function [3], the computation of the maximum LLR can be simplified by rewriting $L_t^{(i)}(b) = \rho_i^2 \lambda_b\left(\hat{x}_t^{(i)}\right)$. For this purpose, by using the reciprocal $\frac{1}{\mu^{(i)}}$ of the effective channel gain, LCU first amplifies or reduce the real and imaginary parts of the equalized time-domain symbols and the effective channel gain. Then, the piecewise linear function $\lambda_b\left(\hat{x}_t^{(i)}\right)$ is estimated, and the result is amplified or reduced by the post-equalization SINR ρ_i^2. Finally, the obtained maximum LLR value is transferred to the output unit. In order to minimize the circuit area, the proposed architecture only uses logic shift and the logic "AND" to access each piecewise linear function. The inverse calculation is performed by looking up a table stored in the B-RAM unit [4]. Each clock cycle LCU processes one symbol, thus the 64-QAM hardware can reach a peak throughput rate of 1.89 Gbit/s at a frequency of 317 MHz.

3.1.5 Inverse Module Based on Cholesky Decomposition

In order to evaluate the performance and complexity of the proposed approximate inverse matrix unit, an exact inverse unit is used to compare the proposed inverse unit, which simply replaces the previous approximate inverse unit. This section first summarizes the inverse algorithm based on the Cholesky decomposition algorithm and then introduces the corresponding VLSI structure design.

3.1.5.1 Inverse Algorithm

In the exact inverse unit, A^{-1} is solved in three steps. First, compute the normalized Gram matrix $A = G + \frac{N_0}{E_s} I$. Then, the Cholesky decomposition is performed according to $A = LL^H$, where L is a lower triangular matrix with a real principal diagonal. Finally, using the effective forward/backward substitution procedure to solve A^{-1}. Specifically, we first solve $Lu_i = e_i$ by forward substitution, where $i = 1, 2, \ldots, N_t$, e_i is the ith unit vector. After that, solve v_i by backward substitution $L^H v_i = u_i$, where $i = 1, 2, \ldots, N_t$, then $A^{-1} = [v_1, v_2, \ldots, v_{N_t}]$.

3.1.5.2 Architecture of Cholesky Decomposition Algorithm

The VLSI architecture based on the Cholesky decomposition algorithm is different from that in Sect. 3.1.2. In particular, three separate units were deployed to compute the normalized Gram matrix, the exact inverse of the matrix, and the forward/backward substitution unit to solve the inverse matrix A^{-1}. The pipeline divides the circuit into several levels, and the details will be explained separately.

The normalized Gram matrix is computed as the sum of the outer products, that is $G = \sum_{i=1}^{N_r} r_i r_i^H$, where r_i represents the row i of the matrix H. Since the Gram matrix is a symmetric matrix, it can be efficiently computed by multiplying and accumulating a systolic array of triangles. The computing unit of the Gram matrix reads the element of one-row matrix H at a time, and outputs the solution of the Gram matrix after N_r clock cycles. To get the normalized Gram matrix A, add each diagonal element of the matrix G to $\frac{N_0}{E_s}$ in the last clock cycle.

Next, use the systolic array to Cholesky decompose matrix A, and get the lower triangular matrix L. The systolic array consists of two different PE: a PE on the main diagonal line and a PE on the non-diagonal line. The data flow is similar to the linear systolic array proposed in Ref. [5]. The difference is that the design uses multiple PEs to process one column of the input matrix A, while only one PE is used in Ref. [5]. Therefore, this design can meet the requirement of LTE-A peak throughput rate. In this design, a pipeline has a depth of 16 levels, and each clock cycle outputs a column of matrix L. Therefore, the data throughput of this design is related to the Cholesky decomposition algorithm per N_t cycles.

3.1.5.3 Architecture of Forward/Backward Substitution Unit

The forward/backward substitution unit inputs a lower triangular matrix L and computes $A^{-1} = (L^H)^{-1} L^{-1}$ as output. There are three parts in the forward/backward substitution unit. The first part solves $Lu_i = e_i$ by forward substitution, where $i = 1, 2, \ldots, N_t$, e_i is the ith unit vector. The second part solves v_i by reverse substitution $L^H v_i = u_i$, where $i = 1, 2, \ldots, N_t$. The third part is the conjugate transpose unit. Because the computation of forward/backward substitution units is symmetrical, you only need to design forward substitution units and reuse them. For convenience, let us say that the forward substitution unit is used to solve the equation $Lx = b$. The equations $Lx_i = b_i$, $i = 1, 2, \ldots, N_t$ solved by the forward substitution element are independent of each other and can be solved simultaneously by N_t PEs. The PE uses pipeline structure and contains N_t-level operational logic units. Each level contains two multiplexers, a complex multiplier, and a complex subtractor. $\Delta_t = b_i - \sum_j L_{i,j} x_j$ and $\frac{\Delta_i}{L_i}$ are computed by controlling signals. Therefore, for a matrix L, there are N_t^2 complex multiplications in the forward substitution unit, and $2 N_t^2$ complex multiplications in the whole forward and backward substitution unit. The conjugate transpose unit makes use of multiplexer and N_t first input first output (FIFO) memory, and the elements in the conjugate transpose matrix L^H are rearranged according to the input sequence of the forward replacement unit.

3.2 Chebyshev Iteration Hardware Architecture

This section describes the VLSI architecture in the massive MIMO detection algorithm for implementing the line soft-output of PCI [6]. Like other latest VLSI architectures for massive MIMO detection, this section also employs independent and distributed Rayleigh fading channel [7, 8]. The VLSI architecture is used to implement the massive MIMO system with 64-QAM and 128×16. Based on the proof and analysis in Chap. 2, the algorithm selects $K = 3$ (including initial solution $K = 0$ and two iterations $K = 1$ and $K = 2$) as the number of iterations number of PCI, so as to achieve a high detection accuracy and a low resource consumption.

3.2.1 VLSI Top-Level Structure

Figure 3.3 is the block diagram of the PCI-based VLSI top-level architecture. To achieve a higher data throughput with limited hardware resources, the top-level architecture is fully pipelined. The MF vector y^{MF} and initial solution $\hat{s}^{(0)}$ are computed in the initial module. In the next three steps, the estimated transmitted vectors ($K = 1, 2, 3$) will be computed in the iterative module 1 and the iterative module 2 (including pre-iterative block and iterative block), and another iteration also must be computed in the initial module. In the iterative module, the $N_t \times N_r$ matrix H^H is multiplied by the $N_t \times 1$ vector $\dot{h}^{(K)}$, and combing subtraction operation to compute the residual vector $r^{(K)}$. The calibration vector $\sigma^{(K)}$ and estimated transmitted signal $\hat{s}^{(K)}$ are also computed therein. Finally, combined with the estimated transmit signal $\hat{s}^{(K)}$, the parameters of β and N_0 are solved respectively and the LLR is output. The architecture memory is used to store the initial data, including the channel matrix H, the received vector y, the parameters N_0 and E_s. In addition, intermediate results such as y^{MF}, $\dot{h}^{(K)}$ and parameter ρ, φ, and β will also be stored in the memory. Four blocks are used to store four different channel matrices H, which meets the requirements of high parallel data access. In each block, there are 32 SRAMs (static random access memory) used to store complex values of the channel matrix. Each clock cycle reads elements of eight channel matrices, and eight elements of the vector y are also read in each clock cycle. Moreover, three blocks are used to store different vectors y^{MF} to prepare data access for two iterative modules, which read one vector y^{MF} element per clock cycle. These modules will be further described later.

3.2.2 Initial Module

There are two main improvements in the design of the initial module, involving a series of new iterative parameter computation methods and user-level pipeline processing mechanism. For the first improvement, according to the PCI and based

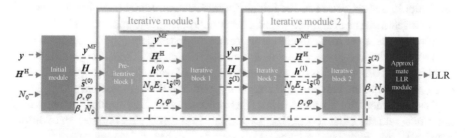

Fig. 3.3 Block diagram of PCI-based VLSI top-level architecture. © [2018] IEEE. Reprinted, with permission, from Ref. [6]

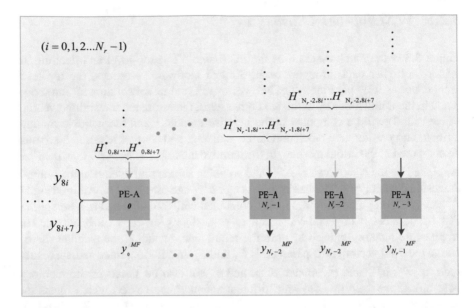

Fig. 3.4 User-level pipeline architecture of the initial module. © [2018] IEEE. Reprinted, with permission, from Ref. [6]

on the properties of the massive MIMO system, the parameters α, β, $\rho^{(K)}$ and $\varphi^{(K)}$ need to be computed. The computation of these parameters is simplified because it only depends on N_r and N_t. To increase the system throughput while limiting resource consumption, the values of these parameters are prestored in the registers to reduce redundant computations. Therefore, these data are converted from the immediate data in the registers and are computed only once (regardless of the number of signal vector groups). For the second improvement, the user-level pipeline mechanism is mainly used to large-scale matrix multiplication in y^{MF} computation, because if the vector y^{MF} has been computed, the calculation of $\hat{s}^{(0)}$ only needs multiplication. Given that y is used every time the product of a row of H^H with vector y is computed, the input of vector y is pipelined to reduce the number of memory accesses. Figure 3.4 is the user-level pipeline architecture of the initial module, which contains N_t PE-As, each of which implements a multiplication and accumulation operations. To achieve a good match with the next level, the structure reads the input data (eight elements) once per clock cycle. The arrangement of input data is highly efficient and parallel in the whole system. Figure 3.5 is the schematic diagram of each PE-A structure, including three main parts: the real part, the imaginary part, and the complex part. Each PE-A has two arithmetic logic unit (ALUs), one implements 8-bit multiplication and seven additions, and the other reorganizes two inputs into a complex number. In addition, matrix H^H and H can be prepared in advance and stored in the register.

In Refs. [7, 8], a similar processing element array is used to compute y^{MF}. In these architectures, each element of vector y is transferred to all the units in the first

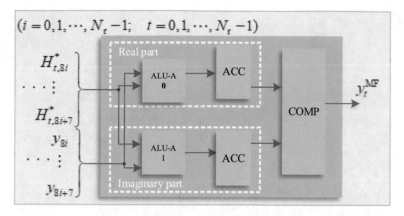

Fig. 3.5 Schematic diagram of the PE-A structure. © [2018] IEEE. Reprinted, with permission, from Ref. [6]

location. Therefore, all outputs from this module are simultaneously transferred to the pre-iterative block. Compared with the proposed architecture, the output end of the architecture in Refs. [7, 8] needs N_t-times registers in the initial block, which means that the proposed architecture saves times area and power consumption of registers. Furthermore, the systolic array architecture consumes more time due to input data. However, time consumption can be negligible and has no effect on the data throughput of the whole system, so the energy and area efficiency are also improved.

3.2.3 Iterative Module

The user-level pipeline processing mechanism is also used in the iterative module. There are two types of units in this module: pre-iterative units and iterative units. Based on the data throughput of the whole system, the large-scale matrix multiplication of $\dot{h}^{(K)}$ needs to be accelerated to match the time consumption of the pipeline. Because the initial solution $\hat{s}^{(0)}$ of each cycle is transferred to the pre-iterative block, the pre-iterative units are arranged into an array structure to implement the user-level pipeline (Fig. 3.6). PE-B has $8 \times (N_t - 1)$ arithmetic units like PE-A. The ALU in PE-B is different. There is no additional module in this ALU because the intermediate data are computed separately. Because of the initial module, the input of the pre-iterative unit is immediately read from the register, including the channel matrix H, the initial solution $\hat{s}^{(0)}$, and N_0. The pre-iterative unit outputs the elements of $\dot{h}^{(K)}$ and transfers them to the iterative unit. Because the pre-iterative unit works in deep pipelining, the input of block matrix H^H and the estimated transmitted signal are utilized simultaneously, which reduces the memory consumption and computation

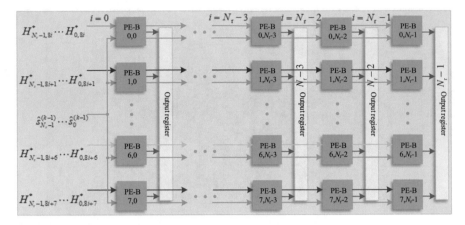

Fig. 3.6 User—level pipeline structure of the pre-iterative module. © [2018] IEEE. Reprinted, with permission, from Ref. [6]

time. The iterative unit in the iterative module is located after the pre-iterative unit. Based on PCI, the iterative unit computes the residual vector $r^{(K)}$ and the estimated transmitted signal $\hat{s}^{(K)}$ of K iterations. In these computation, matrices H^{H} and $\dot{h}^{(K)}$ are the most complex, containing subtraction, small-scale real multiplication and large-scale complex multiplication. As a result, the iterative unit architecture is also designed as a user-level pipelined architecture (Fig. 3.4). This architecture satisfies the time constraints of the entire pipeline, which reduces resource consumption (without affecting the data throughput).

In Refs. [7, 8], similar computations are performed in eight sets of lower triangular systolic arrays because of the large-scale complex matrix multiplication. It takes more resources and energy here (more than 400%). Furthermore, pipeline latency is determined by the time consumption of the systolic array, which means that the data throughput of the entire system is limited by resources. In the proposed architecture, the iterative module maximizes the attributes of the input data and reduces the registers of the intermediate data. However, for Refs. [7, 8], large-scale matrices are required to be stored between iterations.

3.2.4 LLR Module

The LLR module is used to compute the approximate LLR for each transmitted bit. Based on PCI, the $\xi_b(\hat{s}_i)$ function is rewritten as a piecewise linear function of Gray mapping. Figure 3.7 shows the architecture of LLR processing module. There are three identical PE-C in the figure. The input parameters β and N_0 of the unit are multiplied in ALU-B and the reciprocal of their product is output (referenced by the table). To reduce the resource consumption, the coefficient $\xi_b(\hat{s}_i)$ of each linear

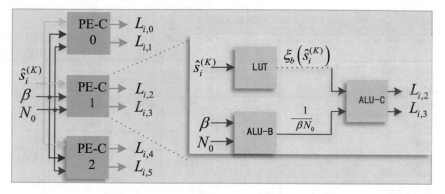

Fig. 3.7 LLR processing module. © [2018] IEEE. Reprinted, with permission, from Ref. [6]

equation is stored in a corrected look-up table (LUT). Next, the final soft-output is computed in ALU-C. The approximate LLR computation module simplifies the soft-output calculation, speeds up the data processing, and reduces the circuit area and power consumption. Although this module increases the number of registers, this increase is negligible.

3.2.5   Experimental Results and Comparison

The VLSI architecture is described in the Verilog Hardware Description Language (HDL) and verified on the FPGA platform (Xilinx Virtex-7). Table 3.1 shows the architecture for MMSE in the massive MIMO systems, and lists the key implementation results of the architecture proposed in this section and other existing technology designs. Compared with the architecture in Ref. [8], the data throughput of the architecture is increased by 2.04 times and the resource consumption is greatly reduced. For example, compared with the Cholesky decomposition algorithm and NSA structure, each unit (LUT + FF resources) decreases by 66.60 and 61.08% respectively. Therefore, the data throughput of each unit (LUT + FF resource) is 6.11 times and 5.23 times that of the two architectures in Ref. [8]. Compared with that described in Ref. [9], the data throughput of each unit is 1.97 times that of Ref. [9]. In addition, considering the high frequency and high resource utilization, the power consumption of these architectures in Ref. [8] is much higher than that of the architecture designed in this book. Note that the detector in Ref. [8] is designed to explicitly realize the signal detection in an advanced LTE system based on single carrier frequency division multiple access. The signal is mapped to the conventional OFDM for transmission. There is a fair comparison between the architecture in this section and the architecture in Ref. [8]. In order to achieve greater fairness, the detectors for OFDM system (including the conjugate gradient least square (CGLS) detector [10] and Gauss–Seidel (GAS) detector [11]) are also selected. The comparison result

Table 3.1 Comparison of the resource consumption of Xilinx virtex-7 FPGA

Comparison item	Reference [8]		Reference [9]	Reference [10]	Reference [11]	This design
Inverse method	CD	NS	OCD	CGLS	GS	PCI
LUT resource	208,161	168,125	23,955	3324	18,976	70,288
FF resource	213,226	193,451	61,335	3878	15,864	70,452
DSP48	1447	1059	771	33	232	1064
Freq/MHz	317	317	262	412	309	205
Throughput rate/(Mbit/s)	603	603	379	20	48	1230
Throughput rate/resources number/[Mbit/(s K slices)]	**1.43**	**1.67**	**4.44**	**2.78**	**1.38**	**8.74**

shows that the data throughput of each unit is 3.14 times and 6.33 times that of Refs. [10, 11] respectively.

Figure 3.8 shows the layout of the ASIC. Table 3.1 lists the proposed PCI architecture and detailed hardware features of other existing technical designs in Ref. [12]. The algorithm in Ref. [12] contains extra processing element. These designs are efficient ASIC architectures for solving massive MIMO system detection problems. Compared with the algorithm in Ref. [12], the proposed architecture achieves 2.04 times data throughput, energy efficiency increases by 2.12 times and area efficiency increases by 1.23 times respectively. The area efficiency and energy efficiency are normalized to 65 nm due to different technologies used in the design. The energy and area efficiency of this structure is obviously better than that of Ref. [12], that is, 4.56 times and 3.79 times. In order to achieve high parallelism, this architecture will consume more memory, and the frequency of memory access will increase. Four blocks (36.86 KB) and three blocks (0.216 KB) are used to store four-channel matrices and three vectors y^{MF} respectively. The remaining data consumes 0.144 kb of memory. The memory bandwidth of the architecture is 6.53 Gbit/s (to support high data throughput). In the massive MIMO system, hundreds of antennas cause considerable computational complexity, and the small MIMO detectors are not suitable for this design. Therefore, the fair comparison between this design and the conventional MIMO detector [3, 13] is difficult. For this reason, the current work is mainly compared with the architecture of the massive MIMO detection algorithm (Table 3.2).

The BER results of PCI and PCI's ASIC implementation as well as the BER results of NSA, GAS, RI, OCD, CG, and MMSE algorithms are shown in Fig. 3.9. To achieve the same SNR, the BER of PCI is lower than that of the NSA method. Compared with that of the floating-point detector, the BER loss caused by ASIC implementation of PCI is less than 1 dB.

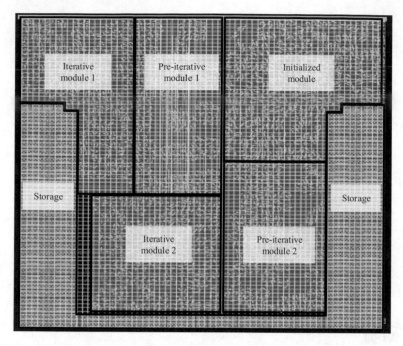

Fig. 3.8 ASIC layout of this design. © [2018] IEEE. Reprinted, with permission, from Ref. [6]

Table 3.2 Comparison of ASIC implementation results for the massive MIMO detector

Comparison item	Reference [21]	This design
Process	45 nm	65 nm 1P9M
MIMO system	128 × 8 64-QAM	128 × 16 64-QAM
Inverse method	NSA	PCI
Logical gates/(M Gates)	6.65	4.39
Storage/KB	15.00	37.22
Area/mm^2	4.65	7.70
Freq./MHz	1000	680
Power/W	1.72(0.81 V)	1.66(1.00 V)
Throughput rate/(Gbit/s)	2.0	4.08
Energy efficiency/[Gbit/(s W)]	1.16	2.46
Area efficiency/[Gbit/(s mm^2)]	0.43	0.53
Normalized[a] energy efficiency/[Gbit/(s W)]	**0.54**	**2.46**
Normalized area efficiency/[Gbit/(s mm^2)]	**0.14**	**0.53**

[a]The process is normalized to the 65 nm CMOS process, assuming: $f \sim s$, $A \sim 1/s^2$, $P_{\text{dyn}} \sim (1/s)(V_{\text{dd}}/V_{\text{dd}}')^2$

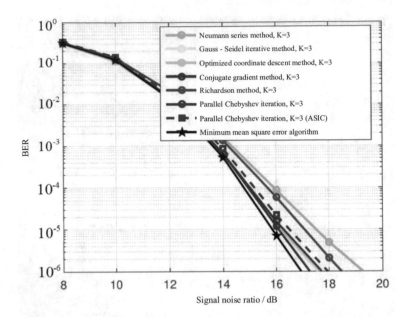

Fig. 3.9 BER performance curve of this design ASIC. © [2018] IEEE. Reprinted, with permission, from Ref. [6]

3.3   Hardware Architecture Based on Weighted Jacobi Iteration

In this section, a WeJi hardware architecture design based on optimized MMSE is described [14], which implements the massive MIMO detection with 64-QAM, 128×8.

3.3.1   VLSI Top-Level Architecture

Figure 3.10 illustrates the block diagram of the top-level structure of the massive MIMO detector designed in this section. In order to implement a higher data throughput with limited hardware resources, the architecture is fully pipelined. The VLSI architecture is divided into three main parts. In the first preprocessing element (a systolic array based on diagonal lines), the Gram matrix G, P^{-1} and the MF vector y^{MF} are computed through the input received vector y, channel matrix H, N_0, and E_s. These input data are stored in different storages of the architecture. All the complex values of the channel matrix H and received vector y are stored in 32 SRAMs. For each clock cycle, altogether eight elements of the channel matrix H and vector y are read. The storage size of the channel matrix H and vector y is 3 KB and about 0.34 KB respectively. In addition, various parameters, such as N0 and Es, are stored in mem-

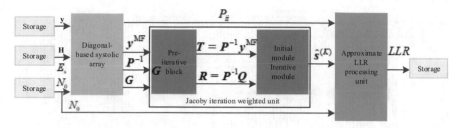

Fig. 3.10 VLSI top-level architecture model. © [2018] IEEE. Reprinted, with permission, from Ref. [14]

ory. Various parameters, such as N_0 and E_s, are also stored in memory. In the second unit, the matrix G/P^{-1} and the vector y^{MF} are used to iterate, and WeJi performs matrix inversion. The WeJi unit includes various modules. The pre-iterative module is used to compute iteration matrix R $(R = P^{-1}Q)$ and vector T $(T = P^{-1}y^{MF})$. The results of the pre-iterative module are output to the initial module and the iteration module to perform the final $\hat{s}^{(K)}$ computation. Based on the simulation results and analysis in Chap. 2, $K = 2$ is selected as the iteration number of WeJi implementation, which can achieve high detection accuracy with low resource consumption. In the third unit, vector $\hat{s}^{(K)}$, the diagonal element P_{ii} and parameter N_0 of the MMSE filter matrix are calculated to obtain the output (LLR).The output is stored in 16 SRAMs, about 0.1 KB.

Figure 3.11 shows the WeJi VLSI sequence. 45 cycles are used to calculate all the results in the diagonal systolic arrays. The 45 clock cycles consist of 32 clock cycles for performing complex multiplication, five clock cycles for performing cumulative computation of matrix P, and eight clock cycles for computing the reciprocal of matrix P. After 38 clock cycles, the results of the diagonal systolic array can be obtained and applied to the pre-iterative module (the first module of the WeJi unit).In the pre-iterative module, the computation of matrix R and vector T requires 15 clock cycles and eight clock cycles, respectively. As long as the initial module and iterative module (the second module of the WeJi cell) is able to start computing the first element of the initial solution, other elements of the initial solution can be computed immediately. After 11 clock cycles, the first iteration can be started. Similar to the first iteration, when the first iteration starts, the second iteration can start after 11 clock cycles. In short, the initial block and the iterative block consume a total of 37 clock cycles. Finally, after 11 clock cycles starting from the second iteration, the approximate LLR processing unit can use the first element of the vector $\hat{s}^{(K)}$ to implement the computation of LLR. After three clock cycles, the value of LLR is solved and stored in the output storage. Then, the remaining 15 LLR values are calculated one after another. The LLR unit consumes 18 clock cycles. In the proposed VLSI architecture, the average utilization of diagonally systolic arrays, initial module and iterative module is close to 100%. These two modules are more complex (compared with pre-iterative module and approximate LLR processing units), and have higher area costs. In order to accurately transmit data, the input and output data

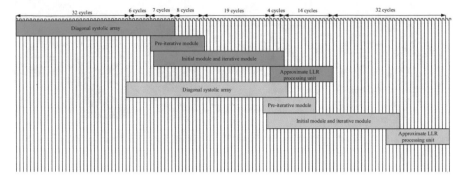

Fig. 3.11 WeJi VLSI sequence diagram. © [2018] IEEE. Reprinted, with permission, from Ref. [14]

of pre-iterative module and approximate LLR processing unit must match the data of the two main models. Therefore, the average utilization rate of the pre-iterative module and the approximate LLR processing unit is about 60%. Each unit will be described in detail below.

3.3.2 Diagonal Systolic Array

In the first preprocessing element, a diagonal-based systolic array with one-sided input is designed to compute Gram matrix and the MF vector. Figure 3.12 illustrates the architecture of the systolic array in detail. Taking into account the scale of the massive MIMO system, the unit contains three different PEs. There are N_t PE-As, $\frac{N_t^2 - N_t}{2}$ PE-Bs, and $N_t - 1$ PE-Cs in deep pipelining. For example, in a 128×8 MIMO system, there are eight PE-As, 28 PE-Bs, and seven PE-Cs. Taking the first PE-A, PE-B, and PE-C for example, their structures are described in detail in Fig. 3.13. PE-A is used to compute the MF vectors y^{MF}, the diagonal elements of the Gram matrix, G matrix and its inversion, and P^{-1}. PE-A contains four groups of ALUs, three accumulators (ACCs) and one RECU. The ALU-A and ALU-B are respectively used to calculate the real and imaginary parts of each element in the input matrix (Fig. 3.13a). $P_{i,i}^{-1}$, the real and imaginary parts of y_i^{MF} are all sent to the next module for the next computation. In RECU, the reciprocal of the diagonal elements of matrix P is obtained by LUT. Since the value of each element of P is close to 128 (the number of antennas at BS), the LUT stores the reciprocal from 72 to 200, which has a minimal impact on detection accuracy. Figure 3.13b is the calculation details of the non-diagonal elements of the matrix A in PE-B. PE-C is used to calculate the conjugate of the input data (Fig. 3.13c). It is noteworthy that the different types of computation in PEs (all PEs in massive MIMO detector) are implemented through multiple pipelines, with pipeline registers between each computation. For example,

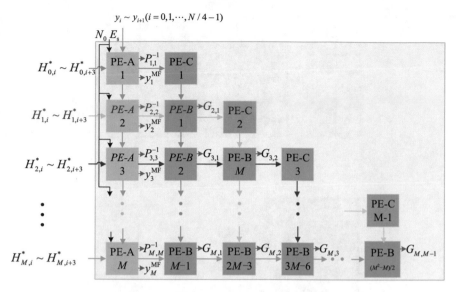

Fig. 3.12 Schematic diagram of diagonal systolic array. © [2018] IEEE. Reprinted, with permission, from Ref. [14]

in ALU-A in Fig. 3.13, the result computed by the multiplier is stored in the pipeline register and is taken as the input to the adder in the next step. In addition computation, the output of ALU-A results require multiple cycles, and the results of each cycle are stored in pipeline registers. All other PEs follow the same multi-cycle pipeline architecture. For this systolic array, the inputs of the transpose matrix $\boldsymbol{H}^{\mathrm{H}}$ and matrix $\boldsymbol{y}$ are simultaneously transmitted to PE-A. To ensure that each PE processes a set of correct operands, the row i of $\boldsymbol{H}^{\mathrm{H}}$ is delayed by $i-1$ clock cycles. First, each value of $\boldsymbol{H}^{\mathrm{H}}$ is transferred from PE-A to PE-B, then to PE-C (by row), and then from PE-C to PE-B (by column). Since the inverse of the matrix $\boldsymbol{P}$ is used to compute the matrix $\boldsymbol{R} = \boldsymbol{P}^{-1}\boldsymbol{Q}$ and the vector $\boldsymbol{T} = \boldsymbol{P}^{-1}\boldsymbol{y}^{\mathrm{MF}}$ in the next unit (WeJi unit), the reciprocal of the diagonal element must be computed as soon as possible. This is where all PE-As is the first processing element on the left side of each row in the array. After the initial delay, the output of PE-A is transmitted to the next unit per clock cycle. Therefore, this diagonal, unilateral input pulse array can achieve a high throughput rate and high hardware utilization.

Similar systolic arrays have been mentioned in Refs. [7, 8, 11]. In these architectures, PE-A is not the first processing element, located in the diagonal of the systolic array. Therefore, the computation of diagonal elements of Gram matrix $\boldsymbol{G}$ is delayed, consuming 15 clock cycles. This structure halves the number of clock cycles required to compute the matrix $\boldsymbol{P}$, even if the throughput rate doubled. In Refs. [7, 8, 11], bilateral input of PE-A was used. However, unilateral input was used in this design. Due to the existence of conjugate PEs, single side input can reduce the number of registers on the input side by half. The cost of this design mainly comes from PE-C, which

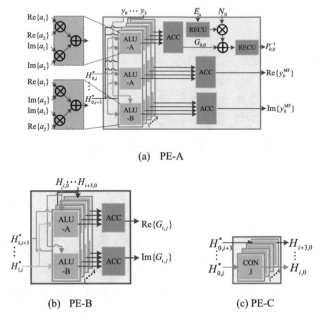

Fig. 3.13 Schematic diagram of **a** PE-A, **b** PE-B and **c** PE-C. © [2018] IEEE. Reprinted, with permission, from Ref. [14]

is acceptable. The architecture of the implicit method does not include the computing units of the Gram matrix G [10, 15, 16], because the Gram matrix is divided into two vector multiplication. These architectures have high data throughput, low hardware resource usage, and power consumption. However, when considering the unique properties of the actual massive MIMO system, the same Gram matrix results as those in the implicit architecture need to be computed many times, so the energy consumption and latency of these implicit architectures are very high in the actual massive MIMO system. In contrast, the systolic array computed by the Gram matrix is less energy intensive due to the reusability of the results of the Gram matrix.

3.3.3 WeJi Module

There are two modules in the WeJi unit: pre-iterative module, initialized module, and iterative module. The pre-iterative module is used to meet the request for input data in the initial module and the iterative module. Figure 3.14 shows the schematic diagram of a pre-iterative module in which $N_t + 1$ PE-Ds are computed in parallel in deep pipelining (nine PE-Ds for 128×8 MIMO systems). This module has two main operations: computation for vector $T = P^{-1}y^{\text{MF}}$ and iterative matrix $R = P^{-1}Q$. The computation of these two parts is carried out at the same time, and the result of one

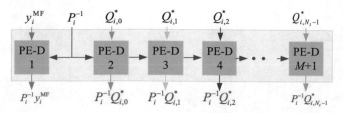

Fig. 3.14 Schematic diagram of the pre-iterative module. © [2018] IEEE. Reprinted, with permission, from Ref. [14]

Fig. 3.15 Schematic diagram of PE-D structure. © [2018] IEEE. Reprinted, with permission, from Ref. [14]

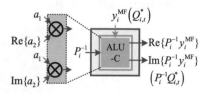

PE-E is computed for each clock cycle, thus achieving high parallelism. Figure 3.15 is the architecture of PE-D, which contains an ALU-C. Here input data P^{-1} is a real matrix, so we can simplify the calculation.

In Refs. [7, 8], after calculating the Gram matrix and the MF vector, the calculation of matrices R and T is performed in a systolic array, so the hardware cost is very low. However, considering the throughput rate of the whole system, the calculation of the second G and y^{MF} is delayed due to matrices T and R, so the throughput rate is reduced. In order to maintain high data throughput, the computation of matrices T and R is performed in another systolic array, which requires more operational units. In this architecture, the pre-iterative module uses pipeline mechanisms to compute matrix T and R within precise time limits. Compared with Refs. [7, 8], this architecture takes into account time constraints and utilizes PEs effectively. The data throughput of the structure proposed in this section will not be reduced, and a lower area and power consumption can be achieved as well.

There is a limited area overhead when the initialized module and iteration module reach high throughput rate and hardware processing speed (Fig. 3.16). Due to the time limitation of the previous module, the module uses the pipelined architecture for iterative computation. In order to adapt to high frequency, the block has N_t PEs, called PE-E. For example, there are eight PE-Es in a 128×8 MIMO system. As shown in Fig. 3.17, the detail of PE-E is illustrated. PE-E consists of two ALU (ALU-D and ALU-E), which is used to compute the real part and imaginary part of $\hat{s}_i^{(K)}$. There are eight pipelined registers at each input of PE-E. Input vectors are sent to the PE-E unit by element (from left to right), and input matrix $R = P^{-1}Q$ is transmitted to each PE-E when the matrix is computed. In the first phase, PE-E is used to compute the initial solution based on WeJi and the total time after receiving the first data. In the second phase, PE-E iteratively computes the transmitted vector $\hat{s}^{(K)}$. The solution $\hat{s}_m^{(K)}$ of PE-E is stored in the input pipeline register of PE-E.

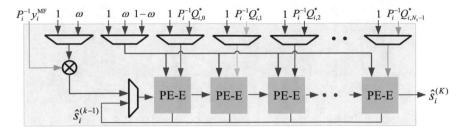

Fig. 3.16 Schematic diagram of initialization and iterative modules. © [2018] IEEE. Reprinted, with permission, from Ref. [14]

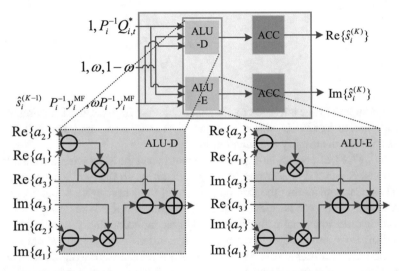

Fig. 3.17 Schematic diagram of PE-E structure. © [2018] IEEE. Reprinted, with permission, from Ref. [14]

In Refs. [7, 8], as matrix multiplication is required, these calculations are also performed in the systolic array. Therefore, this requires additional PEs. And every element of the systolic array is constantly performing calculations, which means a lot of area and power consumption. Compared with Refs. [7, 8], the architecture module for this design can perform vector multiplication with less area overhead and lower power consumption. Compared with Ref. [11], WeJi can achieve eight times higher parallelism than that the GAS algorithm mentioned. User-level parallel units can be used in this module, so this module can be used to obtain a fully pipelined architecture.

3.3.4 LLR Module

The approximate LLR processing unit is used to compute the LLR value of each transmitted bit based on WeJi. The approximate LLR processing method is applied to WeJi and used to design architecture according to Refs. [7, 8, 17]. NPI variance σ_{eq}^2 and SINR ς_i^2 can be found as Eqs. (3.1) and (3.2)

$$\sigma_{eq}^2 = E_s U_{ii} - E_s U_{ii}^2 \tag{3.1}$$

$$\varsigma_i^2 = \frac{1}{E_s}\frac{U_{ii}}{1 - U_{ii}} \approx \frac{1}{E_s}\frac{\frac{P_{ii}}{P_{ii}+N_0/E_s}}{1 - \frac{P_{ii}}{P_{ii}+N_0/E_s}} = \frac{P_{ii}}{N_0} \tag{3.2}$$

Figure 3.18 is a block diagram approximate LLR processing unit, which contains $\frac{1}{2}\log_2 Q$ PE-Fs for Q-QAM modulation. The first step is to calculate SINR ς_i^2 using P_{ii} and N_0. The value of SINR can be used for the same ith user. The linear equation $\varphi_b(\hat{s}_i)$ can be solved by different $\hat{s}_i$, which can be effectively implemented in the hardware structure. In the next computation, the bit LLR value $L_{i,b}$ is calculated using SINR ς_i^2. The details of PE-F are given in Fig. 3.18. The coefficient $\varphi_b(\hat{s}_i)$ of each linear equation is stored in the correction LUT. In addition, the effective channel gain and Pii are transmitted from the Gram matrix and the MF module. In RECU, the reciprocal of N_0 is implemented by LUT. The module facilitates the LLR computation, which improves the processing speed and reduces the area and power consumption. Although this method increases the number of LUTs, it increases very little, so it is acceptable.

3.3.5 Experimental Result and Comparison

The proposed hardware architecture is verified on the FPGA platform (Xilinx Virtex-7) and the tape-out chip is implemented using TSMC 65 nm 1P8 M CMOS technology. We can obtain the detailed hardware parameters from the chip imple-

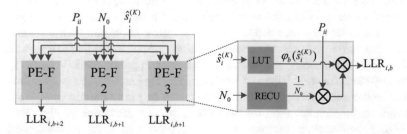

Fig. 3.18 Schematic diagram of approximate LLR processing unit structure. © [2018] IEEE. Reprinted, with permission, from Ref. [14]

mentation and compare them with those of the most advanced designs. Moreover, the design of fixed point and its performance in detecting accuracy will be introduced in this section. The data throughput of detector Θ is expressed as:

$$\Theta = \frac{\log_2 Q \times N_t}{T_s} \times f_{\text{clk}} \tag{3.3}$$

where f_{clk} is the clock frequency, Q is the constellation size, N_t is the number of users, and T_s is the number of clock cycles required for each symbol vector calculation. According to Eq. (3.3), in this architecture, the throughput rate is closely related to clock frequency, number of users, constellation size, and processing cycle. In addition, the iterations number, the size of hardware resources, the number of antennas and users will affect the processing cycle. In this architecture, the number of clock cycles is designed for meeting $T_s = \frac{N_t}{4}$.

3.3.5.1 Fixed-Point Scheme

In order to reduce the consumption of hardware resources, the fixed-point design is used in the whole design. The related fixed-point parameters are determined through several simulations. It should be noted that word width refers to the real or imaginary part of a complex number. The input of the architecture is all quantized to 14 bits, including the received signal v, the flat Rayleigh fading channel matrix H and the power spectral density N_0 of noise. Therefore, the multiplication is quantized to 14 bits and the result is transmitted to the accumulator in the diagonal systolic array, which is set to 20 bits. The LUT unit used to implement the reciprocal of elements in matrix P consists of 128 addresses with 12-bit output. The preprocessing module uses a 14-bit input, which is the input of the initial block and iterative block, indicating that the multiplication is quantized to 14 bits and the result is sent to the accumulator. Besides this, the output is set to 16 bits and sent to the LLR preprocessing module. The multiplication is set to 14 bits, and the output is set to 12 bits in this module. For the 128×8 MIMO system, the fixed-point performance is shown in Fig. 2.8 (marked as "fp"). In this architecture, the SNR loss required to implement a 10^{-2} FER is 0.2 dB, including an error of about 0.11 dB from the algorithm and a fixed-point error of about 0.09 dB from the chip implementation. In hardware implementation, the whole architecture adopts the fixed-point algorithm to reduce hardware resource consumption. Therefore, compared with software simulation, the hardware implementation will increase SNR losses due to errors caused by fixed-point parameters, such as truncation error, approximate RECU, LUT, etc., caused by the finite word length of hardware. During iteration, the fixed-point error should be increased because the proposed architecture is an iterative one. However, according to the simulation results in Fig. 2.8, the more iterations the algorithm has, the higher the detection accuracy is, so the detection accuracy increases with the increase of iterations. Figure 2.4.2 shows the FER performance of the exact MMSE, the proposed algorithm, the fixed-point implementation, and the comparison with other algorithms. Compared with the existing technology, the BER performance loss of WeJi (0.2 dB) is lower than that of NSA [7, 8] (0.36 dB), RI (0.43 dB) [18], intra-iterative interference cancelation (IIC) [16] (0.49 dB), CG [10] (0.66 dB), GAS [11, 17] (0.81 dB) and OCD [15] (1.02 dB).

3.3.5.2   FPGA Verification and Comparison

Table 3.3 summarizes the key experimental results on the FPGA platform (Xilinx Virtex-7). The results are compared with those of the other architectures and the best solution representing large-scale MIMO detectors is implemented with FPGA. Compared with the architecture based on Cholesky decomposition algorithm, the data throughput/unit of the proposed architecture is 4.72 times higher. This architecture reduces the throughput rate to 64.67%. However the unit (LUT + FF) consumption of the NSA-based detector drops 87.40%, so did the DSP consumption. With the same resource, the throughput rate of this design will be 4.05 times higher than that of NSA-based architecture [7], which can be attributed to the proposed algorithm and its VLSI architecture. Compared with WeJi architecture, GAS method architecture [11] achieves lower hardware resource consumption. However, low throughput rate (48 Mbit/s) is a limitation of the GAS methodology architecture due to the low computational parallelism of each element in the estimated vector (as described in Sect. 3.2). Therefore, compared with the GAS method, WeJi-based architecture achieves 4.90 times throughput rate/unit. In addition, the WeJi architecture proposed in this design is compared with the implicit architecture. The CG-based architecture [10] achieves lower hardware overhead, but the throughput rate is only 20 Mbit/s, far less than WeJi. Considering throughput rate/unit, the throughput rate of WeJi method is 2.43 times higher than that of CG-based architecture. Compared with the OCD-based architecture [15], the WeJi architecture achieves 1.20 times the data throughput/unit. IIC-based architecture [19] achieves high data throughput, but consumes many units and DSP. As a result, the WeJi maintains its advantage in data throughput/unit, 1.65 times higher than IIC-based architecture. Finally, the FPGA implementation architecture of massive MIMO detectors with nonlinear algorithms has been developed, such as two designs in Ref. [20]. Nonlinear detection algorithms, such as triangular approximate semi-definite relaxation (TASER), have better detection accuracy than that of the linear detection algorithms based on MMSE (WeJi, Cholesky decomposition algorithm, NSA, CG, IIC, etc.). Similar to the architecture based on GAS and CG, these two TASER-based architectures achieve low throughput rate (38 Mbit/s and 50 Mbit/s). Such low throughput rate limits the use of these architectures. Compared with two TASER-based architectures, the data throughput/unit of the WeJi architecture is increased by 1.23 and 2.79 times.

Table 3.3 Comparison of the resource consumption of various algorithms on the FPGA platform

Compared item	This design	Reference [7]		Reference [11]	Reference [10]		Reference [16]	Reference [20]	
MIMO system	128 × 8 64-QAM	128 × 8 64-QAM	128 × 8 64-QAM	128 × 8 64-QAM	128 × 8 64-QAM	128 × 8 64-QAM	128 × 8 64-QAM	128 × 8 BPSK	128 × 8 QPSK
Inverse method	WeJi	CHD	NS	GS	CG	OCD	IIC	TASER	TASER
Preprocessing	Include (explicit)	Include (explicit)	Include (explicit)	Include (explicit)	Include (implicit)	Include (implicit)	Include (implicit)	Exclusive	Exclusive
LUT resource	20,454	208,161	168,125	18,976	3324	23,914	72,231	4790	13,779
FF resource	25,103	213,226	193,451	15,864	3878	43,008	151,531	2108	6857
DSP48	697	1447	1059	232	33	774	1245	52	168
Freq./MHz	205	317	317	309	412	258	305	232	225
Throughput rate/(Mbit/s)	308	603	603	48	20	376	915	38	50
Throughput rate/resources number/[Mbit/(s K slices)]	6.76	1.43	1.67	1.38	2.78	5.62	4.09	5.51	2.42

3.3.5.3   ASIC Verification and Comparison

This design is realized by using TSMC 65 nm 1P8 M CMOS technology with a chip area of 2. 57 mm^2. Figure 3.19a is the micrograph of the chip. Energy and area efficiency are defined as data throughput/power and data throughput/area respectively.

The detector in Ref. [7] includes additional processing, such as fast Inverse Fourier Transform processing. To ensure a fair comparison, the same architecture is used as a comparison with the design architecture. In addition, for different technologies, energy and area efficiency (Table 3.4) are normalized to 65 nm and 1 V power supply voltage, corresponding to Eq. (3.4):

$$f_{\text{clk}} \sim s, \quad A \sim \frac{1}{s^2}, \quad P_{\text{dyn}} \sim \frac{1}{s}\left(\frac{V_{\text{dd}}}{V'_{\text{dd}}}\right)^2, \tag{3.4}$$

where s, A, P_{dyn} and V_{dd} represent scaling, area, power, and voltage respectively. This scaling method is widely used to compare different architectures of different technologies. The architecture implements the normalized energy efficiency of 0.54 Gbit/(s W) and the normalized area efficiency of 0.14 Gbit/(s mm^2). The comparison shows that the energy and area efficiency are 2.93 times and 2.86 times respectively as shown in Ref. [21].

In Ref. [22], two TASER-based algorithms are described to achieve high detection accuracy. However, their throughput rates are very low (0.099 Gbit/s and 0.125 Gbit/s). The proposed WeJi architecture achieves a throughput of 1.02 Gbit/s, which is 10.3 and 8.16 times that of the two architectures in Ref. [22]. And the architecture in Ref. [22] can only be used for BPSK or QPSK. These architectures are not suitable for high order modulation, which limits their application and development. For comparison, the results are normalized to 65 nm technology, as shown in Table 3.4.

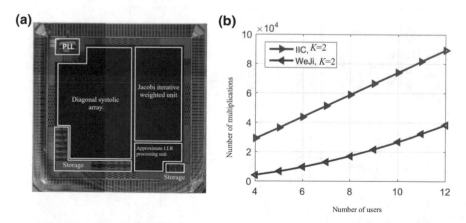

Fig. 3.19 Chip micrograph and number of real multiplications in WeJi and IIC. **a** Chip micrograph, **b** Number of real multiplications in WeJi and IIC © [2018] IEEE. Reprinted, with permission, from Ref. [14]

Table 3.4 Comparison of ASIC results

Compared item	This design	Reference [21]	Reference [16]	Reference [20]		Reference [23]	Reference [22]
Process	65 nm CMOS	45 nm CMOS	65 nm CMOS	40 nm CMOS		40 nm CMOS	28 nm FD-SOI
MIMO system	128 × 8 64-QAM	128 × 8 64-QAM	128 × 8 64-QAM	128 × 8 BPSK	128 × 8 QPSK	128 × 32 256-QAM	128 × 8 256-QAM
Inverse method	WeJi	NSA	IIC	TASER	TASER	MPD	CHD
Silicon verification	Yes	No (layout)	No (layout)	No (layout)	No (layout)	Yes	Yes
Preprocessing	**Inclusive (explicit)**	**Inclusive (explicit)**	**Inclusive (implicit))**	**Exclusive**	**Exclusive**	**Exclusive**	**Exclusive**
Logical gates/(M Gates)	1.07	6.65	4.3	0.142	0.448		0.148
Storage/KB	3.52	15.00					
Area/mm^2	2.57	4.65	9.6	0.15	0.483	0.58	1.1
Freq./GHz	0.68	1.00	0.60	0.598	0.56	0.425	0.30
Power consumption (Voltage)	0.65 (1.00 V)	1.72 (0.81 V)	1.00 ()	0.041 (1.1 V)	0.0087 (1.1 V)	0.221 (0.9 V)	0.018 (0.9 V)
Throughput rate/(Gbit/s)	1.02	2.0	3.6	0.099	0.125	2.76	0.3

(continued)

Table 3.4 (continued)

Compared item	This design	Reference [21]	Reference [16]	Reference [20]		Reference [23]	Reference [22]
Energy efficiency[a] [Gbit/(s W)]	1.58	1.16	3.6	2.41	1.44	12.49	16.67
Area efficiency[a] [Gbit/(s mm^2)]	0.40	0.43	0.375	0.66	0.26	4.76	0.27
Normalized[b] energy efficiency [Gbit/(s W)]	**1.58 (2.93[c] ×)**	**0.54**	**3.6**	**1.11**	**0.66**	**3.83**	**2.51**
Normalized[b] area efficiency [Gbit/(s mm^2)]	**0.40 (2.86[c] ×)**	**0.14**	**0.375**	**0.15**	**0.06**	**1.11**	**0.022**

[a]Energy and area efficiencies are respectively calculated through throughput/power consumption and throughput/area
[b]The process is normalized to the 65 nm CMOS process, assuming: $f \sim s$, $A \sim 1/s^2$, $P_{dyn} \sim (1/s)(V_{dd}/V'_{dd})^2$
[c]The normalized energy and area efficiency of the design is compared with the results in Ref. [21]

Compared with Ref. [22], WeJi architecture performs better in normalized energy and area efficiency. Specifically, compared with the two TASER detectors for BPSK and QPSK, the WeJi detector has 1.42 and 2.39 times normalized energy efficiency, and 2.67 times and 6.67 times normalized area efficiency. However, it should be noted that the preprocessing part is not included in the detector. According to Fig. 3.19a, the preprocessing takes up most of the chip (greater than 50%). Therefore, the preprocessing part will consume a lot of power of the chip. If the preprocessing part is considered in the TASER detector, the proposed WeJi detector should have better performance due to the improvement of energy and area efficiency. Based on the Cholesky decomposition algorithm, a detector design with a relatively low throughput rate (0.3 Gbit/s) is proposed in Ref. [22], thereby limiting its application. The WeJi architecture designed in this paper can achieve a throughput rate of 1.02 Gbit/s (about 3.4 times). The area overhead mentioned in Ref. [22] is 1.1 mm^2, less than WeJi. The area efficiency of WeJi in Ref. [22] is 1.48 times that of WeJi. Considering the use of 28 nm FD-SOI technology in Ref. [22], the results are normalized to 65 nm technology. The normalized area efficiency of WeJi is about 18.18 times that of Ref. [22]. The normalized energy efficiency achieved under the Ref. [22] architecture is 1.58 times that of the WeJi architecture. FD-SOI technology is exploited in the chip developed in Ref. [22]. The power consumption of FD-SOI technology is lower than that of CMOS technology when normalized to 65 nm. Meanwhile, the results in Ref. [22] do not include the preprocessing part (i.e., the power consumed by the preprocess), and the results in WeJi system include the preprocessing part. Therefore, for the above two reasons, the energy efficiency of the architecture in Ref. [22] should be significantly reduced. A message passing detector (MPD) is proposed in Ref. [23], achieving very high throughput rate and normalized energy and area efficiency. The throughput rate of 128×32 MIMO system processed by architecture in Ref. [23] is obviously improved compared with 128×8 MIMO system processing. The architecture in Ref. [23] does not include the preprocessing part. According to the computation complexity analysis, the resource consumption ratio of the preprocessing part in 128×8 MIMO system is greater than that of 128×32 MIMO system. Therefore, taking into account the preprocessing part of the architecture, the area, and power requirements are significant in order to ensure high data throughput at 2.76 Gbit/s. The normalized energy and area efficiency of WeJi architecture are on a par with those in Refs. [22, 23].

The results of ASIC in Table 3.4 come from Ref. [16]. An IIC detector under the implicit architecture is proposed based on ASIC implementation in Ref. [16]. This architecture achieves normalized area efficiency of 0.37 Gbit/(s mm^2), which is lower than the proposed architecture. The energy efficiency of IIC detector is higher than that of WeJi detector. When the channel frequency changes slowly, the channel hardening effect is obvious, and the explicit method can be reused. In practical systems, when the unique properties of massive MIMO systems (i.e., channel enhancement) are considered, the implicit architecture needs to compute the same Gram matrix T_c

times, while the explicit architecture only needs to compute the same Gram matrix once. For example, when considering the typical system parameters in the current LTE-A standard, the channel coherence time satisfies $T_c = 7$. Figure 3.19b is an explicit (WeJi) architecture and an implicit (IIC) architecture. The implicit architecture bears very high computational complexity and energy loss (about Tc times) in actual massive MIMO systems. As a result, the energy cost of the IIC detector increases significantly (about T_c times), and the energy efficiency of the IIC detector (3.6 Gbit/s W) is lower by T_c times. When considering the reusable use of the Gram matrix, the energy efficiency of the WeJi detector is higher than that of the implicit IIC architecture. The slowly changing channel results in T_c times buffer to store channels, which is the limit of the WeJi architecture. According to the design, as the number of antennas or users increases, the number of PE in Figs. 3.12, 3.14, and 3.16 increases by exploiting similar algorithms and architectures. For example, for a $N_r \times N_t$ MIMO system, the number of PE-As, PE-Bs, PE-Cs, PE-Ds, PE-Es, and PE-Fs should be $N_t, \frac{N_t^2 - N_t}{2}, N_t - 1, N_t + 1, N_t$, and $\frac{1}{2} \log_2 Q$ respectively. If this architecture is adopted, the data throughput will satisfy Eq. (3.3). In addition, the area and power consumption will increase as the number of PEs increases. In the latest architecture, such as the NSA, IIC, and the Cholesky decomposition algorithms, the number of PEs will also increase to achieve a scalable MIMO system. Considering the reuse of the chip, chips can be reused as the number of antennas or users increases, because the chip can break down the large-scale channel matrices and received vector into smaller scale matrices and vectors. Other chips are needed for control reasons and storage of intermediate data. The data throughput and efficiency loss of this chip are very small compared to other latest chips. Considering efficiency, time and manpower, there is no need to use chips. Conceptual chip reuse can be implemented under this architecture to increase the number of antennas or users.

3.4 Hardware Architecture Based on Conjugate Gradient Method

3.4.1 VLSI Top-Level Structure

This section introduces the design of VLSI hardware architecture based on TCG. A 128×8 massive MIMO system based on 64-QAM is designed, and the number of iterations K is 3 (including the pre-iterative part $K = 0$, and the two iterations $K = 1$ and $K = 2$). Figure 3.20 is the diagram of the top-level architecture of hardware, which consists of input/output module, multiplication module, initial module, and iterative module. The multiplication module and iterative module can be divided into four levels of pipelined structure, and each level of pipeline needs 16 clock cycles.

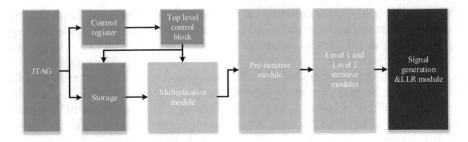

Fig. 3.20 VLSI top-level architecture of TCG

3.4.2   Input/Output Module

The input/output module consists of a joint test action group (JTAG) interface, a storage array, and some registers. In order to get a higher data throughput, only the pipeline mechanism is adopted for the hardware structure. For the massive MIMO system in this section, the scale of the input channel matrix H is 128×8, and the input data volume is very large, which is difficult to match the hardware requirements. Therefore, the JTAG interface is utilized in the design to simplify the port design. The JTAG interface uses JTAG protocol to debug the registers and control the data read, write, input and output of the registers. The storage array is used to store external data. Moreover, registers can control the start, execution, and end of MIMO detection system. JTAG uses the ARM7 ETM module, so the JTAG wiring, software structure and internal scanning structure in the system can be used during chip testing, thus reducing the design load and test risk. The internal storage and registers are connected by advanced microcontroller bus architecture (AMBA) and ARM7 ETM, as shown in Fig. 3.21. The control register is at the top level of the JTAG interface, and JTAG is used to control internal storage. The data is input into the internal storage array by JTAG, and then the control register and execution data are allocated by JTAG. The read/write unit of the MIMO system is used to read and write data in the internal storage, while the MIMO detection system begins to perform computation. In addition, the counter is reduced by one at this point, and reduced by one every 16 cycles.

The storage array consists of nine 32×128 single-port SRAM arrays. One SRAM array is used to store the received vector y, and the other eight SRAM arrays are used to store the values of channel matrix H (each SRAM array to store one column of H). The external port of the storage array is 32 bits, where 10 bits are the address bits, which are composed of {A [9], CS[2:0], A[5:0]}. The internal port is 128 bits, where 10 bits are the address bit, and its composition is {A[5:0]}.

Fig. 3.21 Schematic
diagram of the input/output
module

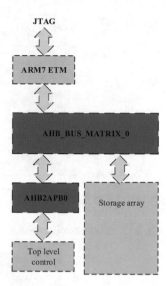

3.4.3 Multiplication Module

The multiplication module computes the matrix multiplication $W = H^\mathrm{H}H$ and the MF vector $y^\mathrm{MF} = H^\mathrm{H}y$, and provides results to the next level. Eight diagonal elements and 28 non-diagonal elements are required to be computed for matrix W (because W is a diagonal matrix). In addition, the multiplication module will compute eight elements in the y^MF vector. All computation of the multiplication module is completed in 16 cycles.

The channel matrix H is a 128×8 matrix. The computation of $W = H^\mathrm{H}H$ and $y^\mathrm{MF} = H^\mathrm{H}y$ are divided into 16 steps in this design. Each step computes $W_k = H_k^\mathrm{H}H_k$ and $y_k^\mathrm{MF} = H_k^\mathrm{H}y_k$, where $k = 0, 1, \ldots, 15$, which is part of the matrix W and vector y^MF. Therefore, there are 16 steps to complete the calculation of the entire matrix W and vector y^MF in this design. Figure 3.22 shows the frame structure of the multiplication module, which contains 8 PE-As and 38 PE-Bs. Figures 3.23 and 3.24 illustrate the structure of PE-A and PE-B respectively. Each PE-A and PE-B are used to compute an 8-bit complex multiplication accumulation. However, the difference between PE-A and PE-B is that the input of PE-A is a set of vectors, each element of the vector is multiplied by itself, and the real and imaginary parts of the vector elements are respectively squared and summed, and then added up. The input of PE-B is two sets of vectors. The real and imaginary parts of the two sets of vectors are multiplied by each other and then the same operation is done in PE-A. In the figure, $i = 0, 2, \ldots, 7$. The reason to distinguish PE-A from PE-B is that PE-A can save multipliers and reduce chip area by designing PE-A separately. Since the matrix W is a symmetric matrix, the structure of the whole multiplication module is a trapezoid structure.

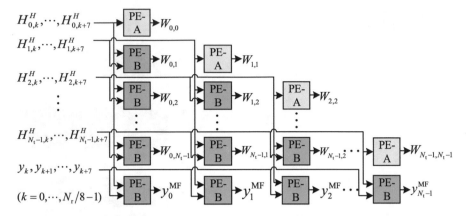

Fig. 3.22   Schematic diagram of the multiplication module based on pipeline mechanism

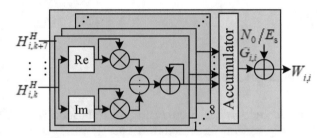

Fig. 3.23   Schematic diagram of the PE-A structure

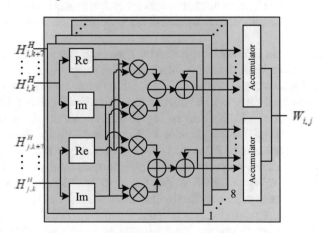

Fig. 3.24   Schematic diagram of the PE-B structure

Because the multiplication module divides the computation of $W = H^H H$ and $y^{MF} = H^H y$ into many steps, and the computation of each step is the same, the module can be reused in hardware, and saves area and power consumption. Although this design increases the time loss of hardware, such design has no effect on the data throughput of the system because the time loss is not the main consideration of the hardware design. Therefore, this design can improve the energy and area efficiency of the system.

3.4.4 Iterative Module

The iterative module adopts the pipeline processing mechanism. The whole iterative module includes the pre-iterative units and the iterative units. The pre-iterative unit completes the initial solution of iteration and prepares for the iterative computation. The iterative unit updates the transmitted signal s and residual z. The whole iterative unit is divided into two levels. The first level unit calculates the parameter γ through the iterative initial solution output by the pre-iterative unit and updates the values of s and z. The second level unit calculates the parameters ρ and γ according

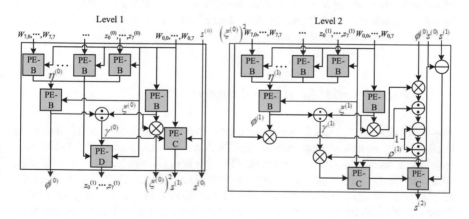

Fig. 3.25 Schematic diagram of iterative module structure based on pipeline mechanism

Fig. 3.26 Schematic diagram of the PE-C structure

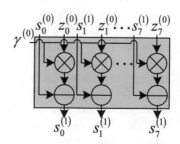

Fig. 3.27 Schematic
diagram of the PE-D
structure

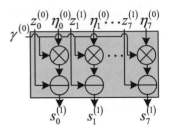

to the output of the first level module, and updates the values of s and z. According to
TCG, the judgment statement in the algorithm will not be executed when the number
of iterations $K = 1$, and ρ will remain 1. Therefore, the operations in step 13 and
step 14 of the algorithm will be simplified to $s_1 = s_0 + \gamma_0 z_0$ and $z_1 = z_0 - \gamma_0 \eta_0$.
After the first iteration, ρ will be updated and no longer equal to 1, then step 13
and step 14 will be restored to the original form. By comparing the two iterations,
it is easy to find that the operations in the second iteration also appear in the first
iteration, so part of the hardware in the two-level module is repeated. Figure 3.25
is the structure of the iterative module, in which PE-C is used for the operation of
constant multiplying a vector and adding a vector, and PE-D is used for the operation
of constant multiplying a vector and subtracting a vector. Figures 3.26 and 3.27 are
respectively schematic diagrams of PE-C and PE-D. The computation of the entire
iteration module is completed in 16 cycles.

3.4.5 Experimental Results and Comparison

The VLSI design of TCG is verified on the FPGA platform and ASIC using the
Verilog HDL. Some key parameters in the FPGA verification are listed in Table 3.5. In
this design, the number of basic units (LUT + FF slices) is relatively small compared
with other designs. For example, compared with GAS [11], OCD [9], and PCI [6], the
number of basic units is reduced by 32.94, 72.60 and 83.40% respectively, so the data
throughput/unit is increased by 19.54 times, 6.07 times, and 3.08 times respectively.
It is also easy to see from the table that the data throughput/unit of TCG hardware
design is much higher than that of other algorithms.

The design also implements the tape-out chip and the micrograph of the ASIC chip
is shown in Fig. 3.28. The partition of the whole chip can be found in the diagram. To
make the chip work at a higher frequency, the PLL part is added to the chip design.
This design adopts TSMC 65 nm CMOS technology with chip area of 1.87 mm ×
1.87 mm. It can work at a frequency of 500 MHz frequency, to a data throughput of 1.5
Gbit/s. The chip power is 557 mW at this point. Some specific parameters in the latest
ASIC hardware design for various massive MIMO detection algorithms are shown
in Table 3.6. In order to obtain a relatively fair comparison result, the normalized
energy and normalized area efficiency of various designs are calculated in the table.

Table 3.5 Comparison of the resource consumption of various algorithms on the FPGA platform

Compared item	This design	Reference [8]		Reference [11]	Reference [10]	Reference [9]	Reference [6]
Inverse method	TCG	CHD	NSA	GAS	CGLS	OCD	PCI
LUT resource	4587	208,161	168,125	18,976	3324	23,914	70,288
FF resource	18,782	213,226	193,451	15,864	3878	43,008	70,452
DSP48	972	1447	1059	232	33	774	1064
Freq./MHz	210	317	317	309	412	258	205
Throughput rate/(Mbit/s)	630	603	603	48	20	376	1230
Throughput/resource number/[Mbit/(s K slices)]	26.96	1.43	1.67	1.38	2.78	5.62	8.74

Table 3.6 Comparison of ASIC results of various latest massive MIMO detectors

Compared item	Reference [14]	Reference [24]	Reference [25]	Reference [22]	Reference [26]	Reference [27]	This design
Arithmetic	MMSE	MMSE	SD	MMSE	MPD	SD	MMSE
MIMO system	128×8	4×4	4×4	128×8	32×8	4×4	128×8
Voltage/V	1.0	1.0	1.2	0.9	0.9	1.2	1.0
Process/nm	6.5	65	65	28	40	65	65
Freq./MHz	680	517	445	300	500	333	500
Throughput rate/(Gbit/s)	1.02	1.379	0.396	0.3	8	0.807	1.5
Preprocessing	Yes	Yes	Yes	No	No	No	Yes
Area (logic gates)/kGE	1070	347	383	148	1167	215	1372
Power/mW	650	26.5	87	18	77.89	38	557
Normalized[a][b] energy efficiency [Mbit/(s mW)]	**1.569**	**1.626**	**0.205**	**0.505**	**7.876**	**0.956**	**2.693**
Normalized[a][b] area efficiency [Mbit/(s kGE)]	**0.953**	**0.083**	**0.022**	**2.027**	**1.055**	**0.078**	**1.093**

[a] The technological process is normalized to the 65 nm CMOS technological process, assuming that $f \sim s$ and $P_{dyn} \sim (1/s)(V_{dd}/V'_{dd})^2$

[b] After the technological process is normalized to the 128×8 MIMO system, the area and the critical path delay respectively increase according to the following rule: $(128/N_r) \times (8/N_t)$, $(\log_2 8 / \log_2 N_t)$

Fig. 3.28 Micrograph of TCG chip

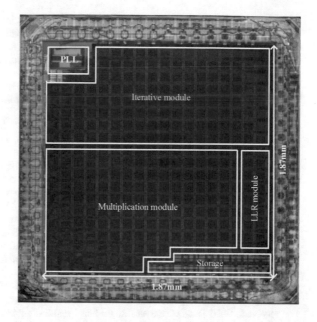

Compared with the chip in Refs. [14, 24, 25], the normalized energy efficiency of this design was increased by 1.72 times, 1.66 times, and 13.14 times, respectively, and the normalized area efficiency was increased by 1.15 times, 13.17 times and 49.68 times respectively. The energy efficiency and area efficiency of this design can reach 12.5 Mbit/(s mW) and 3.79 Mbit/(s kGE) respectively without considering the preprocessing part. Compared with the chips of Cholesky decomposition algorithm [22], MPD [26] and SD [27], the energy efficiency is raised by 4.99 times, 1.59 times and 13.01 times, and the area efficiency is raised by 1.87 times, 3.59 times and 48.56 times, respectively. Based on the data in Table 3.6, we can conclude that the chip can work at a higher frequency and achieve a higher data throughput with a small area and low power.

References

1. Gpp TS (2009) 3rd generation partnership project; Technical specification group radio access network; Evolved Universal Terrestrial Radio Access (E-UTRA); Physical Channels and Modulation (Release 8). 3GPP TS 36.211 V.8.6.0
2. Bartlett MS, Gower JC, Leslie PH (1960) The characteristic function of Hermitian quadratic forms in complex normal variables. Biometrika 47(1/2):199–201
3. Studer C, Fateh S, Seethaler D (2011) ASIC implementation of soft-input soft-output MIMO detection using MMSE parallel interference cancellation. IEEE J Solid-State Circuits 46(7):1754–1765

4. Wu M, Yin B, Vosoughi A et al (2013) Approximate matrix inversion for high-throughput data detection in the large-scale MIMO uplink. In: IEEE international symposium on circuits and systems, pp 2155–2158

5. Schreiber R, Tang WP (1986) On systolic arrays for updating the Cholesky factorization. BIT 26(4):451–466

6. Peng G, Liu L, Zhang P et al (2017) Low-computing-load, high-parallelism detection method based on Chebyshev iteration for massive MIMO systems with VLSI architecture. IEEE Trans Signal Process 65(14):3775–3788

7. Yin B, Wu M, Wang G et al (2014) A 3.8 Gb/s large-scale MIMO detector for 3GPP LTE-Advanced. In: IEEE international conference on acoustics, speech and signal processing, pp 3879–3883

8. Wu M, Yin B, Wang G et al (2014) Large-scale MIMO detection for 3GPP LTE: algorithms and FPGA implementations. IEEE J Sel Top Sign Process 8(5):916–929

9. Yin B, Wu M, Cavallaro JR et al (2015) VLSI design of large-scale soft-output MIMO detection using conjugate gradients. IEEE Int Symp Circ Syst 1498–1501

10. Wu Z, Zhang C, Xue Y et al (2016) Efficient architecture for soft-output massive MIMO detection with Gauss-Seidel method. IEEE Int Symp Circ Syst 1886–1889

11. Choi JW, Lee B, Shim B et al (2013) Low complexity detection and precoding for massive MIMO systems. In: Wireless communications and NETWORKING conference, pp 2857–2861

12. Wu M, Dick C, Cavallaro JR et al (2016) FPGA design of a coordinate descent data detector for large-scale MU-MIMO. IEEE Int Symp Circ Syst 1894–1897

13. Liu L (2014) Energy-efficient soft-input soft-output signal detector for iterative MIMO receivers. IEEE Trans Circ Syst I Regul Pap 61(8):2422–2432

14. Peng G, Liu L, Zhou S et al (2017) A 1.58 Gbps/W 0.40 Gbps/mm? ASIC implementation of MMSE detection for 128×8 64-QAM massive MIMO in 65 nm CMOS. IEEE Trans Circ Syst I Regul Pap PP(99):1–14

15. Wu M, Dick C, Cavallaro JR et al (2016) High-throughput data detection for massive MU-MIMO-OFDM using coordinate descent. IEEE Trans Circuits Syst I Regul Pap 63(12):2357–2367

16. Chen J, Zhang Z, Lu H et al (2016) An intra-iterative interference cancellation detector for large-scale MIMO communications based on convex optimization. IEEE Trans Circuits Syst I Regul Pap 63(11):2062–2072

17. Dai L, Gao X, Su X et al (2015) Low-complexity soft-output signal detection based on gauss–C-Seidel method for uplink multiuser large-scale MIMO systems. IEEE Trans Veh Technol 64(10):4839–4845

18. Gao X, Dai L, Ma Y et al (2015) Low-complexity near-optimal signal detection for uplink large-scale MIMO systems. Electron Lett 50(18):1326–1328

19. Kincaid D, Cheney W (2009) Numerical analysis: mathematics of scientific computing. vol 2. Am Math Soc

20. Casta eda O, Goldstein T, Studer C (2016) Data detection in large multi-antenna wireless systems via approximate semidefinite relaxation. IEEE Trans Circ Syst I Regul Pap PP(99):1–13

21. Yin B (2014) Low complexity detection and precoding for massive MIMO systems: algorithm, architecture, and application. Diss. Rice University

22. Prabhu H, Rodrigues JN, Liu L et al (2017) A 60 pJ/b 300 Mb/s 128×8 massive MIMO precoder-detector in 28 nm FD-SOI[C]. Solid-State Circuits Conference (ISSCC), 2017 IEEE International. IEEE, pp 60–61

23. Tang W, Chen CH, Zhang Z (2016) A 0.58mm2 2.76 Gb/s 79.8pJ/b 256-QAM massive MIMO message-passing detector. Vlsi Circ 1–5

24. Chen C, Tang W, Zhang Z (2015) 18.7 A 2.4 mm 2 130mW MMSE-nonbinary-LDPC iterative detector-decoder for 4×4 256-QAM MIMO in 65 nm CMOS. In: Solid-state circuits conference, pp 1–3

25. Noethen B, Arnold O, Perez Adeva E et al (2014) 10.7 A 105GOPS 36 mm 2 heterogeneous SDR MPSoC with energy-aware dynamic scheduling and iterative detection-decoding for 4G in 65 nm CMOS. In: Solid-state circuits conference digest of technical papers, pp 188–189
26. Chen YT, Cheng CC, Tsai TL et al (2017) A 501 mW 7.6l Gb/s integrated message-passing detector and decoder for polar-coded massive MIMO systems[C]. VLSI Circuits, 2017 Symposium on IEEE, pp C330-C331
27. Winter M, Kunze S, Adeva EP et al (2012) A 335 Mb/s 3.9 mm^2 65 nm CMOS flexible MIMO detection-decoding engine achieving 4G wireless data rates[J]

Chapter 4
Nonlinear Massive MIMO Signal Detection Algorithm

Currently, there are two kinds of signal detectors for MIMO systems: linear signal detectors and nonlinear signal detectors [1]. Linear signal detection algorithms include the conventional ZF algorithm and MMSE algorithm [2], as well as some recently proposed linear signal detection algorithms [3–5]. Although these linear signal detection algorithms have the advantages of low complexity, their deficiency in detection accuracy cannot be ignored, especially when the number of user antennas is close to or equal to the number of base station antennas [3]. The optimal signal detector is a nonlinear ML signal detector, but its complexity increases exponentially with the increase of the number of the transmitting antennas, so it cannot be implemented for massive MIMO systems [6]. The SD detector [7] and the K-Best detector [8] are two different variations of the ML detector. They can balance the computational complexity and performance by controlling the number of nodes in each search stage. However, the QR decomposition in these nonlinear signal detectors can lead to high computational complexity and low parallelism because of the inclusion of matrix operations such as element elimination. Therefore, people urgently need a detector with low complexity, high precision and high processing parallelism.

This chapter first introduces several conventional nonlinear MIMO signal detection algorithms in the Sect. 4.1. Section 4.2 presents a K-best signal detection and preprocessing algorithm in high-order MIMO systems, combining the Cholesky sorted QR decomposition and partial iterative lattice reduction (CHOSLAR) [9]. Section 4.3 presents another new signal detection algorithm, TASER algorithm [10].

© Springer Nature Singapore Pte Ltd. and Science Press, Beijing, China 2019
L. Liu et al., *Massive MIMO Detection Algorithm and VLSI Architecture*,
https://doi.org/10.1007/978-981-13-6362-7_4

4.1 Conventional Nonlinear MIMO Signal Detection Algorithm

4.1.1 ML Signal Detection Algorithm

The ML signal detection algorithm can achieve the optimal estimation of the transmitted signal. The detection process is to find the nearest constellation point in all the constellation points set as the estimation of the transmitted signal. The detailed analysis is as follows. Considering the MIMO system of N_t root transmitting antenna and Nr root received antenna, the symbols received by all receiving antennas are represented by vector $y \in C^{N_r}$, then

$$y = Hs + n, \tag{4.1}$$

where, $s \in \Omega^{N_t}$, is the transmitted signal vector containing all user data symbols (Ω denotes the set of constellation points); $H \in C^{N_r \times N_t}$ is Rayleigh flat-fading channel matrix, and its element $h_{j,i}$ is the channel gain from transmitting antenna $i(i = 1, 2, \ldots, N_t)$ to receiving antenna $j(j = 1, 2, \ldots, N_r)$. $n \in C^{N_r}$ is the additive Gaussian white noise vector with independent components and obeying $N(0, \sigma^2)$ distribution.

The conditional probability density of the receiving signal can be expressed as

$$P(y|H, s) = \frac{1}{(\pi\sigma^2)^M} \exp\left(-\frac{1}{\sigma^2}\|y - Hs\|^2\right) \tag{4.2}$$

As the optimal signal detection algorithm, ML signal detection algorithm solves s by the finite set constrained least mean square optimization, as shown in Eq. (4.3).

$$\tilde{s} = \arg\max_{s \in \Omega} P(y|H, s) = \arg\min_{s \in \Omega} \|y - Hs\|^2 \tag{4.3}$$

Perform QR decomposition on the channel matrix H, and we can get

$$
\begin{aligned}
\|y - Hs\|^2 &= \left\|QQ^H(y - Hs) + (I_{N_r} - QQ^H)(y - Hs)\right\|^2 \\
&= \left\|QQ^H(y - Hs)\right\|^2 + \left\|(I_{N_r} - QQ^H)(y - Hs)\right\|^2 \\
&= \left\|Q^H y - Rs\right\|^2 + \left\|(I_{N_r} - QQ^H)y\right\|^2 \\
&= \left\|Q^H y - Rs\right\|^2 \\
&= \left\|y' - Rs\right\|^2,
\end{aligned}
\tag{4.4}
$$

where $y' = Q^H y$, according to the upper triangular property of matrix R,

$$\tilde{x} = \arg\min_{s\in\Omega} \left\| y' - Rs \right\|^2$$

$$= \arg\min_{s\in\Omega} \left(\sum_{i=1}^{N_t} \left| y'_i - \sum_{j=i}^{N_t} R_{i,j}s_j \right|^2 + \sum_{i=N_t+1}^{N_r} \left| y'_i \right|^2 \right)$$

$$= \arg\min_{s\in\Omega} \left(\sum_{i=1}^{N_t} \left| y'_i - \sum_{j=i}^{N_t} R_{i,j}s_j \right|^2 \right)$$

$$= \arg\min_{s\in\Omega} \left[f_{N_t}(s_{N_t}) + f_{N_t-1}(s_{N_t}, s_{N_t-1}) + \cdots + f_1(s_{N_t}, s_{N_t-1}, \ldots, s_1) \right], \quad (4.5)$$

where the function in $f_k(s_{N_t}, s_{N_t-1}, \ldots, s_k)$ can be expressed as

$$f_k(s_{N_t}, s_{N_t-1}, \ldots, s_k) = \left| y'_k - \sum_{j=k}^{N_t} R_{k,j}s_j \right|^2 \quad (4.6)$$

Here we constructed a search tree to seek the optimal solution for the set of all constellation points, as shown in Fig. 4.1.

There are S nodes (S is the number of possible values of each point in the modulation mode) in the first stage expanded by the root node. Each node has a value of $f_{N_t}(s_{N_t})(s_{N_t} \in \Omega)$. In the first stage, each node expands S child nodes to get the structure of the second stage, a total of S^2 nodes. The value of second stage node is $f_{N_t}(s_{N_t}) + f_{N_t-1}(s_{N_t}, s_{N_t-1})$, and so on, the last generated N_t stage has S^{N_t} child nodes. The value of the child node is $f_{N_t}(s_{N_t}) + f_{N_t-1}(s_{N_t}, s_{N_t-1}) + \cdots + f_1(s_{N_t}, s_{N_t-1}, \cdots, s_1)$. We can find the optimal solution by looking for all the nodes.

ML signal detection algorithm searches all nodes to find the optimal node, and then estimates the transmitted signal, which is obviously the optimal estimation algo-

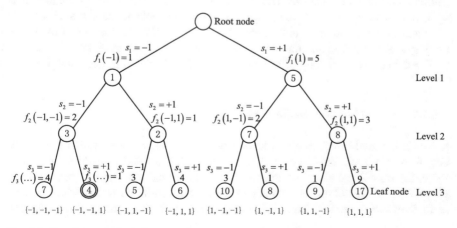

Fig. 4.1 Search tree in the ML signal detection algorithm

rithm. Although ML signal detection algorithm can achieve the best performance, its computational complexity increases exponentially with the increase of the number of the transmitting antennas, the number of bits per symbol after modulation and the length of processing data. The computational complexity is $O(M^{N_t})$ (M is the number of constellation points, N_t is the number of the transmitting antennas). For example, in a 4×4 MIMO system with 16QAM modulation, the search amount for each symbol block is as high as $4^{16} = 65,536$. Therefore, it is difficult for the ML signal detection algorithm to apply to the actual communication system in the application scenarios of high-order modulation (M is large) and large number of transmitting antennas (N_t is large). In order to reduce the computational complexity of the algorithm, we need some approximate detection algorithm [11].

4.1.2 SD Signal Detection Algorithm and K-Best Signal Detection Algorithm

To achieve near-optimal performance and reduce computational complexity, several nonlinear signal detectors are proposed. One of the typical algorithms is the tree-based search algorithm [12, 13]. So far, many near-ML signal detection algorithms have been presented, including tree-based search algorithms with low complexity. K-best signal detection algorithm [14] searches K nodes in each stage to find the possible transmitted signals, while SD signal detection algorithm [15] searches the hypersphere near the receiving signal vector to find the possible transmitted signals. However, no other signal detection algorithms can achieve full diversity gain [16] except the ML signal detection algorithm. The fixed-complexity SD (fixed-complexity sphere decoding, FSD) signal detection algorithm [17] utilizes the potential grid structure of the receiving signal, and is considered the most promising algorithm to achieve the ML detection performance and reduce the computational complexity. In the conventional small-scale MIMO system, the algorithm performs well, but its computational complexity is still unbearable when the antenna size increases or the modulation order increases (for example, the number of the transmitting antennas is 128 and the modulation order is 64 QAM modulation) [6].

The above-mentioned algorithm will be described in detail in the following.

4.1.2.1 K-Best Signal Detection Algorithm

K-best [14] signal detection algorithm is a depth-first search algorithm that performs only forward search. This algorithm only keeps K paths with optimal metric in each stage. Figure 4.2 shows the search tree of K-best signal detection algorithm when $N_t = 2$ [11]. This algorithm expands all possible candidate nodes from the root node, then sorts them by metric, selects the first K paths with the smallest metric as

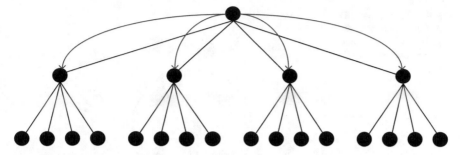

Fig. 4.2 Search tree *in the* K-best signal detection algorithm. © [2018] IEEE. Reprinted, with permission, from Ref. [11]

the next stage, and so on, until the leaf node. The specific implementation of K-best signal detection algorithm is shown in Algorithm 4.1.

Algorithm 4.1 K-best signal detection algorithm

1. For number of stages $i = N_t$ to 1,Do
2. Extend $\sqrt{2^Q}$ possible paths out of each survival path;
3. Update the partial Euclidean distance (PED) of each path;
4. Sort all paths according to PED;
5. Screen K best paths and update the history paths;
6. If number of stages $i = 1$, end the algorithm. Otherwise, jump to Step 2.

The K-best signal detection algorithm can achieve near-optimal performance within a fixed complexity and moderate parallelism. The fixed complexity depends on the number of reserved candidate nodes K, the modulation order and the number of the transmitting antennas. In the search number of the K-best signal detection algorithm, the total number of nodes searched is $2^Q + (N_t - 1)K2^Q$. Although the K-best signal detection algorithm has the above advantages, it does not take into account the noise variance and channel conditions. In addition, the K-best signal detection algorithm extends all the K reserved paths for each stage to 2^Q possible child nodes. Therefore, great complexity is required to enumerate these child nodes, especially when the number of high-order modulation and survivor paths is large. The algorithm also needs to compute and sort $2^Q K$ paths of each stage, where $K\left(2^Q - 1\right)$ is the path of the cropped tree. This sort is also very time-consuming. At the same time, the algorithm has a large bit error rate at a low K value.

4.1.2.2 SD Signal Detection Algorithm

The SD signal detection algorithm is applicable to a wide SNR range, which can achieve near-optimal performance while maintaining the average computational

Fig. 4.3 Principle of SD signal detection algorithm. © [2018] IEEE. Reprinted, with permission, from Ref. [11]

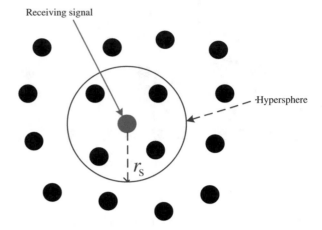

complexity of the polynomial stage [7]. The SD signal detection algorithm was origi-nally used to calculate the minimum length grid vector, which was further developed to solve the short grid vector, and finally used for the ML estimation [15]. The basic principle of the SD signal detection algorithm is to limit the search space of the optimal ML solution to the hypersphere with a radius of r_s near the receiving vector, as shown in Fig. 4.3 [11], and formulate it as an expression (4.7). Therefore, the computational complexity is reduced by simply verifying the grid points within the hypersphere rather than all possible points of the transmitting signals.

$$\hat{s}_{SD} = \underset{s \in 2^{QN_t}}{\arg\min}\left\{\|\boldsymbol{y} - \boldsymbol{H}\boldsymbol{s}\|^2 \leq r_s^2\right\} \tag{4.7}$$

The channel matrix $\boldsymbol{H}$ can be decomposed into matrices $\boldsymbol{Q}$ and $\boldsymbol{R}$, i.e., $\boldsymbol{H} = \boldsymbol{Q}\boldsymbol{R}$, Eq. (4.7) is equivalent.

$$\hat{s}_{SD} = \underset{s \in 2^{QN_t}}{\arg\min}\left\{\|\tilde{\boldsymbol{y}} - \boldsymbol{R}\boldsymbol{s}\|^2 \leq r_s^2\right\}, \tag{4.8}$$

where $\boldsymbol{R}$ is the upper triangular matrix. The Euclidean distance is defined as $d_1 = \|\tilde{\boldsymbol{y}} - \boldsymbol{R}\boldsymbol{s}\|^2$, PED is shown in Eq. (4.9)

$$d_i = d_{i+1} + \left|\tilde{y}_i - \sum_{j=i}^{N_t} R_{i,j}s_j\right|^2 = d_{i+1} + |e_i|^2 \tag{4.9}$$

The number search process with $N_t + 1$ nodes can be expressed by Fig. 4.4, where Stage i denotes the i(th) transmitting antenna.

The search algorithm starts with the root node or the first child node in Stage N_t, where Stage N_t represents the N_t(th) antenna symbol at the transmitter. Then compute PED. If PED (i.e., d_{N_t}) is less than the sphere radius r_s, then the search

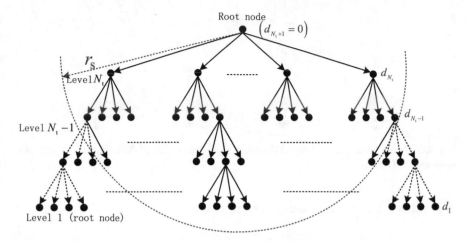

Fig. 4.4 Search tree in the SD signal detection algorithm. © [2018] IEEE. Reprinted, with permission, from Ref. [11]

process will continue until it reaches the stage $N_t - 1$. Otherwise, if the d_{N_t} is greater than the sphere radius r_s, then the search has exceeded the set hypersphere, no longer search for all the child nodes below this node, but continue to search in another path. Thus, the step-by-step search is carried out until a valid leaf node in the first stage is estimated.

The selection of D in the SD signal detection algorithm has a certain influence on the search process. If the value of D is too small, it will result in the value of the first stage node exceeding D or the value of all nodes exceeding D when finding a middle stage, thus the optimal solution cannot be obtained. The SD-pruning signal detection algorithm can effectively solve this problem. It first defines the preset value as infinity, then updates the preset value when the last stage searched, and then keeps updating when smaller values searched, so that the performance of SD-pruning signal detection algorithm can reach that of ML signal detection algorithm.

4.1.2.3 FSD Signal Detection Algorithm

The SD signal detection algorithm is another suboptimal MIMO signal detection algorithm, which can further reduce the computational complexity of the K-best detection algorithm [13]. The detection process of the FSD signal detection algorithm is based on a two-stage tree search, as shown in Fig. 4.5 [11]. The algorithm implementation is shown in Algorithm 4.2.

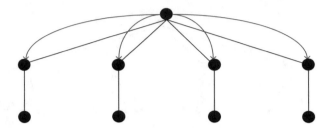

Fig. 4.5 Search tree in the FSD signal detection algorithm. © [2018] IEEE. Reprinted, with permission, from Ref. [11]

Algorithm 4.2 FSD signal detection algorithm

Full extension: First, the first p stages are fully extended, i.e. all the possible candidate nodes are reserved.

Single extension: A single extension of the remaining $N_t - p$ stages, that is, each node holds only candidate nodes with the smallest measure.

A conventional FSD signal detection algorithm has a fixed complexity but without consideration of noise and channel conditions. A simplified version of FSD signal detection algorithm is described in Ref. [18], which introduces the path selection into the remaining stages, so that the FSD signal detection algorithm can be highly parallel and fully pipelined [19].

4.2   CHOSLAR Algorithm

4.2.1   System Model

The CHOSLAR algorithm exploits a complex domain system with the same number of transmitting antennas and the number of receiving antennas, i.e., $N_r = N_t = N$, and assumes that the channel matrix has been estimated [8, 10, 20–22]. The channel model is shown in Eq. (4.1). The ML signal detection is shown in Eq. (4.3). After the channel matrix $\boldsymbol{H}$ is decomposed by QR, Eq. (4.3) can be written as

$$\hat{\boldsymbol{s}}_{\mathrm{ML}} = \arg\min_{s \in \Omega} \left\| \boldsymbol{Q}^{\mathrm{H}} \boldsymbol{y} - \boldsymbol{R}\boldsymbol{s} \right\|^2 = \arg\min_{s \in \Omega} \left\| \hat{\boldsymbol{y}} - \boldsymbol{R}\boldsymbol{s} \right\|^2, \tag{4.10}$$

where $\boldsymbol{H} = \boldsymbol{Q}\boldsymbol{R}$, $\boldsymbol{Q}$ is unitary matrix, and $\boldsymbol{R}$ is upper triangular matrix. ML detector generates soft-output through LLR, that is, seeking other symbol vectors closest to $\hat{\boldsymbol{y}}$. According to the Ref. [23], LLR of the b(th) bit of the i(th) symbol can be computed by Eq. (4.11).

$$\mathrm{LLR}\big(s_{i,b}\big) = \frac{1}{\sigma^2}\bigg(\min_{s\in s_{i,b}^0}\big\|\hat{\pmb{y}} - \pmb{R}\pmb{s}\big\|^2 - \min_{s\in s_{i,b}^1}\big\|\hat{\pmb{y}} - \pmb{R}\pmb{s}\big\|^2\bigg), \qquad (4.11)$$

where $s_{i,b}^0$ and $\mathrm{s}_{i,b}^1$ represents the set of symbols from the modulation constellation point Ω when $s_{i,b}$ equals to 0 and 1 respectively.

Matrix $\pmb{R}$ is an upper triangular matrix, so the tree search starts with the last element of s. The scale of the channel matrix $\pmb{H}$ increases as the scale of the MIMO system increases (e.g., from 2×2 to 16×16). Due to the high complexity and data dependency, QR decomposition is difficult to implement on hardware, especially for high-order MIMO systems. Therefore, the computational complexity and parallelism of QR decomposition are the main factors limiting the throughput of high-order MIMO systems at present.

4.2.2 QR Decomposition

QR decomposition is the process of decomposing the estimated channel matrix $\pmb{H}$ into a unitary matrix $\pmb{Q}$ and an upper triangular matrix $\pmb{R}$ [8]. Gram–Schmidt (GS) [21, 24, 25], Householder transformation (HT) [26] and GR [22, 27] are three widely used QR decomposition algorithms. A simplified GS algorithm is proposed to achieve stable permutation QR decomposition in Ref. [28]. In GR algorithm, the computation of matrix $\pmb{Q}$ can be replaced by the same rotation operation as that of upper triangular matrix $\pmb{R}$. However, in high-order MIMO systems, with the increase of the scale of matrix $\pmb{H}$, it is inefficient to eliminate only one element at a time. Another disadvantage of GR algorithm is that the rotation operation cannot be executed until the two elements on the left are eliminated. Therefore GR algorithm parallelism is constrained, resulting in lower data throughput in higher order MIMO systems. In order to reduce the computational complexity while maintaining the hierarchical gain, a lattice reduction (LR) algorithm proposed in Ref. [20] as a preprocessing method to adjust the estimated channel matrix in polynomial time according to Lenstra–Lenstra–Lovasz (LLL) algorithm [20]. In addition, the LR-based MMSE signal detection algorithm proposed in Ref. [21] and the LR-based zero-forcing decision feedback (ZF-DF) algorithm proposed in Ref. [29] have better performance than conventional linear detection algorithms. However, LR requires huge quantity of conditional validation and column exchanging, resulting in uncertain data throughput, low parallelism, and long latency in hardware implementation, especially when the size of MIMO systems increases. QR decomposition of channel matrix is also required in LR-based ZF-DF algorithm, and QR decomposition is very sensitive to the order of search stage. The scale increases gradually with the increase of system dimensions. Therefore, with the increase of the number of users and antennas, the preprocessing part (including QR decomposition and LR) as part of the K-best signal detection algorithm has high complexity and low parallelism [20], which becomes one of the main factors hindering the performance of the whole detector, especially when the channel matrix is not slowly changing. After preprocess, the detection program itself is also a rather complicated part of the whole process. The number of child nodes in each stage determines the complexity of K-best signal detection algorithm. Seeking the optimal solution in a smaller vector space can greatly reduce the number of child nodes. Therefore, when the channel matrix obtained by preprocess has better performance (when the preprocess is more asymptotically orthogonal), the complexity of the detection program itself can be greatly reduced.

In Ref. [8], there is a detailed comparison of the computational complexity of GS, HT and GR algorithms based on real number calculation quantity. For complex domain system, GS algorithm and GR algorithm approximate need $O(4N^3)$ multiplication on the real domain, while HT algorithm needs more computation d it needs to compute the unitary matrix Q [26]. Therefore, although HT algorithm can eliminate elements by column, its hardware consumption is also unsustainable. An improved GS algorithm for 4×4 QR decomposition is proposed in Ref. [25]. Matrices Q and R as well as the median of computation are expanded for parallel design. However, compared with the conventional GS algorithm, the number of multiplication has not decreased. In Ref. [23], a simplified GS algorithm for matrix decomposition is proposed. This algorithm can realize low complexity, especially QR decomposition for stable replacement of slowly varying channels. However, when the channel is not slowly varying, this algorithm has a nonnegligible loss in detection accuracy. A time-dependent channel matrix is proposed in Ref. [30], in which only the columns of partial matrices will be updated with time. Meanwhile, an algorithm combining an approximate Q matrix retention scheme and a precise QR decomposition update scheme is proposed to reduce the computational complexity. In Ref. [8], GR algorithm using coordinate rotation digital computer (CORDIC) unit simplifies complex arithmetic units in QR decomposition by iterative shift and addition operation. A new decomposition scheme based on real-domain system is also proposed in Ref. [8] to further reduce the number of arithmetic operation. However, the simplified QR decomposition combined with K-best signal detection has a serious loss of precision compared with ML signal detection algorithm. Other nonlinear signal detection algorithms, such as TASER [10] and probabilistic data association (PDA) algorithm [31] are also not suitable for high-order modulation systems. There will be serious precision loss, and hardware implementation is very difficult.

The sorted QR decomposition can reduce the search range and further improve the accuracy of K-best signal detection algorithm while keeping low complexity. In the execution of K-best signal detection algorithm, each element of $\hat{s}^{ML}$ is estimated as shown in Eq. (4.12).

$$\hat{s}_i = \left(\hat{y}_i - \sum_{j=i+1}^{N} R_{i,j} s_j \right) \bigg/ R_{i,i} \qquad (4.12)$$

Although $\hat{y}_i$ contains noise, the principal diagonal element $R_{i,i}$ can avoid the influence of noise and signal interference. QR decomposition is performed by column, which means that the matrix R is generated by row. Because the absolute value of the matrix H is fixed, the product of all diagonal elements of R is also a constant. When we decompose the i(th) column, there is

$$R_{i,i} = \text{norm}(H(i : N, i)) = \sqrt{\sum_{j=i}^{N} H_{j,i}^2},$$ (4.13)

where $R_{i,i}$ decreases as i increases. Sort and screen out the columns with the smallest norm as the next column to be decomposed, ensuring that the product of the remaining diagonal elements is as large as possible.

4.2.3 Lattice Reduction

The LR algorithm can achieve more efficient detection performance. For example, the LR algorithm is combined with the MMSE signal detection algorithm in Ref. [30] and LR algorithm is combined with ZF-DF algorithm in Ref. [29]. In the LR-based K-best signal detection algorithm and ZF-DF algorithm, the preprocessing part (QR decomposition and LR) plays an important role in reducing the computational complexity, especially when the size of MIMO system increases. When LR is used for channel matrix H, an approximately orthogonal channel matrix can be obtained, that is

$$\bar{H} = HT,$$ (4.14)

where T is a unimodular matrix. $\bar{H}$ is a matrix whose condition is much better than H, thus $\bar{H}$ contains less noise. LLL algorithm is an LR algorithm known for its polynomial time computational complexity [20]. This algorithm checks and corrects the matrix R, making it satisfy two conditions. In this section, different LLL algorithms are adopted, which need to satisfy the Siegel condition [20] expressed in Eq. (4.15) and the size reduction conditions expressed in Eq. (4.16). The parameter δ satisfies $0.25 < \delta < 1$. The Siegel condition keeps the difference between the two adjacent diagonal elements within a small range to prevent the generation of excessively small diagonal elements. The size reduction condition ensures that diagonal elements are slightly dominant so as to achieve approximate orthogonality. The adjusted channel matrix H or upper triangular matrix R can inhibit interference between different antennas.

$$\delta |R_{k-1,k-1}| > |R_{k,k}|, \quad k = 2, 3, \ldots, N$$ (4.15)

$$\frac{1}{2}\left|R_{k-1,k-1}\right| > \left|R_{k-1,r}\right|, \ 2 \leq k \leq r \leq N \qquad (4.16)$$

However, LR algorithm requires multiple conditional checks and column exchanges, resulting in uncertain data throughput, low complexity, and long latency. The sorted QR decomposition is essentially similar to the LR algorithm, both of which achieve better properties by adjusting the matrix R. Therefore, K-best signal detection algorithm can achieve the near-ML accuracy when the number of branch extensions is small. However, as the scale of the MIMO system increases, the computational complexity of preprocessing becomes uncontrollable. At the same time, sorted QR decomposition and LR will result in higher computational complexity, especially LR, which requires an uncertain number of low parallelism operations.

4.2.4 Cholesky Preprocessing

4.2.4.1 Cholesky-Sorted QR Decomposition

In the K-best signal detector, the matrix preprocessing requires high detection accuracy and computational efficiency. The first step of the K-best signal detector preprocessing is QR decomposition of channel matrix H. In the decomposition process, the computation of unitary matrix Q will lead to high computational complexity. The QR decomposition algorithm in this section uses the properties of matrices Q and R to avoid the computation of matrix Q. QR decomposition is $H = QR$. There is

$$H^{H}H = (QR)^{H}QR = R^{H}\left(Q^{H}Q\right)R = R^{H}R \qquad (4.17)$$

The elements of the channel matrix H are independent and identically distributed complex Gaussian variables with a mean value of 0 and a variance of 1. So matrix H is nonsingular, that is, $A = H^{H}H = R^{H}R$ is positive definite. Therefore, A is a Hermite positive semi-definite matrix. Cholesky decomposition is to decompose Hermite positive semi-definite matrix into the product of a lower triangular matrix and its conjugate transpose matrix, namely $A = LL^{T}$. When using Cholesky decomposition, the matrix R is equivalent to the upper triangular matrix L^{T}, and the calculation of the special unitary matrix Q is avoided. Although matrix R is obtained by shortcut, it is known from Eq. (4.10) that matrix Q still needs to be calculated to solve $Q^{H}y$. Since $H = QR$, so you can solve Q by $Q = HR^{-1}$. Therefore, the calculation for $Q^{H}y$ is converted to

$$Q^{H}y = \left(HR^{-1}\right)^{H}y = \left(R^{-1}\right)^{H}H^{H}y \qquad (4.18)$$

In Eq. (4.18), the direct solution of $\boldsymbol{Q}^H \boldsymbol{y}$ is replaced by the inversion of the upper triangular matrix $\boldsymbol{R}$ and two matrix-vector multiplications. The computation complexity of the upper triangular matrix is $O(N^3)$, which is significantly lower than that of the directly computing matrix $\boldsymbol{Q}$. In the GR algorithm, when eliminating the elements of matrix $\boldsymbol{H}$, do the same conversion for vector $\boldsymbol{y}$ to solve $\boldsymbol{Q}^H \boldsymbol{y}$ instead of directly solving matrix $\boldsymbol{Q}$. However, the computation matrix $\boldsymbol{R}$ takes the multiplication of $O(4N^3)$ complexity. The computational complexity of these algorithms will be compared in the subsequent chapters.

The next question is how to combine sorting operations with QR decomposition. In the Cholesky-based algorithm, the QR decomposition is realized through the Gram matrix $\boldsymbol{A} = \boldsymbol{H}^H \boldsymbol{H}$. Based on the Cholesky decomposition, the matrix $\boldsymbol{R}$ is generated by row, and the pseudo codes of Cholesky decomposition are shown in Algorithm 4.3.

Algorithm 4.3 Cholesky decomposition algorithm

1: **for** $i = 1; i \leq N - 1; i + 1$ **do**

2:     $R_{i,i} = \sqrt{A_{i,i}}$;

3:     **for** $j = i + 1; j \leq N; j + 1$ **do**

4:         $R_{i,j} = A_{i,j} / R_{i,i}$;

5:     **end for**

6:     **for** $m = i + 1; m \leq N; m + 1$ **do**

7:         **for** $n = m; n \leq N; n + 1$ **do**

8:             $A_{m,n} \leftarrow A_{m,n} - R_{i,m}^H R_{i,n}$;

9:         **end for**

10:     **end for**

11: **end for**

12: $R_{N,N} = \sqrt{A_{N,N}}$;

Here, a 4×4 matrix is used as an example. When $i = 1$ (the first round of sorting), $\boldsymbol{A} = \boldsymbol{H}^H \boldsymbol{H}$, and the values to be compared in the sorting are shown in Eq. (4.19).

$$V_k = \boldsymbol{h}_k^H \boldsymbol{h}_k = \text{norm}^2(\boldsymbol{h}_k) = A_{k,k}, \; k = 1, 2, 3, 4 \tag{4.19}$$

That is, each diagonal element $A_{k,k}$ is the square of the norm of the column corresponding to the matrix $\boldsymbol{H}$, so the elements of the row and column corresponding to the smallest diagonal element in $\boldsymbol{A}$ are exchanged with elements of the first row and first column, and then the first round of decomposition can begin. At $i = 2$ (the second round of sorting), the value of V_k satisfies

$$
\begin{aligned}
V_k &= \text{norm}^2(\boldsymbol{h}_k) - R_{i-1,k}^{\text{H}} R_{i-1,k} \\
&= A_{k,k} - R_{i-1,k}^{\text{H}} R_{i-1,k}, \ k = 2, 3, 4
\end{aligned}
\tag{4.20}
$$

According to the Eq. (4.20) and the eighth row of the Algorithm 4.3, V_k is the diagonal element of matrix $\boldsymbol{A}$ (updated in the previous decomposition), so when $i = 2$, the diagonal element of updated matrix $\boldsymbol{A}$ can be reused as the base of the sorting. When $i = 3, 4$, the analysis process is similar. Therefore, the diagonal elements of Gram matrix $\boldsymbol{A}$ can always be used as the norm for each column of the sorting operation. In the conventional GR algorithm, when $i = 2$ (after the first round of decomposition), the matrix $\boldsymbol{H}$ can be written as

$$
\begin{bmatrix}
H_{1,1}, & H_{1,2}, & H_{1,3}, & H_{1,4} \\
H_{2,1}, & H_{2,2}, & H_{2,3}, & H_{2,4} \\
H_{3,1}, & H_{3,2}, & H_{3,3}, & H_{3,4} \\
H_{4,1}, & H_{4,2}, & H_{4,3}, & H_{4,4}
\end{bmatrix}
\rightarrow
\begin{bmatrix}
R_{1,1}, & R_{1,2}, & R_{1,3}, & R_{1,4} \\
0, & H_{2,2}', & H_{2,3}', & H_{2,4}' \\
0, & H_{3,2}', & H_{3,3}', & H_{3,4}' \\
0, & H_{4,2}', & H_{4,3}', & H_{4,4}'
\end{bmatrix},
\tag{4.21}
$$

where $H_{i,j}'$ is the updated element of row i, column j of the matrix $\boldsymbol{H}$. While the GR algorithm does not change the norm of each column, so when the decomposition of column 2 is performed, the value to be compared when the sorting column k is

$$
V_k = \sum_{j=i}^{4} \left(H_{j,k}'\right)^{\text{H}} H_{j,k}', \ k = 2, 3, 4
\tag{4.22}
$$

To get the correct sort, the value to be compared in each round of sort must be computed according to the updated matrix $\boldsymbol{H}$. This computation takes $\frac{2}{3}N^3$ real number multiplications. Figure 4.6 exhibits the difference between the Cholesky-sorted QR decomposition algorithm in this section and the conventional algorithm, assuming that the fourth column has the smallest norm value at the second round of decomposition. In the proposed algorithm, sorting is achieved by exchanging rows and columns after matrix updates that only on diagonal elements. In conventional algorithms, the square value of the norm of all three columns of vectors must be obtained before sorting, and then the column is exchanged.

Compared with other Cholesky decomposition algorithms for MIMO detection [29, 32, 33], the proposed algorithm in this section is slightly different in terms of purpose and implementation details. In ZF-DF algorithm and successive interference cancelation (SIC) detector [29, 33], the diagonal elements of $\boldsymbol{R}$ are calculated by the QR decomposition. The Gram matrix is decomposed by the LDL algorithm, which

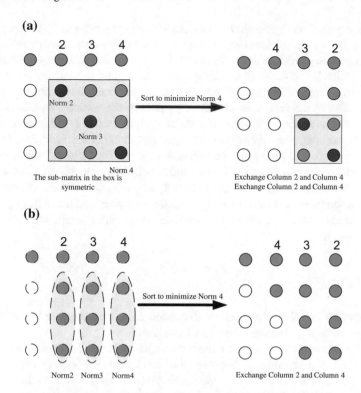

Fig. 4.6 Difference between Cholesky sorted QR decomposition algorithm and a conventional algorithm. **a** Use this algorithm to sort QRD, **b** use a conventional algorithm to sort QRD. © [2018] IEEE. Reprinted, with permission, from Ref. [9]

is broken down into a lower triangular matrix (L) with diagonal elements of 1 and a diagonal matrix (D) with real numbers. We can compute the diagonal elements of R after decomposing the matrix D. However, in the K-best signal detection algorithm, the whole upper triangular matrix R needs to be calculated. Therefore, the matrix R cannot be calculated simply by using the LDL algorithm, but by using other algorithms, then the subsequent K-best program will be affected. In the CHOSLAR algorithm in this section, the matrices R and Q can be solved directly by matrix decomposition, which makes the subsequent K-best program execution can start in a very short time. The Cholesky algorithm is used to decompose and invert the Gram matrix in the linear MMSE signal detection algorithm in Ref. [32]. Compared with the above algorithms, in CHOSLAR algorithm, Cholesky QR decomposition is performed first, then LR algorithm and K-best algorithm are executed, so the algorithm can be applied to nonlinear algorithm with high performance. In a word , the algorithm in this section is more favorable to subsequent K-best search than other algorithms. The algorithm also includes sorting operations in the decomposition process, and the detector accuracy is greatly improved compared with the Cholesky decomposition algorithm in Refs. [29, 32, 33]. The obtained matrix has a flat diagonal element, which is conducive to the computation of LR and K-best, and increases the accuracy of detection. In conventional QR decomposition, the sorting operation is used to optimize the matrix R [21, 24, 25, 27]. In the sorted QR decomposition algorithm in this section, the norm of each column of matrix H has been computed when matrix A is adjusted. As a result, the algorithm does not need additional addition operation, because the adjustment of matrix A as part of the Cholesky decomposition process has been implemented. The conventional QR decomposition, whether using GR [27] or GS algorithm [21, 24, 25], is directly implemented with the matrix H. The decomposition process is performed by column operation, and the sorting operation requires an extra step to compute the norm of all remaining columns that have not been decomposed.

4.2.4.2 PILR

In this section, a PILR algorithm is described to reduce the number of column exchanges while maintaining a constant throughput.

The algorithm runs T times from the countdown to the $N/2+1$ row in an iterative way. In Step k of each iteration, the algorithm first detects the Siegel condition of the Eq. (4.15). If the condition does not meet, this algorithm will calculate the parameter μ according to Eq. (4.23), where round() is a rounding function.

$$\mu = \text{round}\left(R_{k-1,k}/R_{k-1,k-1}\right) \tag{4.23}$$

Then, the $k - 1$ column of R is multiplied by μ, and the result is subtracted by the k column, as shown in Eq. (4.24):

$$R_{1:k,k} \leftarrow R_{1:k,k} - \mu R_{1:k,k-1} \tag{4.24}$$

The channel matrix H performs the operation shown in Eq. (4.25).

$$H_{:,k} \leftarrow H_{:,k} - \mu H_{:,k-1} \tag{4.25}$$

The operation described in Eq. (4.25) is a single size reduction to ensure that the Siegel condition can be executed correctly. After a single size reduction, the elements of column k and column $k - 1$ in R and H are exchanged to get $\hat{R}$ and $\hat{H}$. This

(a) **(b)**

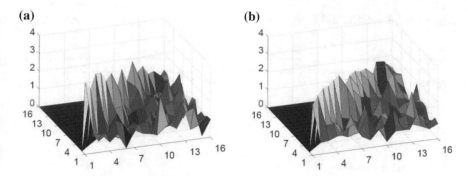

Fig. 4.7 a Element values of unsorted matrix R, **b** sorted matrix R. © [2018] IEEE. Reprinted, with permission, from Ref. [9]

exchange operation changes the triangular form of the matrix $\boldsymbol{R}$. Therefore, a 2×2 GR matrix θ is needed, that is

$$\theta = \begin{bmatrix} a^{\mathrm{H}}, & b^{\mathrm{H}} \\ -b, & a \end{bmatrix}, \tag{4.26}$$

where the value of a and b can be expressed as

$$a = \frac{\hat{R}_{k-1,k-1}}{\sqrt{\hat{R}_{k-1,k-1}^2 + \hat{R}_{k,k-1}^2}}, \quad b = \frac{\hat{R}_{k,k-1}}{\sqrt{\hat{R}_{k-1,k-1}^2 + \hat{R}_{k,k-1}^2}} \tag{4.27}$$

Finally, Row k and Row $k - 1$ are updated by the left multiplying matrix θ, to restore the triangular form of matrix $\boldsymbol{R}$, i.e.,

$$\begin{bmatrix} \hat{R}_{k-1,k-1}, & \hat{R}_{k-1,k}, & \cdots \\ 0, & \hat{R}_{k,k}, & \cdots \end{bmatrix} \leftarrow \theta \times \begin{bmatrix} \hat{R}_{k-1,k}, & \hat{R}_{k-1,k-1}, & \cdots \\ \hat{R}_{k,k}, & 0, & \cdots \end{bmatrix} \tag{4.28}$$

The full size reduction process is similar to the single size reduction step but is performed for the entire matrix $\hat{\boldsymbol{R}}$. When performing the full size reduction operation, the algorithm iterates elements from $\hat{R}_{k,k}$ to $\hat{R}_{1,1}$ one by one. For example, Fig. 4.7 shows the values of each element of the sorted and unsorted matrix $\boldsymbol{R}$ for a 16×16 MIMO system. The values of diagonal elements after sorted QR decomposition are flat, so the LR algorithm can be applied to the 8×8 submatrix in the lower right corner of $\boldsymbol{R}$. The algorithm is iterated from the sixteenth column to the ninth column. The complete preprocessing algorithm includes the Cholesky sorted QR decomposition and PILR, as shown in Algorithm 4.4.

Algorithm 4.4 CHOSLAR algorithm

1: Input: $N \times N$ channel matrix $\boldsymbol{H}$, $N \times 1$ received vector; Each element of $N \times 1$ is $1 + 1j$ vector; Parameter δ

2: Output: Matrix $\boldsymbol{R}$; Vector $\hat{\boldsymbol{y}} = \boldsymbol{Q}^H \boldsymbol{y}$;

3: //Initialization:

4: $\boldsymbol{A} = \boldsymbol{H}^H \boldsymbol{H}$; $\dot{\boldsymbol{y}} = 0.5 \times (\boldsymbol{y} - \boldsymbol{H}\boldsymbol{v})$;

5: //Sorted QR decomposition:

6: **for** $k = 1; k \leq N - 1; k + 1$ **do**

7:      i is sorted by $A_{i,i} = \underset{d=k,k+1,\ldots N}{\arg\min} \left(A_{d,d} \right)$;

8:      Exchange the column i and column k of the matrix $\boldsymbol{H}$ to get the matrix $\dot{\boldsymbol{H}}$;

         Exchange the row i and row k, column i and column k of the matrix $\boldsymbol{A}$

9:      $R_{k,k} = \sqrt{A_{k,k}}$;

10:     **for** $j = k; j \leq N; j + 1$ **do**

11:          $R_{k,j} = A_{k,j}/R_{k,k}$;

12:     **end for**

13:     **for** $m = k + 1; m \leq N; m + 1$ **do**

14:          **for** $n = m; n \leq N; n + 1$ **do**

15:               $A_{m,n} \leftarrow A_{m,n} - R_{k,m}^H R_{k,n}$;

16:          **end for**

17:     **end for**

18: **end for**

19: $R_{N,N} = \sqrt{A_{N,N}}$; //Obtain matrix $\boldsymbol{R}$

20: //PILR:

21: **for** $t = 1; t \leq T; t + 1$ **do**

22:     **for** $k = N; k \geq N/2 + 1; k - 1$ **do**

23:          **if** $\delta \left| R_{k-1,k-1} \right| > \left| R_{k,k} \right|$ **then**

24:               $\mu = R_{k-1,k}/R_{k-1,k-1}$;

25:               $R_{1:k,k} \leftarrow R_{1:k,k} - \mu R_{1:k,k-1}$; $\dot{H}_{:,k} \leftarrow \dot{H}_{:,k} - \mu \dot{H}_{:,k-1}$;

26:               Exchange the column k and column $k-1$ of the matrix $\boldsymbol{R}$ and $\dot{\boldsymbol{H}}$ to get the matrix $\hat{\boldsymbol{R}}$ and $\hat{\boldsymbol{H}}$;

27:               $a = \dfrac{\hat{R}_{k-1,k-1}}{\sqrt{\hat{R}_{k-1,k-1}^2 + \hat{R}_{k,k-1}^2}}$; $b = \dfrac{\hat{R}_{k,k-1}}{\sqrt{\hat{R}_{k-1,k-1}^2 + \hat{R}_{k,k-1}^2}}$;

28: $$\theta = \begin{bmatrix} a^{\mathrm{H}} & b^{\mathrm{H}} \\ -b & a \end{bmatrix};$$

29:            Update the row k and row $k-1$ of matrix $\hat{\mathbf{R}}$,

$$\begin{bmatrix} \hat{R}_{k-1,k-1} & \hat{R}_{k-1,k} & \cdots \\ 0 & \hat{R}_{k,k} & \cdots \end{bmatrix} \leftarrow \theta \times \begin{bmatrix} \hat{R}_{k-1,k} & \hat{R}_{k-1,k-1} & \cdots \\ \hat{R}_{k,k} & 0 & \cdots \end{bmatrix};$$

30:        **end if**

31:      **end for**

32: **end for**

33: //**Full size reduction:**

34: **for** $m = N-1; m \geq 2; m-1$ **do**

35:      **for** $n = m+1; n \leq N; n+1$ **do**

36:          $\mu = round\left(\hat{R}_{k-1,k} \middle/ \hat{R}_{k-1,k-1}\right);$

37:          $\hat{R}_{1:k,k} \leftarrow \hat{R}_{1:k,k} - \mu \hat{R}_{1:k,k-1};$

38:          $\hat{H}_{:,k} \leftarrow \hat{H}_{:,k} - \mu \hat{H}_{:,k-1};$

39:      **end for**

40: **end for**

41: //**Inversion of matrix** $\hat{R}$:

42: **for** $i = N; i \geq 2; i-1$ **do**

43:      $R^{inv}_{i,i} = 1\middle/ \hat{R}_{i,i};$

44:      **for** $j = i+1; j \leq N; j+1$ **do**

45:          $R^{inv}_{i,j} = -\left(\sum_{k=j}^{N} \hat{R}_{i,k} R^{inv}_{k,j}\right) R^{inv}_{i,i};$

46:      **end for**

47: **end for**

48: //**Post-vector calculation:**

49: $\hat{y} = \left(R^{inv}\right)^{\mathrm{H}} \left(\hat{H}^{\mathrm{H}} \hat{y}\right).$

In Ref. [34], conventional LR and partial LR operations are used to optimize matrix $\mathbf{R}$. First, a conventional algorithm uses the unit matrix T as the trace of all column adjustments, where the channel matrix H can be used directly. Then, when the full size reduction is executed, the original LR [29] and partial LR [34] algorithms iterate elements from $\hat{R}_{1,1}$ to $\hat{R}_{k,k}$ one by one, whereas the algorithm proposed in this section operates in the opposite direction, which makes it more possible to operate in line to improve parallelism. As will be mentioned in Chap. 5, the algorithm can achieve high throughput ASIC hardware architecture design, thus proving its high parallelism. Finally, in the conventional LR algorithm, the entire matrix $\hat{R}$ needs to be adjusted, and the algorithm in this section combines LR with sorted QR decomposition, both of them need to adjust the matrix $\hat{R}$ and make it owns more advantageous features. The PILR algorithm runs T times from the last row to the $N/2 + 1th$ row iteratively. This algorithm makes use of this characteristic and combines it in order to reduce the total number of column exchanges.

4.2.5 Improved K-Best Detector and Its Performance Simulation

Based on the proposed preprocessing algorithm, an improved K-best detector with $K = 10$ is adopted in this section. The detector is divided into N stages to solve the output signal $\hat{s}$. Taking the 16×16 MIMO system as an example, the search tree is shown in Fig. 4.8. An approximate solution $\hat{s}_N$ is computed in Stage 1, as shown in Eq. (4.29).

$$\hat{s}_N = \hat{y}_N / \hat{R}_{N,N} \tag{4.29}$$

Then, four Gaussian integers closest to $\hat{s}_N$ in the 64QAM constellation are obtained in a certain order, as shown in Fig. 4.9. These four nodes based on the PED ascending order are as the parent node of Stage 2. In Stage 2, four parent nodes perform interference cancelation in turn. Then, the parent node with the minimum

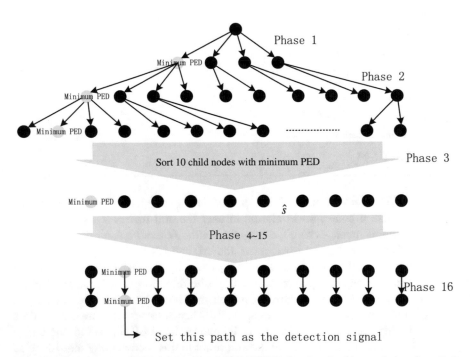

Fig. 4.8 Tree extension of K-best detector. © [2018] IEEE. Reprinted, with permission, from Ref. [9]

Fig. 4.9 Enumeration of constellation points. © [2018] IEEE. Reprinted, with permission, from Ref. [9]

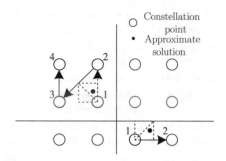

PED extends four child nodes in the same way as Stage 1, and the other three parent nodes extend two child nodes in a certain order respectively. The 10 child nodes in their PEDs ascending order serve as the parent nodes of Stage 3, where Stages 3–15 are structurally similar. First, 10 parent nodes perform interference cancelation successively. Then, the parent node with the minimum PED expands to four child nodes, and the other parent nodes expand to two child nodes respectively. Then 10 child nodes with the smallest PED are selected as the parent node of the next stage. In the last stage, after interference cancelation, each parent node expands only one child node. Finally, the child node with minimum PED is selected as the final solution, and the path corresponding to the child node is the output signal $\hat{s}$.

BER is simulated below to estimate the performance of the algorithm. All algorithms consider 64QAM modulation high-order MIMO systems and employ a rate-1/2 random interleaved convolutional codes [1, 5]. The number of symbols encoded is 120 and the number of frames is 100,000. The channel is an independent and identically distributed Rayleigh fading channel, and SNR is defined at the receiver. Figure 4.10a compares the BER performance of full-matrix LR and PILR algorithms in 16×16 MIMO systems with different iterations. It can be seen that K-best signal detection algorithm with reduced sequencing at $K = 10$ can achieve near-optimal performance when the number of iterations is 3. Therefore, when the 8×8 PILR algorithm in the lower right corner mentioned in this section is specially customized for the 16×16 MIMO system, the algorithm iteratively traverses three times from the last row to the ninth row. Figure 4.10b compares the BER performance of the K-best signal detection algorithm with different K values ($K = 8, 10, 12$) using CHOSLAR preprocess. It can be seen that compared with ML and CHOSLAR ($K = 12$) algorithms, the performance loss of CHOSLAR ($K = 10$) algorithm is 1.44 dB and 0.53 dB (BER is 10^{-5}) respectively, which is in the acceptable range. However, when $K = 8$, there is a performance loss of 3.46 dB compared to the ML algorithm, so $K = 10$ is used in this section. Figure 4.10c compares BER performance of different detection algorithms ($K = 10$) for 16×16 MIMO systems with 64QAM modulation. Compared with ML algorithm, the performance loss of the K-best algorithm with CHOSLAR preprocessing in this section is 1.44 dB when BER is 10^{-5}, which is very close to that of K-best algorithm (0.89 dB) with sorted QR decomposition and LR and K-best algorithm (1.2 dB) with QR decomposition and LR [20].

Furthermore, the performance of the algorithms used in this section and the K-best algorithm (3.37 dB) with sorted QR decomposition in Refs. [22, 24], the K-best algorithm (3.88 dB) with QR decomposition in Ref. [25], the K-best algorithm (4.96 dB) with simplified QR decomposition in Ref. [35], the MMSE-LR algorithm (more than 8 dB) and the MMSE algorithm (more than 8 dB) in Ref. [21], are comparable. At the same time, Fig. 4.10c shows that the performance of the algorithm combining sorted QR decomposition and LR is significantly better than that of the algorithm only using sorted QR decomposition. It should be noted that the simulation results are based on 64QAM modulation, so TASER algorithm is not supported [10].

The above simulation results are all based on 16×16 MIMO system. It can be seen that the CHOSLAR algorithm in this section has great advantages in the BER performance. To demonstrate the advantages of the proposed algorithm in both higher order MIMO systems and different modulation types, Fig. 4.11 shows the BER performance of the algorithm in different simulation types. Figure 4.11a, b compare the BER performance under the 64QAM modulation of the 64×64 MIMO system and the 128×128 MIMO system respectively. Because the complexity of ML signal detection algorithm is very high in 64×64 MIMO systems and 128×128 MIMO system, the simulation results of ML signal detection algorithm are not shown in the figure. As we can see from Fig. 4.11a, the proposed algorithm has a performance loss of 0.77 dB compared with the K-best signal detection algorithm using sorted QR decomposition and LR in 64×64 MIMO systems. While Fig. 4.11b shows a performance loss of 1.41 dB in 128×128 MIMO system. Moreover, the BER performance of the proposed algorithm is better than that of the algorithm mentioned in Refs. [21, 22, 24, 35]. Therefore, the CHOSLAR algorithm adopted maintains its advantages in the higher order MIMO system. Figure 4.11c shows BER performance of higher order modulation (256QAM). The K-best signal detection algorithm uses $K = 14$. According to Fig. 4.11c, the CHOSLAR algorithm retains its advantages while keeping a performance loss of only 1.01 dB.

Although only symmetrical MIMO systems which have equal receiving antennas and transmitting antennas are discussed for the present algorithms, this algorithm is also applicable to asymmetrical MIMO systems which have more receiving antennas than transmitting antennas. Throughout the algorithm, QR decomposition is converted to the Cholesky decomposition, and the number of the receiving antennas only affects the initialization stage (row 4 in Algorithm 4.4) and the updating and column exchange of matrix H (rows 8, 25, 26, 38 of Algorithm 4.4). With the increase of the number of the receiving antennas (more than the number of the transmitting antennas), these elements of the process would be affected. Initialization is still possible with simple matrix and vector multiplication, while the update and column exchange process of matrix H is only based on a single column of matrix H. Therefore, the CHOSLAR algorithm is also applicable to the asymmetric MIMO system. Figure 4.12a illustrates the BER performance of an asymmetric (16×32) MIMO system. The results show that the CHOSLAR algorithm is also applicable to the asymmetric MIMO system, which still maintains its advantages.

To better reflect how different channel characteristics affect the algorithm and simulation results, the Kronecker channel model is used to estimate the performance [5].

Fig. 4.10 Comparison of
BER performance, **a** PILR
and full-matrix LR,
b different *K* values,
c different algorithms. ©
[2018] IEEE. Reprinted, with
permission, from Ref. [9]

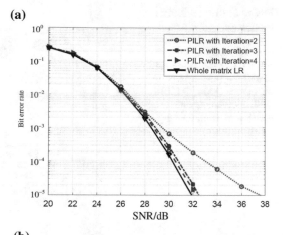

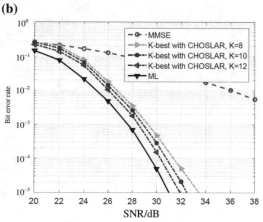

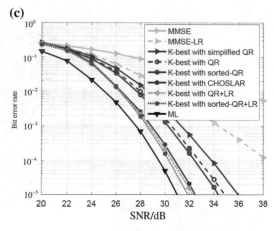

Fig. 4.11 BER performance comparison with different configurations and different modulation. **a** 64 × 64 MIMO 64QAM, **b** 128 × 128 MIMO 64QAM, **c** 16 × 16 MIMO 256QAM. © [2018] IEEE. Reprinted, with permission, from Ref. [9]

(a)

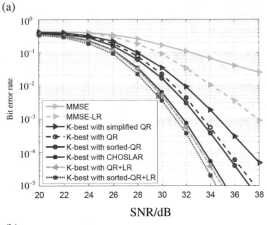

(b)

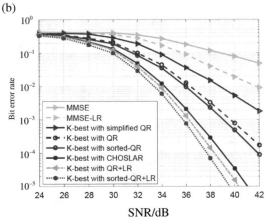

(c)

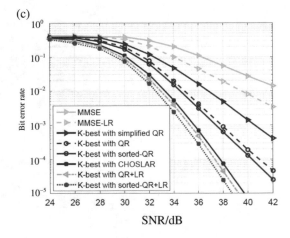

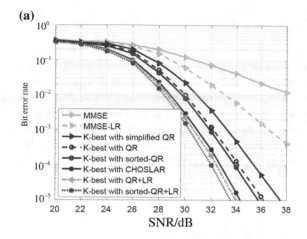

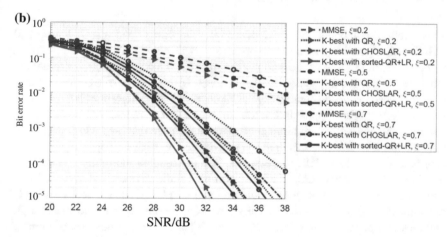

Fig. 4.12 Comparison of BER performance. **a** Asymmetric (16 × 32) MIMO system, **b** Kronecker channel (16 × 16 MMIMO, 64QAM). © [2018] IEEE. Reprinted, with permission, from Ref. [9]

The elements of the channel matrix of the channel model follow the distribution in the form of $N(0, d(z)\mathbf{I}_B)$, where $d(z)$ denotes channel attenuation (such as path attenuation and shielding). The classic path attenuation model is adopted and channel attenuation variable is $d(z) = C/\|z - b\|^\kappa$, where $b \in \mathbf{R}^2$, κ and $\|\cdot\|$ denote the location of base station, path attenuation index and Euclidean norm respectively. The independent shielding attenuation represented by C satisfies $10\lg C \sim N(0, \sigma_{sf}^2)$. Another distinctive feature of the Kronecker channel model is its consideration of channel correlation. $\mathbf{R}_r$ and $\mathbf{R}_t$ respectively denote the channel correlation parameters of receiving antennas and the channel correlation parameters of transmit antennas. The exponential correlation model [5] is adopted, and ξ stands for the correlation factor. Therefore, Channel matrix $\mathbf{H}$ can be expressed as

$$H = R_{\mathrm{r}}^{1/2} H_{\mathrm{i.i.d.}} \sqrt{d(z)} R_{\mathrm{t}}^{1/2}, \tag{4.30}$$

where $H_{\mathrm{i.i.d.}}$ is a random matrix of independent identically distributed complex Gaussian distribution with an average of 0 and a variance of 1 for each element. During the simulation process, the radius of each hexagonal element is $r = 500\mathrm{m}$, and the user location $z \in R^2$ is independent and random. The simulation also adopts the following assumptions: $\kappa = 3.7$, $\sigma_{\mathrm{sf}}^2 = 5$, and transmitting antennas power $\rho = r^{\kappa}/2$. Figure 4.12b compares the BER performance of the algorithm using the Kronecker channel model with three different correlation factors ($\xi = 0.2, 0.5, 0.7$). According to Fig. 4.12b, CHOSLAR algorithm still maintains its advantages in this actual model.

4.2.6 Summary and Analysis

This section compares the computational complexity and parallelism of the CHOSLAR algorithm with other algorithms (GS, GR, and HT). A specific summary is made in Ref. [8]. The analysis shows that most of the computational complexity is attributed to QR decomposition and LR process. In this section, the complex domain system is used, while the complexity of QR decomposition algorithm is expressed by the required real-field operands. For the computational complexity of the QR decomposition, the real-valued multiplication (RMUL) and real-valued addition (RADD) play a main role. It is assumed that the real-valued division and operation of square root is equivalent to RMUL in Refs. [23, 35]. A complex multiplication operation requires 4 RMUL and 2 RADD, while a complex addition needs 2 RADD. Table 4.1 lists the number of real operations required by GS, GR, HT, and Cholesky-sorted QR decomposition algorithm. The Cholesky sorted QR decomposition algorithm adopted includes two parts: matrix multiplication of $H^{\mathrm{H}} H$ and decomposition of Gram A. The matrix multiplication of $H^{\mathrm{H}} H$ requires $2N^3 + 2N^2$ RMUL and $N^3 - N$ RADD, because it requires conjugate symmetric matrix multiplication. The Cholesky decomposition of matrix A requires $N^3 - 2N^2 + 5N$ RMUL and $N^3 - 3N^2 + 3N$ RADD. Because direct computation for matrix Q in GR does not need, its computational complexity is omitted in Table 4.1. In order to perform sorting operations in each algorithm, additional $\frac{2}{3}N^3$ RMUL [27] is required. As can be seen from Table 4.1, the computational complexity of the Cholesky sorted QR decomposition algorithm is lower than that of other algorithms. For example, when $N = 16$, compared with that of GS, GR, and HT algorithms, the number of RMULs required by the algorithm is decreased by 25.1, 44.6, and 93.2%, respectively, and the number of RADDs required by the algorithm is decreased by 55.1, 58.9, and 95.2%, respectively.

Figure 4.13 shows the simulation results of the average column exchange times and maximum column exchange times of LR algorithm with non-sorted QR decomposition and of LR algorithm with sorted QR decomposition in a 16×16 MIMO system. The results show that the number of columns exchanged by sorted QR decomposition is reduced. The PILR algorithm with constant throughput needs three iterations,

Table 4.1 Computational complexity of different detection algorithms in a higher order MIMO system

Algorithm	Real addition	Real multiplication
GS	$4N^3 + N^2 - 2N$	$\frac{14}{3}N^3 + 4N^2 + N$
GR	$4N^3 + \frac{15}{2}N^2 - \frac{23}{2}N$	$\frac{18}{3}N^3 + \frac{23}{2}N^2 - \frac{107}{6}N$
HT	$2N^4 + 5N^3 + \frac{21}{2}N^2$	$\frac{8}{3}N^4 + \frac{22}{3}N^3 + 14N^2$
Cholesky sorted QR decomposition	$2N^3 - 3N^2 + 2N$	$\frac{11}{3}N^3 + 5N$

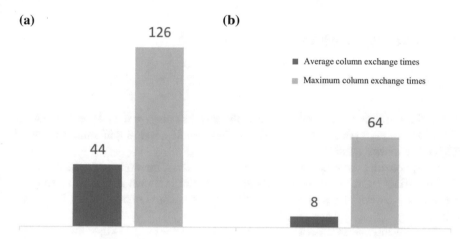

Fig. 4.13 Column exchange comparison between LR for non-sorted QR decomposition and LR for sorted QR decomposition. **a** LR for non-sorted QRD, **b** LR for sorted QRD. © [2018] IEEE. Reprinted, with permission, from Ref. [9]

and each iteration needs eight matrix exchanges. Therefore, a total of 24 matrix exchanges is required, less than the 44 matrix exchanges required for whole-matrix LR (reduced by 45.5%). The number of multiplications required for row updates is also reduced, as the number of matrix exchanges is reduced, so only the PILR needs to be executed in the lower right corner of the triangular matrix. In 16×16 MIMO system, the average number of multiplications required in LR and PILR algorithm is 3960 and 1296 respectively, that is, the PILR algorithm adopted can reduce the number of multiplications by 67.3%. In the calculation of K-best, in order to achieve the same detection accuracy, the K-best signal detection algorithm adopted only needs 33 comparators in each stage, while the unsorted algorithm needs 216 comparators in each stage, that is, the number of comparators in each stage is reduced by 84.7%.

Considering the parallelism of hardware implementation, the Cholesky sorted QR decomposition algorithm can eliminate an entire column of matrix A at one time, while the conventional GR algorithm based on paired CORDIC can only eliminate one element of matrix H at one time. Also, the elements in the left column must have been eliminated before removing a new column of elements. The correlation between these elimination limits its parallelism, especially when the user number and antennas number increase in higher order MIMO systems. For example, for a 16×16 matrix, the GR algorithm based on paired CORDIC takes four rounds to eliminate all elements in the first column. Therefore, compared with the conventional QR decomposition, the Cholesky-sorted QR decomposition algorithm can achieve higher parallelism and lower preprocessing delay. In addition, the parallelism of the size reduction process in the LR algorithm is also improved, because the column updates of matrices H and R are implemented by row rather than by element.

4.3 TASER Algorithm

4.3.1 System Model

TASER algorithm is applicable to two scenarios: the coherent data detection in MU-MIMO wireless systems and the JED in large-scale single-input multiple-output (SIMO) wireless systems.

First, consider the first application scenario, that is, the coherent data detection in MU-MIMO wireless system. The channel model is shown in Eq. (4.1), the ML detection is shown in Eq. (4.3), assuming that the channel matrix H has been estimated.

For conventional small-scale MIMO systems, a series of SD algorithms are presented in Ref. [15]. However, the computational complexity of these algorithms increases exponentially with the increase of the number of the transmitting antennas N_t, so they are not suitable for massive MIMO system. When the ratio of the number of receiving antennas to the transmitting antennas is more than 2 in the massive MU-MIMO system, some newly linear algorithms can achieve the near-ML performance [3]. However, when the number of the receiving antennas increases, the ratio of the number of receiving antennas to the transmitting antennas is close to 1, the BER performance of these linear algorithms is too poor to be accepted [3].

In order to ensure low complexity and near-optimal performance in this application scenario, the Eq. (4.3) in the ML problem is relaxed into a semi-definite program (SDP) [36], and the ML detection problem needs to be reformulated in the relaxation process. Assume that the modulation mode is fixed to QAM modulation, such as BPSK and QPSK. First, the system model is transformed into the real-valued decomposition, as shown in Eq. (4.31).

$$\bar{y} = \bar{H}\bar{s} + \bar{n}, \tag{4.31}$$

where the elements are defined as follows:

$$\bar{y} = \begin{bmatrix} \mathrm{Re}\{y\} \\ \mathrm{Im}\{y\} \end{bmatrix}, \quad \bar{H} = \begin{bmatrix} \mathrm{Re}\{H\} & -\mathrm{Im}\{H\} \\ \mathrm{Im}\{H\} & \mathrm{Re}\{H\} \end{bmatrix}$$

$$\bar{s} = \begin{bmatrix} \mathrm{Re}\{s\} \\ \mathrm{Im}\{s\} \end{bmatrix}, \quad \bar{n} = \begin{bmatrix} \mathrm{Re}\{n\} \\ \mathrm{Im}\{n\} \end{bmatrix} \tag{4.32}$$

The decomposition of Eq. (4.31) causes the ML problem to be formulated as

$$\bar{s}^{\mathrm{ML}} = \arg\min_{\tilde{s} \in \chi^N} \mathrm{Tr}\left(\tilde{s}^{\mathrm{H}} T \tilde{s}\right) \tag{4.33}$$

For QPSK, $T = \left[\bar{H}^{\mathrm{H}} \bar{H}, -\bar{H}^{\mathrm{H}} \bar{y}; -\bar{y}^{\mathrm{H}} \bar{H}, \bar{y}^{\mathrm{H}} \bar{y} \right]$ is the matrix of $N \times N$ ($N = 2N_t + 1$), $\tilde{s} = [\mathrm{Re}\{s\}; \mathrm{Im}\{s\}; 1]$. The range of its element value is $\mathcal{X} \in \{-1, +1\}$. In this way, the solution $\bar{s}^{\mathrm{ML}}$ can be converted back to the solution $\left[\hat{s}^{\mathrm{ML}}\right]_i = \left[\bar{s}^{\mathrm{ML}}\right]_i + j\left[\bar{s}^{\mathrm{ML}}\right]_{i+U}, i = 1, \ldots, U$ in the complex domain. For BPSK, $T = \left[\underline{H}^{\mathrm{H}} \underline{H}, -\underline{H}^{\mathrm{H}} \bar{y}; -\bar{y}^{\mathrm{H}} \underline{H}, \bar{y}^{\mathrm{H}} \bar{y} \right]$ is the matrix of $N \times N$ ($N = N_t + 1$), $\tilde{s} = [\mathrm{Re}\{s\}; 1]$. The matrix $\underline{H} = [\mathrm{Re}\{H\}; Im\{H\}]$ of $2N_r \times N_t$ is defined. At this time $Im\{s\} = 0$, so $\left[\hat{s}^{\mathrm{ML}}\right]_i = \left[\bar{s}^{\mathrm{ML}}\right]_i, i = 1, \ldots, U$.

The second application scenario is JED in the massive SIMO wireless system. Suppose that the transmitting time slot of a single user is $K + 1$, the number of receiving antennas is N_r, and the system model of SIMO wireless channel with narrowband flat-fading attenuation is [37].

$$Y = hs^{\mathrm{H}} + N, \tag{4.34}$$

where $Y \in C^{N_r \times (K+1)}$ is the received vector obtained in the $K + 1$ time slot. $h \in C^{N_r}$ is the unknown SIMO channel vector and is assumed to remain constant in the $K + 1$ time slot. $s^{\mathrm{H}} \in O^{1 \times (K+1)}$ is the transmitting antennas vector containing all the data symbols in the $K + 1$ time slot, $N \in C^{N_r \times (K+1)}$ is the cyclic symmetric Gaussian noise with independent identical distribution, and the variance is N_0. The ML JED problem can be expressed as [37].

$$\left\{\hat{s}^{\mathrm{JED}}, \hat{h}\right\} = \arg\min_{s \in O^{K+1}, h \in C^B} \left\| Y - hs^{\mathrm{H}} \right\|_{\mathrm{F}} \tag{4.35}$$

Note that both two outputs of JED have stage ambiguity that is, for a certain stage ϕ, if $\hat{s}^{\mathrm{ML}} e^{j\phi} \in O^{K+1}$, then $\hat{h} e^{j\phi}$ is also one of the solutions. To avoid this problem, assume that the first transmitted entry has been known to the receiver.

Since s is assumed to be a vector modulated in a constant modulus (such as BPSK and QPSK), the ML JED estimation of the transmitting antennas vector can be expressed as [37].

$$\hat{s}^{\mathrm{JED}} = \arg\max_{s \in O^{K+1}} \| Y s \|_2 \tag{4.36}$$

$\hat{h} = Y\hat{s}^{\text{JED}}$ is the estimation for channel vectors. When the time slot $K + 1$ is very small, Eq. (4.36) can be accurately solved [37] with a low-complexity SD algorithm. However, the complexity of the SD algorithm will become very high when the time slot is very large. Compared with the coherent ML detection algorithm in the above Eq. (4.33), it is not recommendable to approximate the solution of Eq. (4.36) by the linear algorithm, because the possible value of every item of s is infinite after the constraint $s \in O^{K+1}$ is relaxed to $s \in C^{K+1}$.

The ML JED problem of Eq. (4.36) is relaxed to the SDR of the same structure as the coherent ML problem of Eq. (4.33). Previously, we have assumed that the symbol s_0 at the first transmitting end is known, then $\|Ys\|_2 = \|y_0 s_0 + Y_r s_r\|_2$, where $Y_r = [y_1, \ldots, y_K]$, $s_r = [s_1, \ldots, s_K]^{\text{H}}$. Similar to the coherent ML problem, we transform it into a real-valued decomposition. First, we define

$$\bar{y} = \begin{bmatrix} \text{Re}\{y_0 s_0\} \\ \text{Im}\{y_0 s_0\} \end{bmatrix}, \quad \bar{H} = \begin{bmatrix} \text{Re}\{Y_r\} & -\text{Im}\{Y_r\} \\ \text{Im}\{Y_r\} & \text{Re}\{Y_r\} \end{bmatrix}, \quad \bar{s} = \begin{bmatrix} \text{Re}\{s_r\} \\ \text{Im}\{s_r\} \end{bmatrix} \quad (4.37)$$

$\|y_0 s_0 + Y_r s_r\|_2 = \|\bar{y} + \bar{H}\bar{s}\|_2$ can be obtained from Eq. (4.37). Equation (4.36) can be written in the same form as Eq. (4.33), i.e.

$$\bar{s}^{\text{JED}} = \underset{\bar{s} \in \chi N}{\arg \min} \, \text{Tr}\left(\bar{s}^{\text{T}} T \bar{s}\right) \quad (4.38)$$

For QPSK, $T = -\left[\bar{H}^{\text{H}}\bar{H}, \bar{H}^{\text{H}}\bar{y}; \bar{y}^{\text{H}}\bar{H}, \bar{y}^{\text{H}}\bar{y}\right]$ is the matrix of $N \times N$ ($N = 2K + 1$), $\bar{s} = [\text{Re}\{s_r\}; \text{Im}\{s_r\}; 1]$. The range of its element value is $\mathcal{X} \in \{-1, +1\}$. For BPSK, $T = -\left[\underline{H}^{\text{H}}\underline{H}, \underline{H}^{\text{H}}\bar{y}; \bar{y}^{\text{H}}\underline{H}, \bar{y}^{\text{H}}\bar{y}\right]$ is the matrix of $N \times N$ ($N = K + 1$), $\bar{s} = [\text{Re}\{s_r\}; 1]$. The matrix $\underline{H} = [\text{Re}\{Y_r\}; \text{Im}\{Y_r\}]$ of $2N \times K$ is defined. Similar to coherent ML problem, the solution $\bar{s}^{\text{JED}}$ can be converted to the solution in the complex domain.

Next, we will introduce how to find the approximate solutions of Eqs. (4.33) and (4.38) with the same SDRS-based algorithm.

4.3.2 Semi-definite Relaxation

SDR is known to all because it can be used to solve the coherent ML problem [36]. It can significantly reduce the computational complexity of BPSK and QPSK modulation systems. SDR cannot only provide near-ML performance, but also achieve the same diversity order of magnitude as ML detector [38]. Meanwhile, SDR is rarely used to solve the ML JED problem.

The data detection based on SDR first reformulates Eqs. (4.33) and (4.38) into the form of Eq. (4.39) [36], i.e.,

$$\hat{S} = \underset{S \in \mathbf{R}^{N \times N}}{\arg \min} \operatorname{Tr}(TS), \text{ s.t. } \operatorname{diag}(S) = 1, \operatorname{rank}(S) = 1, \qquad (4.39)$$

where $\operatorname{Tr}(s^H T s) = \operatorname{Tr}(T s s^H) = \operatorname{Tr}(TS)$ and $S = ss^H$ are the matrices with rank 1, $s \in \mathcal{X}^N$, and dimension is N. However, the constraint with the rank 1 in Eq. (4.39) makes Eq. (4.39) as complex as Eqs. (4.33) and (4.38). Therefore, the key of SDR is to relax the constraint of rank, so that SDP can be solved in polynomial time. By applying the SDR to Eq. (4.39), we get the famous optimization problem shown in Eq. (4.40) [36].

$$\hat{S} = \underset{S \in \mathbf{R}^{N \times N}}{\arg \min} \operatorname{Tr}(TS), \text{ s.t. } \operatorname{diag}(S) = 1, S \succeq 0 \qquad (4.40)$$

Constraint $S \geq 0$ makes S a positive semi-definite (PSD) matrix. If the rank of the result from Eq. (4.40) is 1, then $\hat{s}$ in $\hat{S} = \hat{s}\hat{s}^H$ is an accurate estimation of Eqs. (4.33) and (4.38), that means SDR solves the original problem in an optimal way. If the rank of $\hat{S}$ is greater than 1, the ML solution can be estimated by taking the symbol of the leading eigenvector of $\hat{S}$ or adopting a random scheme [36].

4.3.3 Algorithm Analysis

In this section, we will introduce a new algorithm, TASER, which is used to approximate the SDP problem presented in Eq. (4.40).

TASER algorithm is based on the fact that PSD matrix $S \geq 0$ in real domain can be factorized by Cholesky decomposition $S = L^H L$, where L is a lower triangular matrix whose main diagonal of $N \times N$ is a nonnegative term. Therefore, the SDP problem of Eq. (4.40) can be transformed into.

$$\hat{L} = \underset{L}{\arg \min} \operatorname{Tr}(L T L^H), \text{ s.t. } \|l_k\|_2 = 1, \quad \forall k \qquad (4.41)$$

Equation (4.41) uses two-norm constraint to replace $\operatorname{diag}(L^H L) = 1$ in Eq. (4.40), where $l_k = [L]_k$. The symbolic bit of the last row of solution $\hat{L}$ from Eq. (4.41) is taken as the solution to the problem in Eqs. (4.33) and (4.38), because if the rank of $\hat{S} = \hat{L}^H \hat{L}$ is 1, then the last row of $\hat{L}$, as the unique vector, must contain the relevant eigenvectors. If the rank of $\hat{S} = \hat{L}^H \hat{L}$ is greater than 1, an near-ML solution needs to be extracted. In Ref. [39], it is proposed that the last row of the result from Cholesky decomposition can be used as an approximation of PSD matrix with rank 1. In Chap. 5, the simulation results all prove that the performance achieved by this approximation is close to that of the exact SDR detector from eigenvalue decomposition. This approach avoids eigenvalue decomposition and stochastic solution in conventional schemes, thus reducing the complexity.

An effective algorithm is proposed to directly solve the triangular SDP of Eq. (4.41). However, the matrix L of Eq. (4.41) is non-convex, so it becomes difficult to find the optimal solution. For the TASER algorithm, a special method to solve the convex optimization problem, that is FBS method, it is applied to solve the non-convex problem of Eq. (4.41) [40]. This method cannot guarantee that the non-convex problem of Eq. (4.41) can converge to the optimal solution. Therefore, in Chap. 5, TASER algorithm will be proved to converge to a key point, and simulation results also prove that this method can guarantee the BER performance similar to that of the ML algorithm.

FBS is an effective iterative method for solving convex optimization problems in the form of $\hat{x} = \arg\min_{x} f(x) + g(x)$, where f is a smooth convex function, g is a convex function, but not necessarily smooth or bounded. The equation for solving FBS is [40, 41]

$$x^{(t)} = \text{prox}_g\left(x^{(t-1)} - \tau^{(t-1)}\nabla f\left(x^{(t-1)}\right); \tau^{(t-1)}\right), \quad t = 1, 2, \ldots \qquad (4.42)$$

The stopping condition of Eq. (4.42) is to achieve convergence or the maximum iterations number $t_{\max}$. $\left\{\tau^{(t)}\right\}$ is a sequence of step parameters and $\nabla f(x)$ is a gradient function of function f. The proximity operator of function g is defined as [40, 41].

$$\text{prox}_g(z; \tau) = \arg\min_{x}\left\{\tau g(x) + \frac{1}{2}\|x - z\|_2^2\right\} \qquad (4.43)$$

In order to approximately solve the Eq. (4.42) with FBS, we define $f(L) = \text{Tr}\left(LTL^H\right)$ and $g(L) = \chi(\|l_k\|_2 = 1, \forall k)$, where χ is the eigenfunction (the value is 0 when the constraint is satisfied, otherwise the value is infinite). The gradient function expressed as $\nabla f(L) = \text{tril}(2LT)$, where $\text{tril}(\cdot)$ means extracting its lower triangular part. Although the function g is non-convex, the approximation operator still has a closed solution $\text{prox}_g(l_k; \tau) = l_k/\|l_k\|_2, \forall k$.

In order to be friendly in hardware implementation, the complex step rule proposed in Ref. [40] is not used here, but a fixed step is used to improve the convergence rate of TASER algorithm [41]. The value of the fixed step is proportional to the reciprocal of Lipschitz constant $\tau = \alpha/\|T\|_2$ of the gradient function $\nabla f(L)$, where $\|T\|_2$ is the spectral norm of matrix T and $0 < \alpha < 1$ is dependent on the system's regulation parameters, used to improve the convergence rate of TASER algorithm [41]. To further improve the convergence rate of FBS, we need to preprocess the problem in Eq. (4.41). First, the diagonal scaling matrix $D = \text{diag}\left(\sqrt{T_{1,1}}, \ldots, \sqrt{T_{M,M}}\right)$ is computed, which is used to scale the matrix T to get a matrix with $\tilde{T} = D^{-1}TD^{-1}$. $\tilde{T}$ is a matrix with a main diagonal of 1. The processor that implements this operation, called the Jacobian preprocessor, is used to increase the conditional number of the original PSD matrix T [42]. Then run FBS to get the lower triangular matrix $\hat{L} = LD$. In the process of preprocessing, we also need to correct the proximity operators. In this case, $\text{prox}_{\tilde{g}}\left(\tilde{l}_k\right) = D_{k,k}\tilde{l}_k/\left\|\tilde{l}_k\right\|_2$, where $\tilde{l}_k$ is the kth column of $\tilde{L}$. Because we only take the symbol bits of the last row of $\hat{L}$ to estimate the ML problem, here we can take only the symbols of the normalized triangular matrix $\tilde{L}$.

The pseudocode of the TASER algorithm is shown in Algorithm 4.5. Input is the preprocessing matrix $\tilde{T}$, scaling matrix D and step size τ. $\tilde{L}^{(0)} = D$ is used for initialization. The main loop body of TASER algorithm is to run gradient function and proximity operator until the final iterations number $t_{\max}$ is obtained. In most cases, only a few iterations are required to obtain BER performance approximate to near-ML.

Algorithm 4.5 TASER algorithm

1: **Input:** $\tilde{T}$, D 和 $\tau = \alpha / \|\tilde{T}\|_2$

2: **Initialization:** $\tilde{L}^{(0)} = D$

3: **for** $t = 1, \ldots, t_{\max}$ **do**

4:      $V^{(t)} = \tilde{L}^{(t-1)} - \mathrm{tril}\left(2\tau \tilde{L}^{(t-1)} \tilde{T}\right)$

5:      $\tilde{L}^{(t)} = \mathrm{prox}_{\tilde{g}}\left(V^{(t)}\right)$

6: **end for**

7: **Output:** $\bar{s}_k = \mathrm{sign}\left(\tilde{L}_{N,k}^{(t_{\max})}\right), k = 1, \ldots, N-1$

The TASER algorithm tries to use FBS to solve a non-convex problem, which will result in two problems. One is whether the algorithm converges to the minimum; the other is whether the local minimum of the non-convex problem corresponds to the minimum of SDP for the convex optimization problem. Next, we will solve these two problems.

For the first problem, although it is still novel to use FBS to solve the minimum value of the positive semi-definite, there has been a lot of research on the convergence of FBS to solve non-convex problems. In Ref. [43], the condition of solving convergence of non-convex problems with FBS is proposed, and the problem to be solved must be semi-algebraic. Equation (4.41) meets this condition exactly. A strict proof of this solution will be given below.

Proposition 4.3.1 *If FBS (Algorithm 4.5) is used to solve the Eq. (4.41) and the step is $\tau = \alpha / \|T\|_2 (0 < \alpha < 1)$, then the iterative sequence $\{L^{(t)}\}$ will converge to a key point.*

Proof Function $\|\ell_k\|_2^2$ is a polynomial, and the constraint set of the Eq. (4.41) is the solution of the polynomial system $\|l_k\|_2^2 = 0$, $(\forall k)$, so it is semi-algebraic. Theorem 5.3 in Ref. [43] shows that if the upper bound of the step is the inverse of the Lipschitz constant of the gradient function of the objective function, then the iterative sequence $\{L^{(t)}\}$ is convergent. The Lipchitz constant is the spectral radius (two norm) of the matrix T. $\qquad\square$

Jacobi preprocessor leads to a problem in the same form as Eq. (4.41), except that the constraint is $\left\|\tilde{l}_k\right\|_2^2 = D_{k,k}^2$ and the step is $\tau = \alpha / \left\|\tilde{T}\right\|_2$, so Proposition 4.3.1 is applicable as well. However, this proposition does not guarantee that a minimum point is found, but only a stable point (in fact, a minimum point is often found). Nevertheless, this proposition is much better than other known low-complexity SDP algorithms in guaranteeing convergence. For example, non-convex enhanced Lagrange scheme adopted in Ref. [44] cannot guarantee convergence.

The second question is whether the local minimum of Eq. (4.41) corresponds to the minimum of convex SDP of Eq. (4.40). In Ref. [44], it is proposed that the local minimum in Eq. (4.41) is the minimum of SDP in Eq. (4.40) when the factors L and L^H are not constrained into triangular form. Nevertheless, L and L^H are constrained to a triangular form, as this simplifies the hardware architecture designed in Chap. 5.

4.3.4 Performance Analysis

Figure 4.14a, b show the respective simulation results of vector error rate (VER) of TASER algorithm modulated by BPSK and QPSK. For the massive MU-MIMO system of $128 \times 8(N_r \times N_t)$, 64×16 and 32×32, coherent data detection is used and the channel is flat Rayleigh flat-fading channel. Meanwhile, the performance of ML detection (using SD algorithm in Ref. [15]), exact SDR detection in Eq. (4.39), linear MMSE detection, and K-best detection ($K = 5$) in Ref. [45] are given, and the performance of SIMO is given as the lower reference bound.

For 128×8 massive MIMO systems, we can see that the performance of all detectors is close to the optimal performance, including the SIMO lower bound. This result is obvious [46]. For the 64×16 massive MIMO system, only linear detection has rather small performance loss, and other detectors have better performance. As for 32×32 massive MIMO system, we can see that that TASER algorithm can still achieve near-ML performance, which is obviously superior to MMSE algorithm and K-best algorithm (however, even with SD algorithm, ML detection complexity is still very high). Figure 4.14a, b also show the fixed-point performance of the TASER algorithm, showing only a small performance loss (SNR below 0.2 dB at 1% VER point).

Figure 4.15a, b show the BER simulation results of the TASER algorithm with BPSK and QPSK modulation in the SIMO system respectively, where the number of receiving antennas is $N_r = 16$, the time slot $K + 1 = 16$ at the transmitter, the maximum number of iterations $t_{\max} = 20$, and the independent identically distributed flat Rayleigh block fading channel model is adopted. The simulation includes SIMO detection, exact SDR detection and ML JED detection, which use perfect receiver channel state information (CSIR) and CHEST [44] respectively. We can see that TASER algorithm can achieve not only near-optimal performance close

(a)

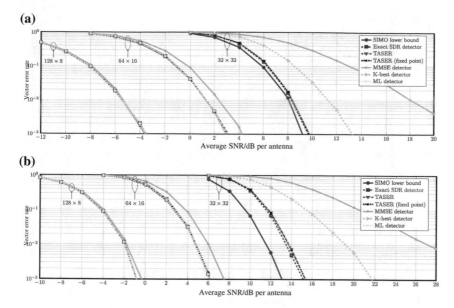

(b)

Fig. 4.14 VER performance comparison of MIMOs with different configurations **a** BPSK, **b** QPSK. © [2018] IEEE. Reprinted, with permission, from Ref. [10]

to perfect CSIR and detection superior to SIMO CHEST, but also performance similar to ML JED algorithm and exact SDR algorithm within controllable complexity.

4.3.5 Computational Complexity

Next, let us compare the computational complexity of the TASER algorithm with other massive MIMO data detection algorithms, including the CGLS algorithm [47], the NSA algorithm [48], the OCD algorithm [49] and the GAS algorithm [50]. Table 4.2 is the number of real number multiplication with the maximum number of iterations t_{max} by different algorithms. As we can see, the complexity of the TASER algorithm (BPSK and QPSK) and NSA is $t_{max} N_t^3$. The complexity of the TASER algorithm is slightly higher. The complexity of CGLS and GAS are both at the $t_{max} N_t^2$ stage, of which GAS is slightly higher. The complexity of OCD is at the $t_{max} N_r N_t$ stage. Obviously, TASER algorithm can achieve near-ML performance with the highest computational complexity, while CGLS, OCD, and GAS have lower computational complexity, but their performance is poor in 32×32 system. So only TASER algorithm can be used for JED, and other linear algorithms cannot be applied in this scenario.

(a)

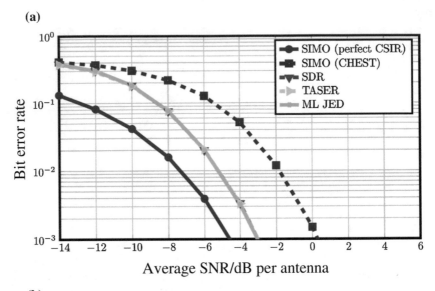

(b)

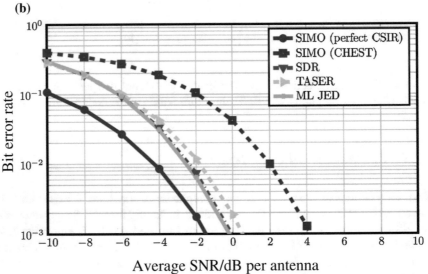

Fig. 4.15 BER performance comparison of SIMO system. **a** BPSK, **b** QPSK. © [2018] IEEE. Reprinted, with permission, from Ref. [10]

Table 4.2 Computational complexity of different data detection algorithms in the massive MIMO system

Algorithm	Computational complexity[a]
BPSK TASER	$t_{\max}\left(\frac{1}{3}N_t^3 + \frac{5}{2}N_t^2 + \frac{37}{6}N_t + 4\right)$
QPSK TASER	$t_{\max}\left(\frac{8}{3}N_t^3 + 10N_t^2 + \frac{37}{3}N_t + 4\right)$
CGLS [47]	$(t_{\max} + 1)\left(4N_t^2 + 20N_t\right)$
NSA [48]	$(t_{\max} - 1)2N_t^3 + 2N_t^2 - 2N_t$
OCD [49]	$t_{\max}(8N_rN_t + 4N_t)$
GAS [50]	$t_{\max}6N_t^2$

[a]The complexity is expressed by the number of the RMUL under the number of iterations of $t_{\max}$. The complex number multiplication requires four RMULs. All the results ignore the preprocessing complexity

References

1. Dai L, Gao X, Su X et al (2015) Low-complexity soft-output signal detection based on Gauss-Seidel method for uplink multiuser large-scale MIMO systems. IEEE Trans Veh Technol 64(10):4839–4845
2. Studer C, Fateh S, Seethaler D (2011) ASIC implementation of soft-input soft-output MIMO detection using MMSE parallel interference cancellation. IEEE J Solid-State Circuits 46(7):1754–1765
3. Wu M, Yin B, Wang G et al (2014) Large-scale MIMO detection for 3GPP LTE: algorithms and FPGA implementations. IEEE J Sel Top Sign Proces 8(5):916–929
4. Peng G, Liu L, Zhou S et al (2017) A 1.58 Gbps/W 0.40 Gbps/mm^2 ASIC implementation of MMSE detection for $128x8$ 64-QAM massive MIMO in 65 nm CMOS. IEEE Trans Circuits Syst I Regul Pap PP(99):1–14
5. Peng G, Liu L, Zhang P et al (2017) Low-computing-load, high-parallelism detection method based on Chebyshev iteration for massive MIMO systems with VLSI architecture. IEEE Trans Signal Process 65(14):3775–3788
6. Gao X, Dai L, Hu Y et al (2015) Low-complexity signal detection for large-scale MIMO in optical wireless communications. IEEE J Sel Areas Commun 33(9):1903–1912
7. Chu X, Mcallister J (2012) Software-defined sphere decoding for FPGA-based MIMO detection. IEEE Trans Signal Process 60(11):6017–6026
8. Huang ZY, Tsai PY (2011) Efficient implementation of QR decomposition for gigabit MIMO-OFDM systems. IEEE Trans Circuits Syst I Regul Pap 58(10):2531–2542
9. Peng G, Liu L, Zhou S et al (2018). Algorithm and architecture of a low-complexity and high-parallelism preprocessing-based K-best detector for large-scale MIMO systems. IEEE Trans Sig Process PP(99):1
10. Castañeda O, Goldstein T, Studer C (2016) Data detection in large multi-antenna wireless systems via approximate semidefinite relaxation. IEEE Trans Circuits Syst I Reg Pap PP(99):1–13
11. Soma U, Tipparti AK, Kunupalli SR Improved performance of low complexity K-best sphere decoder algorithm. In: International Conference on Inventive Communication and Computational Technologies, pp 490–495
12. Fincke U, Pohst M (1985) Improved methods for calculating vectors of short length in a lattice, including a complexity analysis. Math Comput 44(170):463–471

13. Barbero LG, Thompson JS (2006) Performance analysis of a fixed-complexity sphere decoder in high-dimensional mimo systems. In: Proceedings of the IEEE International Conference on Acoustics Speech and Signal Processing, p IV
14. Shen CA, Eltawil AM (2010) A radius adaptive K-best decoder with early termination: algorithm and VLSI architecture. IEEE Trans Circuits Syst I Regul Pap 57(9):2476–2486
15. Burg A, Borgmann M, Wenk M et al (2005) VLSI implementation of MIMO detection using the sphere decoding algorithm. IEEE J Solid-State Circuits 40(7):1566–1577
16. Taherzadeh M, Mobasher A, Khandani AK (2006) LLL reduction achieves the receive diversity in MIMO decoding. IEEE Trans Inf Theory 53(12):4801–4805
17. Barbero LG, Thompson JS (2008) Fixing the complexity of the sphere decoder for MIMO detection. IEEE Trans Wireless Commun 7(6):2131–2142
18. Xiong C, Zhang X, Wu K et al (2009) A simplified fixed-complexity sphere decoder for V-BLAST systems. IEEE Commun Lett 13(8):582–584
19. Khairy MS, Abdallah MM, Habib ED (2009) Efficient FPGA implementation of MIMO decoder for mobile WiMAX system. In: IEEE International Conference on Communications, pp 2871–2875
20. Liao CF, Wang JY, Huang YH (2014) A 3.1 Gb/s 8*8 sorting reduced K-best detector with lattice reduction and QR decomposition. IEEE Trans Very Large Scale Integr Syst 22(12):2675–2688
21. Fujino T, Wakazono S, Sasaki Y (2009) A gram-schmidt based lattice-reduction aided MMSE detection in MIMO Systems. 1–8
22. Yan Z, He G, Ren Y et al (2015) Design and implementation of flexible dual-mode soft-output MIMO detector with channel preprocessing. IEEE Trans Circuits Syst I Regul Pap 62(11):2706–2717
23. Sarieddeen H, Mansour MM, Jalloul L et al (2017) High order multi-user MIMO subspace detection. J Sign Process Syst 1:1–17
24. Zhang C, Liu L, Marković D et al (2015) A heterogeneous reconfigurable cell array for MIMO signal processing. IEEE Trans Circuits Syst I Regul Pap 62(3):733–742
25. Chiu PL, Huang LZ, Chai LW et al (2011) A 684Mbps 57mW joint QR decomposition and MIMO processor for 4×4 MIMO-OFDM systems. In: Solid State Circuits Conference, pp 309–312
26. Kurniawan IH, Yoon JH, Park J (2013) Multidimensional householder based high-speed QR decomposition architecture for MIMO receivers. In: IEEE International Symposium on Circuits and Systems, pp 2159–2162
27. Wang JY, Lai RH, Chen CM et al (2010) A 2x2—8x8 sorted QR decomposition processor for MIMO detection. Inst Electr Electron Eng
28. Sarieddeen H, Mansour MM, Chehab A (2016) Efficient subspace detection for high-order MIMO systems. In: The IEEE International Conference on Acoustics, Speech and Signal Processing
29. Liu T, Zhang JK, Wong KM (2009) Optimal precoder design for correlated MIMO communication systems using zero-forcing decision feedback equalization. IEEE Trans Signal Process 57(9):3600–3612
30. Zhang C, Prabhu H, Liu Y et al (2015) Energy efficient group-sort QRD processor with on-line update for MIMO channel pre-processing. IEEE Trans Circuits Syst I Regul Pap 62(5):1220–1229
31. Yang S, Hanzo L (2013) Exact Bayes' theorem based probabilistic data association for iterative MIMO detection and decoding. In: Global Communications Conference, pp 1891–1896
32. Chen Y, Halbauer H, Jeschke M et al (2010) An efficient Cholesky Decomposition based multiuser MIMO detection algorithm. In: IEEE International Symposium on Personal Indoor and Mobile Radio Communications, pp 499–503
33. Xue Y, Zhang C, Zhang S et al (2016) Steepest descent method based soft-output detection for massive MIMO uplink. In: IEEE International Workshop on Signal Processing Systems, pp 273–278
34. Jiang W, Asai Y, Kubota S (2015) A novel detection scheme for MIMO spatial multiplexing systems with partial lattice reduction. In: IEEE International Symposium on Personal, Indoor and Mobile Radio Communications, pp 2524–2528

35. Mansour MM, Jalloul LMA (2015) Optimized configurable architectures for scalable soft-input soft-output MIMO detectors with 256-QAM. IEEE Trans Signal Process 63(18):4969–4984
36. Luo ZQ, Ma WK, So MC et al (2010) Semidefinite relaxation of quadratic optimization problems. IEEE Signal Process Mag 27(3):20–34
37. Alshamary HAJ, Anjum MF, Alnaffouri T et al (2015) Optimal non-coherent data detection for massive SIMO wireless systems with general constellations: a polynomial complexity solution. In: Signal Processing and Signal Processing Education Workshop, pp 172–177
38. Jalden J, Ottersten B (2008) The diversity order of the semidefinite relaxation detector. IEEE Trans Inf Theory 54(4):1406–1422
39. Harbrecht H, Peters M, Schneider R (2012) On the low-rank approximation by the pivoted Cholesky decomposition. Appl Numer Math 62(4):428–440
40. Goldstein T, Studer C, Baraniuk R (2014) A field guide to forward-backward splitting with a FASTA implementation. Computer Science
41. Beck A, Teboulle M (2009) A fast iterative shrinkage-thresholding algorithm for linear inverse problems. Siam J Imaging Sci 2(1):183–202
42. Benzi M (2002) Preconditioning techniques for large linear systems: a survey. J Comput Phys 182(2):418–477
43. Attouch H, Bolte J, Svaiter BF (2013) Convergence of descent methods for semi-algebraic and tame problems: proximal algorithms, forward–backward splitting, and regularized Gauss-Seidel methods. Math Program 137(1–2):91–129
44. Boumal N (2015) A Riemannian low-rank method for optimization over semidefinite matrices with block-diagonal constraints. Mathematics 1001–1005
45. Wenk M, Zellweger M, Burg A et al (2006) K-best MIMO detection VLSI architectures achieving up to 424 Mbps. In: Proceedings of the IEEE International Symposium on Circuits and Systems, 2006. ISCAS 2006, pp 4–1154
46. Rusek F, Persson D, Lau BK et al (2012) Scaling up MIMO: opportunities and challenges with very large arrays. Sig Process Mag IEEE 30(1):40–60
47. Yin B, Wu M, Cavallaro JR et al (2015) VLSI design of large-scale soft-output MIMO detection using conjugate gradients. In: IEEE International Symposium on Circuits and Systems, pp 1498–1501
48. Wong KW, Tsui CY, Cheng SK et al (2002) A VLSI architecture of a K-best lattice decoding algorithm for MIMO channels. IEEE Int Symp Circuits Syst 3:273–276
49. Wu M, Dick C, Cavallaro JR et al (2016) FPGA design of a coordinate descent data detector for large-scale MU-MIMO. In: IEEE International Symposium on Circuits and Systems, pp 1894–1897
50. Wu Z, Zhang C, Xue Y et al (2016) Efficient architecture for soft-output massive MIMO detection with Gauss-Seidel method. In: IEEE International Symposium on Circuits and Systems, pp 1886–1889

Chapter 5
Architecture for Nonlinear Massive MIMO Detection

When the algorithm is mapped onto the corresponding hardware architecture design, people need to evaluate the performance of the hardware architecture such as data throughput, area, power consumption and delay, and research the resources reuse, the sub-module design, and the whole module pipeline of the hardware architecture, so as to obtain the innovative method with practical application values. Up to now, only the suboptimal linear data detection algorithm has been implemented on FPGA [1] or ASIC [2, 3]. The results of the linear algorithm in the hardware design are not ideal due to the characteristics of the algorithm itself, so it is necessary to try to design the hardware architecture for the nonlinear algorithm.

This chapter first introduces a VLSI architecture designed by us based on the CHOSLAR algorithm in Sect. 4.2. It is implemented with K-best detection preprocessor for the 64 QAM modulation and the 16×16 MIMO system [4]. In order to achieve an optimal trade-off among throughput, area, and power consumption, here, we will present three types of systolic arrays with diagonal priority to perform the initial matrix computation, LR and matrix inversion. An antenna-**level** is also proposed to realize high data throughput and low latency in sorted QRD and post-vector computation. Experimental results show that this architecture has great advantages over the existing designs in data throughput, latency, energy efficiency (throughput/power) and area (throughput/gate number).

Then, the corresponding systolic array is designed according to the TASER algorithm. The systolic array can realize high-throughput data detection with a lower silicon area [5]. VLSI is implemented with Xilinx virtex-7 FPGA and the 40 nm CMOS technology, and the performance and computational complexity are compared in detail with that of other data detectors recently proposed for the massive MU-MIMO wireless system [1, 6–9].

© Springer Nature Singapore Pte Ltd. and Science Press, Beijing, China 2019
L. Liu et al., *Massive MIMO Detection Algorithm and VLSI Architecture*,
https://doi.org/10.1007/978-981-13-6362-7_5

5.1   CHOSLAR Hardware Architecture

5.1.1   VLSI Architecture

This section describes the VLSI architecture implementation based on the CHOSLAR algorithm [4] used in Sect. 4.2. This architecture is designed for 64 QAM 16×16 MIMO systems. The circuit design method is similar to that of other larger scale MIMO systems. From the BER simulation results in Sect. 4.2.5, we know that three iterations are enough to achieve near-optimal detection accuracy and lower resource consumption, so we take 3 as the number of iterations for LR.

Figure 5.1 is the top-level module diagram of the CHOSLAR algorithm. It is composed of five parts: the initialization unit, the sorted QRD unit, the PILR unit, the inversion unit, and the post-vector unit. These units are fully pipelined to achieve high data throughput. First, the initialization results (Gram matrix A and vector $\dot{y}$) in Line 4 of Algorithm 4.2.2 is solved in the initialization unit. As the initialization unit output, Gram matrix A is used to perform sorted QRD in Line 5–18 of Arithmetic 4.2.2, channel matrix H perform exchange operation to obtain $\dot{H}$ at the same time as shown in Line 8. Next, matrix R performs PILR to get matrix $\widehat{R}$ and $\widehat{H}$ in the Line 20–40 of Algorithm 4.2.2. Matrix $\widehat{R}$ is one of the outputs of the CHOSLAR algorithm, and matrix $\widehat{H}$ is transferred to the post-vector unit. Then, the matrix $\widehat{R}$ is inversed in the inversion unit, i.e., Line 41–447 of Algorithm 4.2.2. Finally, in the post-vector unit, the final output $\hat{y}$ of the CHOLSAR algorithm is obtained by matrix–vector multiplication using the outputs (matrix $\widehat{H}$ and R^{inv}, vector $\dot{y}$) of the previous steps.

5.1.1.1   The Initialization Unit

The ultimate purpose of the initialization unit is to calculate the Gram matrix A and vector $\dot{y}$ that will be used in the subsequent units. To achieve high data throughput, this unit designs a systolic array including two types of PEs. In a systolic array, there

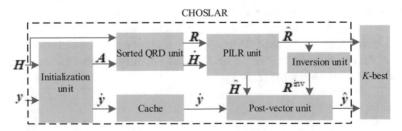

Fig. 5.1 Top-level modules of the CHOSLAR algorithm. © [2018] IEEE. Reprinted, with permission, from Ref. [4]

are N PE-As and $1/2N^2 - 1/2N$ PE-Bs (that is, there are 16 PE-As and 120 PE-Bs in 16×16 MIMO systems). The next unit (the sorted QRD unit) needs to compare the diagonal elements of matrix A, so the PE-As that have obtained these diagonal elements constitute the first block of each row, as shown in Fig. 5.2. In addition, $N - 1$ registers (REG) are used to store the elements of the channel matrix and get the conjugate of each element at the output time. These REGs are used to balance the timing of the pipeline, which balances the different latencies of multiple PE-Bs computation. The first type of the processing element PE-A is used for the computation of vector $\dot{y}$ and diagonal elements that make up matrix A. Every PE-A contains two types of ALU (two ALU-As and one ALU-B), three accumulators, two subtractors, and two shifters. One ALU-A performs complex number multiplication (CM) of $H_{i,j}^*$ and $H_{i,j}^*$, whose results of each cycle are accumulated. Others combined with ALU-B performs CM of matrix H and vector v. The results of ALU-A/ALU-B are accumulated like computation of the elements of matrix A. To execute the computation of Line 4 of Algorithm 4.2.2, subtract the above results from the elements of y, and the shifter computes of the real and imaginary parts of the vector $\dot{y}$. The subsequent processing element PE-B is used to compute the off-diagonal elements of matrix A. Each PE-B contains a CM unit consisting of an ALU-A and an ALU-B. To ensure that each processing element correctly processes operands, the values of Column i in H^{H} delay $i - 1$ clock cycles. First, each value of H^{H} is transferred from PE-A to the subsequent PE-B, then to REG (operate by row), and then the corresponding conjugate value is transferred from REG to PE-B (operate by column).

Similar systolic arrays are used to compute Gram matrix in linear detection algorithms in Refs. [1, 3], the computations in those architectures do not begin with computing the PEs of the diagonal elements of matrix A [1, 3] unlike the systolic arrays designed in this section. Therefore, computation of the diagonal elements of Gram matrix G is postponed. Then, the subsequent sorted QRD algorithm has to wait for more time to receive the input data, data throughput of the overall architecture decreases and the latency increases. The PE-A in Ref. [1, 3] uses the bilateral input while the architecture in this section adopts the unilateral input. Therefore, the number of ports of the systolic array used in Refs. [1, 3] has doubled (at the input end).

5.1.1.2 Sorted QRD Unit

After the unit is initialized, the output matrix A is transferred to the next unit to perform the sorted QRD operation based on the Cholesky decomposition, to obtain the matrix R, as shown in Fig. 5.3. The channel matrix H is also updated in this unit. To achieve higher parallelism, the unit adopts a deep pipeline architecture including N similar processing elements PE-Cs (for example, 16 PE-Cs in a 16×16 MIMO system). All PE-Cs are similar, but the number of ALU-Cs in each PE-C is different, decreasing successively from the first PE-C to the last PE-C in each column. Take the k(th) PE-C as an example to illustrate the architecture. First, use a comparator to compare all diagonal elements of matrix A to find the smallest $A_{i,i}$ and its location

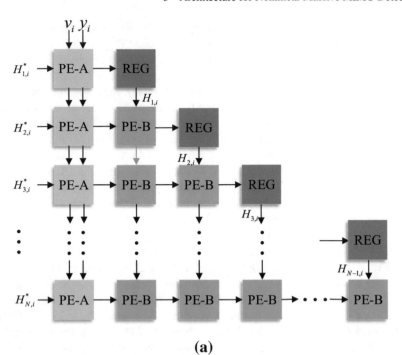

(a)

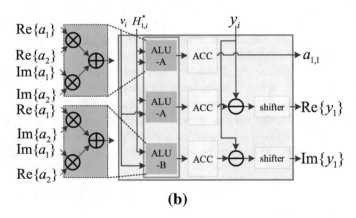

(b)

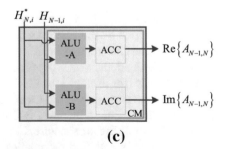

(c)

◀**Fig. 5.2** Architecture of the initialization unit. **a** architecture of the initialization unit, **b** internal architecture of PE-A, **c** internal architecture of PE-B. © [2018] IEEE. Reprinted, with permission, from Ref. [4]

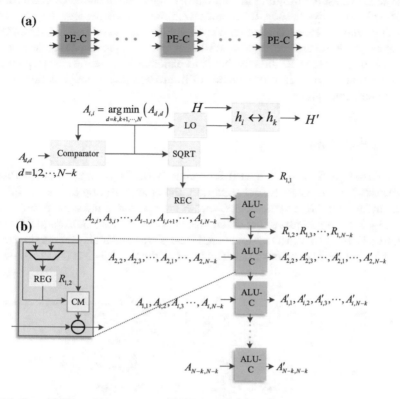

Fig. 5.3 Sorted QRD architecture. **a** sorted QRD architecture, **b** internal architecture of PE-C. © [2018] IEEE. Reprinted, with permission, from Ref. [4]

(LO) in A. Then, $A_{i,i}$ is used to compute $R_{1,1}$ in the square root (SQRT) element. According to the location of $A_{i,i}$, get matrix $\dot{H}$ by exchanging Column i and Column k of matrix H, as shown in Line 8 of Algorithm 4.2.2. Next, the reciprocal (REC) unit is used to compute the REC of $R_{1,1}$, and the result is transferred to the first ALU-C and later used to get the k(th) column of R, which is shown as Line 9–12 of Algorithm 4.2.2. Finally, the elements of A are updated by operating the matrices R and A, and then transferred to the k+1(th) PE-C, as shown in Line 15 of Algorithm 4.2.2. The multiplier and subtractor in ALU-C are used to perform the computation of the matrix A elements in Line 13–17 of Algorithm 4.2.2. Note that the diagonal elements of matrix A are solved first in this framework and can be directly used for the next PE-C, thus reducing the latency in the sorted QRD unit.

The VLSI architecture based on the GR algorithm and the sorting algorithm is proposed in Refs. [10, 11]. A flexible architecture for a 64 QAM 1×1–4×4 MIMO system is proposed in Ref. [10], and a 8×8 K-best signal detector with less sorting operations combined with the LR and the QRD is proposed in Ref. [11]. The sorted QRD units in these architectures are built into a long chain of paired CORDIC arrays, resulting in excessive latency. It becomes even more problematic as the size of the MIMO system increases. The increased computing time for computing sorted QRD, in turn, affects the overall detector's data throughput. The proposed architecture does not need to compute QRD directly, but is implemented by a series of deep pipeline multiplications to meet the requirements of future wireless communication systems for high data throughput.

5.1.1.3   PILR Unit

As shown in Figs. 5.4 and 5.5, the PILR unit has two main functions. One is to update the matrix $\boldsymbol{R}$ based on the Siegel condition, the other is to implement the full-size reduction of matrix $\boldsymbol{R}$, and matrices $\boldsymbol{H}$ and $\boldsymbol{R}$ are updated at the same time.

In Fig. 5.4, the first part of the unit consists of $3N$ PE-Ds (for example, there are 48 PE-Ds in 16×16 MIMO systems). All PE-Ds are similar and operate in parallel. Take the k(th) PE-D as an example. The input of the first PE-D is the k(th) row and

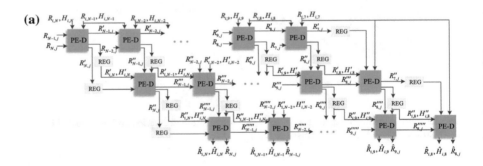

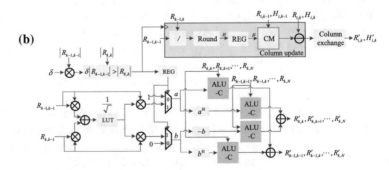

Fig. 5.4 **a** Architecture for updating matrix R based on Seagal condition, **b** internal architecture of PE-D. © [2018] IEEE. Reprinted, with permission, from Ref. [4]

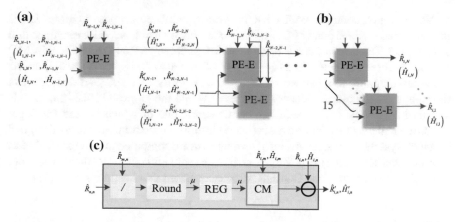

Fig. 5.5 **a** Architecture of full size reduction for matrix R, **b** internal architecture of PE-D, **c** internal architecture of PE-E. © [2018] IEEE. Reprinted, with permission, from Ref. [4]

$k - 1$(th) row of matrix R. PE-D updates the two rows. Then these two rows are used as the input of the next PE-D, and REG can ensure the timing of the pipeline. The framework of the PE-D array is shown as Fig. 5.4. Each PE-D consists of three parts: the column updates for matrices $\dot{H}$ and R, the column exchanges for matrices $\dot{H}$ and R, and the updates for matrix R. Prior to this, there was a unit dedicated to processing the comparison of Line 23 of Algorithm 4.2.2, and the comparison results were used as the enabling signals of the subsequent unit. First, compute the parameter μ and divide it by LUT. Then, the CM unit performs the multiplication of Line 25 of Algorithm 4.2.2. And the resulting matrices R and H are exchanged in columns to obtain matrices $\widehat{R}$ and $\widehat{H}$. Next, compute the parameters a and b in matrix θ to update $\widehat{R}$. During this process, the real multiplication, real number addition, and a LUT are used to realize multiplication, SQRT and REC operations of the elements in matrix $\widehat{R}$. The multiplier, conjugate and negative units are used to obtain θ and then the k(th) and the $k - 1$(th) rows of the matrix $\widehat{R}$ are updated.

In the second part of the PILR unit, there are $1/2N^2 - 1/2N$ identical processing elements PE-Es (e.g., there are 120 PE-Es in a 16×16 MIMO system). Figure 5.5 shows the framework of the size reduction for matrix $\widehat{R}$ and $\widehat{H}$. Take a single PE-E as an example. PE-E first computes the parameter μ, then multiplies each element of the $N - 2$(th) column of the matrices $\widehat{R}$ and $\widehat{H}$ by μ, and then subtract the result from each element of the $N - 1$(th) column. In the size reduction framework, the purpose of the first stage is to update the elements of the N(th) column of matrices $\widehat{R}$ and $\widehat{H}$ (except the elements of Row $N - 1$, Column N). In the second stage, using the results of the first stage to update the elements of the N(th) column of matrices $\widehat{R}$ and $\widehat{H}$ (except the elements of Row $N - 1$, Column N and Row $N - 2$, Column N). In addition, the elements of the $N - 1$(th) and $N - 2$(th) columns are input to PE-E to update the $N - 1$(th) column. The size reduction framework includes $N - 1$ stages, and the computation method is the same in all subsequent stages.

VLSI, which contains similar LR programs, has been proposed before. LR is implemented on three pairs of CORDIC processors by using an odd–even algorithm in Ref. [11]. The detector can achieve near-ML accuracy for 64 QAM 8 × 8 MIMO system. Several CORDIC pairs for QR and LR are also designed. According to the sequence diagram, this part accounts for the majority of latency so that the data throughput of the overall detector decreases. While the proposed PILR unit in this section implements LR through two frameworks, one of which is used for Segal condition and the other for size reduction condition. These frameworks are designed based on systolic arrays. All the intermediate data are computed in the next PE-D and all computations are deeply pipelined, so its hardware utilization and data throughput are higher than those of the CORDIC processor-based architecture in Ref. [11].

5.1.1.4 Inversion Unit

A systolic array is designed for the inverse part of matrix $\widehat{\boldsymbol{R}}$ in the CHOSLAR architecture, as shown in Fig. 5.6. The systolic array has two types of PEs, N PE-Fs and $1/2N^2 - 1/2N$ PE-Gs (for example, there are 16 PE-Fs and 120 PE-Gs in 16 × 16 MIMO systems). PE-Fs and PE-Gs are used to compute the diagonal and non-diagonal elements of matrix $\boldsymbol{R}^{\text{inv}}$, respectively. PE-F makes up the first PE in each

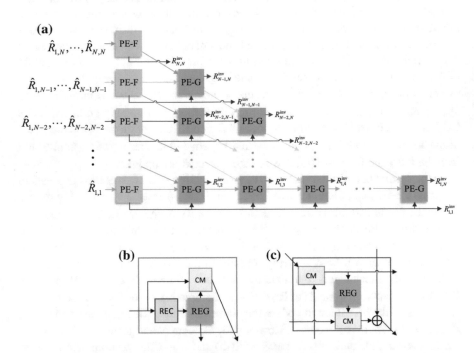

Fig. 5.6 a Architecture of the inversion unit, **b** internal architecture of PE-F, **c** internal architecture of PE-G. © [2018] IEEE. Reprinted, with permission, from Ref. [4]

row because computing the off-diagonal elements of matrix R^{inv} needs the diagonal elements. To ensure that each processing element processes the operands correctly, the values of each column of $\widehat{R}$ are delayed by $i - 2$ clock cycles, and each value of the matrix R is passed from the PE-F to the subsequent PE-G (row by row). Take the third PE-F and PE-G for illustration. PE-F consists of a REC, a REG, and a CM unit and PE-G consists of two CM units, a REG and an adder. Diagonal elements are computed by REC and then output to REG, which is used many times in PE-G. In the next cycle, the off-diagonal elements of matrix $\widehat{R}$ are transferred to the PE-F, the diagonal elements of $\widehat{R}$ is transferred to PE-G on the same row, the PE-F performs the multiplication of the diagonal element of $\widehat{R}$ and off-diagonal element of R^{inv}, and then passes the solution to the PE-G at the lower right. PE-G computes the off-diagonal elements of R^{inv} using the solution from the upper left PE, the diagonal elements of $\widehat{R}$ from the left PE, and the diagonal elements of R^{inv}, which is shown as Line 45 of Algorithm 4.2.2.

In some previously proposed linear detectors' VLSI architectures, such as Refs. [1, 3], they also contain inversion units. These inversion units are approximately implemented based on NSA, similar to the inversion units mentioned in this section (based on systolic arrays). However, the proposed architecture can accurately invert the matrix R, while there is an approximate error with architecture in Refs. [1, 3]. In addition, the systolic array of this design first inverts the diagonal element of R in PE–F of PE in the first column of the unit. Thus the result can be utilized by PE-G. whereas diagonal elements can only be computed after a long delay in Refs. [1, 3] because it does not start with the computation of the PE of these elements. Moreover, the architecture in Refs. [1, 3] requires more ports than that of the architecture in this section.

5.1.1.5 Post-Vector Unit

The post-vector unit performs multiplication operations on the matrix $(R^{inv})^H$, matrix $\widehat{H}^H$ and vector $\dot{y}$ in the 49th row of the Algorithm 4.2.2. Matrix $\widehat{H}^H$ is the output of the PILR unit, and matrix $(R^{inv})^H$ is obtained from the inversion unit. First, matrix $\widehat{H}^H$ is multiplied by vector $\dot{y}$, then the result is multiplied by matrix $(R^{inv})^H$ to get the final solution $\hat{y}$. In addition, since the matrices $(R^{inv})^H$ and $\widehat{H}^H$ are obtained successively by computation, the two matrix–vector multiplications can also be computed in turn, the resources for matrix–vector multiplications can be reused. The framework of the post-vector vector unit is shown in Fig. 5.7. The unit contains N PE-Hs. Each PE-H computes an element of the result vector. Each PE-H contains a CM unit for CM and an ACC unit for accumulation.

5.1.2 Implementation Results and Comparison

The VLSI architecture layout uses TSMC 65 nm 1P8M technology. This section presents the ASIC implementation results and compares them with those of the other

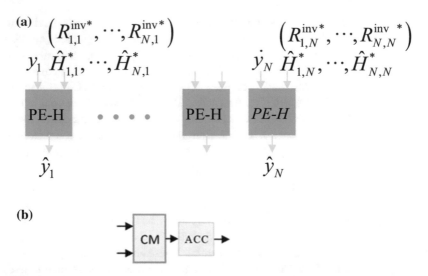

(a)

$$\left(R_{1,1}^{\mathrm{inv}*}, \cdots, R_{N,1}^{\mathrm{inv}*}\right)$$
$$y_1 \ \hat{H}_{1,1}^*, \cdots, \hat{H}_{N,1}^*$$

PE-H • • • • PE-H *PE-H*

$$\hat{y}_1 \qquad\qquad \hat{y}_N$$

$$\left(R_{1,N}^{\mathrm{inv}*}, \cdots, R_{N,N}^{\mathrm{inv}*}\right)$$
$$\dot{y}_N \ \hat{H}_{1,N}^*, \cdots, \hat{H}_{N,N}^*$$

(b)

CM → ACC →

Fig. 5.7 **a** Architecture of the post-vector unit, **b** internal architecture of PE-H. © [2018] IEEE. Reprinted, with permission, from Ref. [4]

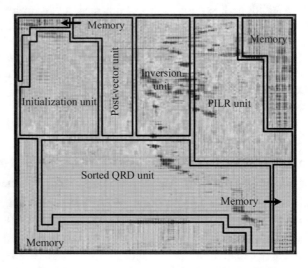

Fig. 5.8 ASIC layout of CHOSLAR-based architecture. © [2018] IEEE. Reprinted, with permission, from Ref. [4]

existing nonlinear detectors. Figure 5.8 is the ASIC layout. Table 5.1 is the detailed comparison of the hardware features based on the CHOSLAR architecture with the post-layout simulation results in Refs. [5, 10–15]. The latter is the existing ASIC architectures, which are effective for nonlinear detection preprocessing of small-scale or high-order MIMO systems.

Table 5.1 Comparison of ASIC implementation results with other MIMO detectors

Parameter	Reference [11]	Reference [10]	Reference [15]	Reference [12]	Reference [14]	Reference [5]			This section	
Antenna size	Complex 8×8	Complex 4×4	Complex 4×4	Complex 4×4	Complex 4×4	Complex 4×4	Complex 8×8	Complex 16×6	Complex 16×6	
Modulation mode	64 QAM	64 QAM	64 QAM	64 QAM	256 QAM	QPSK			64 QAM	
Algorithm	GR+LR	Sorted GR	GS	Sorted QR	WLD	TASER			Sorted QR	Sorted QR+LR
SNR loss[1]/dB	1.2	3.37	3.88	3.37	4.96	#			3.37	1.44
Process/nm	90	65	90	65	90	40			65	
Voltage/V	1.1	1.2	1	1.2	#	1.1			1.2	
Frequency/MHz	65	550	114	500	275	598	560	454	588	
Throughput/(Mbit/s)	585	2640	684	367.88	733	598	374	363	3528	
Delay/µs	2.1	1.26	0.28	#	#	#	#	#	0.7	1.2
Number of gates/kG	612	943	505	1055	1580	148	471	1428	3720	5681
Power/mW	37.1	184	56.8	315.36	320.56	41	87	216	1831	2513
Energy efficiency[2]/[Gbit/(s W)]	15.77	14.34	12.04	1.17	2.29	7.27	4.30	1.68	1.93	1.40
Area efficiency[2]/[Mbit/(s kG)]	0.96	2.80	0.73	0.348	0.46	2.01	0.79	0.25	0.950	0.62
Normalized energy efficiency[3][4]/[Gbit/(s W)]	6.35[3][4]	0.89[4]	1.00[3][4]	0.07[3][4]	0.27[3][4]	0.14[3][4]	0.34[3][4]	0.54[3]	1.93	1.40
Normalized area efficiency[3][4]/[Mbit/(s kG)]	0.33[3][4]	0.18[4]	0.12[3][4]	0.02[4]	0.04[3][4]	0.08[3][4]	0.12[3][4]	0.16[3]	0.95	0.62

① The SNR loss (BER target is 10^{-5}) compared with ML detection in 64 QAM 16 × 16 MIMO system

② Energy efficiency is defined as throughput/power, and area efficiency is defined as throughput/gate number

③ The process is normalized to 65 nm CMOS technology, following $f \sim s$, $P_{dyn} \sim (1/s)\left(V_{dd}/V'_{dd}\right)^2$

④ Zoom to 16 × 16 MIMO configuration: energy efficiency × $(N \times N)/(16 \times 16)$, area efficiency × $(N \times N)/(16 \times 16)$

The architecture for this design can achieve a data throughput of 3.528 Gbit/s, which is, respectively, 5.16 times, 9.59 times, 1.34 times, and 4.81 times of that in the Refs. [10, 12, 14, 15] for the small-scale MIMO systems. One of the requirements of future wireless communication systems is higher throughput. However, high data throughput generally suffers a huge amount of hardware resources and power consumption, so we also compare area and power consumption with that of the recent designs. It should be noted that the CHOSLAR algorithm is designed for high-order MIMO systems, while the architectures in Refs. [10–15] have higher resource consumption and power consumption in high-order systems. Besides this, different technologies and MIMO configurations are used for these architectures. To ensure fairness of comparison, the energy efficiency and area efficiency are normalized under 65 nm technology and 16×16 MIMO configurations, as shown in Table 5.1. This normalization method is widely used when comparing different technologies and hardware implementations for MIMO configurations, such as that in Refs. [5, 11, 12, 15–17]. The CHOSLAR algorithm can achieve the energy efficiency of 1.40 Gbit/(s W), which is 1.40 times, 20.00 times, 1.57 times, and 5.19 times, respectively, of that in Refs. [10, 12, 14, 15]. At the same time, the area efficiency of CHOSLAR algorithm is 0.62 Mbit/(s kG), which is, respectively, 5.17 times, 31.00 times, 3.44 times, and 15.5 times of that in Refs. [10, 12, 14, 15]. In addition, the architectures in Refs. [10, 12, 14, 15] do not perform LR, while the architecture for this design with LR can achieve 1.93 Gbit/(s W) energy efficiency and 0.95 Mbit/(s kG) area efficiency, respectively. We can see that the architecture proposed has greater advantages in both energy efficiency and area efficiency. In terms of latency, CHOSLAR can realize 0.7 μs latency without LR, which is 55.56% of the latency in Ref. [10]. The latency in Ref. [15] is slightly lower than that of CHOSLAR, but CHOSLAR has significant advantages in energy efficiency and area efficiency. The architecture in Ref. [11] has higher energy efficiency than that of CHOSLAR, but lower efficiency area, and the data throughput in Ref. [11] is computed under the assumption that the channel conditions remain constant, so only one MIMO detection preprocessing is performed. To be fair to compare, data throughput and energy efficiency should be reduced accordingly. Furthermore, compared with Ref. [11], the architecture for this design achieves 6.03 times of data throughput, and only 57.14% latency. Meanwhile, CHOSLAR is designed for 16×16 MIMO configuration, whereas it is for 4×4 or 8×8 MIMO configuration in Refs. [10, 11, 14, 15]. For higher order MIMO configurations, the latency in Refs. [10, 11, 14, 15] will increase significantly.

The architecture in Ref. [5] is suitable for nonlinear detection of high-order MIMO systems. Compared with the different MIMO configurations in Ref. [5], the data throughput of CHOSLAR is 11.84 times, 9.43 times, and 9.72 times, respectively, while the low data throughput in Ref. [5] may not meet the data rate requirements of future wireless communication systems. The normalized energy efficiency of CHOSLAR is 10.00 times, 4.12, times and 2.59 times of that of different MIMO configurations in Ref. [5], and the area efficiency of CHOSLAR is 7.75 times, 5.17 times, and 3.88 times higher than that in Ref. [5]. At the same time, the architectures in Ref. [5] only support BPSK and QPSK, and does not support higher level modulation. This limitation is another disadvantage of these architectures in future wireless systems. The architecture in Refs. [1–3] is designed for linear detection algorithms, which can achieve the performance of near-MMSE. These linear detectors suffer nonnegligible loss in detection accuracy, especially when the number of user antennas in MIMO systems is comparable to the number of base station antennas. This is why these linear detectors are not included in Table 5.1.

5.2 TASER-Based Hardware Architecture

5.2.1 Architecture Overview

We present a systolic VLSI architecture with low hardware complexity and high data throughput for TASER algorithm in this section [5]. Figure 5.9 shows a triangular systolic array composed of $N(N + 1)/2$ PEs, mainly used for MAC operation. Each PE is connected to $\tilde{L}_{i,j}^{(t-1)}$ and stores $\tilde{L}_{i,j}^{(t-1)}$ and $V_{i,j}^{(t)}$. All PEs receive data from the column-broadcast unit (CBU) and the row-broadcast unit (RBU).

In the k(th) cycle of the t(th) iteration of TASER, the i(th) RBU sends $\tilde{L}_{i,k}^{(t-1)}$ to all PEs in the i(th) row, while the j(th) RBU sends $\widehat{T}_{k,j}$ to all PEs in the i(th) column. Assume that matrix $\widehat{T} = 2\tau\tilde{T}$ has been solved in the preprocessing stage and the

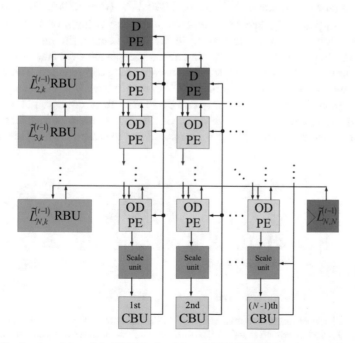

Fig. 5.9 Top-level modules of the TASER algorithm. © [2018] IEEE. Reprinted, with permission, from Ref. [5]

result has been stored in the storage. Take $\tilde{L}_{i,k}^{(t-1)}$ from the PE in Row i, Column k((i,k)) and send it to other PEs in the same row. After receiving data from CBU and RBU, each PE begins to perform MAC operations until $\tilde{L}^{(t-1)}\widehat{T}$ in Row 4 of Algorithm 4.3.1 is finished being computed. To include the subtraction of Row 4, $\tilde{L}_{i,j}^{(t-1)} - \tilde{L}_{i,1}^{(t-1)}\widehat{T}_{1,j}$ operation is performed in the first cycle of each iteration of TASER and the results are stored in the accumulator. In subsequent cycles, $\tilde{L}_{i,k}^{(t-1)}\widehat{T}_{k,j}(2 \leq k \leq N)$ is subtracted from the accumulator in turn. The matrix $\tilde{L}$ is the lower triangular matrix. There is $\tilde{L}_{i,k'} = 0$ when $i < k'$, thereby subtraction of $\tilde{L}_{i,k'}^{(t-1)}\widehat{T}_{k',j}$ can be avoided. The $V_{i,j}^{(t)}$ of PE in the i(th) row of systolic array is solved after i cycles, so the matrix $V^{(t)}$ in the fourth line of Algorithm 4.3.1 can be solved after N cycles.

Figure 5.10 shows an example of the TASER array when $N = 3$. In the first cycle of the t(th) iteration, PE(1,1) inputs $\tilde{L}_{1,1}^{(t-1)}$ and $\widehat{T}_{1,1}$ for computing $V_{1,1}^{(t)} = \tilde{L}_{1,1}^{(t-1)} - \tilde{L}_{1,1}^{(t-1)}\widehat{T}_{1,1}$. At the same time, the PE in the second row performs the first MAC operation and stores the values of $\tilde{L}_{2,j}^{(t-1)} - \tilde{L}_{2,1}^{(t-1)}\widehat{T}_{1,j}$ in their accumulators. In the second cycle, the PE of the second row receives $\tilde{L}_{2,2}^{(t-1)}$ through RBU and $\widehat{T}_{2,j}$ through CBU, thus completing the computation of $V_{2,j}^{(t)} = \tilde{L}_{2,j}^{(t-1)} - \tilde{L}_{2,1}^{(t-1)}\widehat{T}_{1,j} - \tilde{L}_{2,2}^{(t-1)}\widehat{T}_{2,j}$. Meanwhile, PE (1,1) utilizes MAC units to compute the square of $V_{1,1}^{(t)}$ and transfers the solution to the next PE in the same column in the next cycle. In the third cycle, PE(2,1) can utilize $V_{1,1}^{(t)^2}$ from PE(1,1) and $V_{2,1}^{(t)}$ stored internally. PE(2,1) use MAC unit to compute the square of $V_{2,1}^{(t)}$ and add it to $V_{1,1}^{(t)^2}$ (Fig. 5.10c). The result is the sum of the square of the first two elements in the first column of $V^{(t)}$. In the next cycle, the results are sent to the next PE in the same column [in this case, PE(3,1)], so the same steps are repeated. The procedure is repeated over and over in all the columns, until all PEs has completed the corresponding computation. Therefore, the square of two norms of each column of $V^{(t)}$ is solved over $N + 1$ clock cycles, that is, the computation is completed only one cycle after the completion of $V^{(t)}$. In the $N + 2$(th) cycle, the square of two norm of the j(th) column is transferred to the scale unit, where the inverse SQRT is computed and the result is multiplied by $D_{j,j}$.

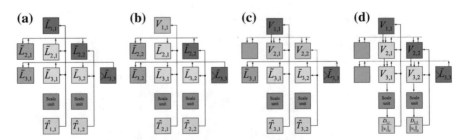

Fig. 5.10 Different cycles of the i(th) iteration of TASER array when $N = 3$. **a** first round, **b** second round, **c** third round, **d** seventh round. © [2018] IEEE. Reprinted, with permission, from Ref. [5]

This operation takes two cycles so that the result is finally solved in the N + 4(th) cycle. In the N + 4(th) cycle, the scaling factor $D_{j,j}/\|v_j\|_2$ (v_j is the j(th) column of $V^{(t)}$) is transferred to all PEs in the same column through the CBU, as shown in Fig. 5.10d. Then, in the N + 5(th) and the last cycles of the iteration, all PEs will multiply the received scaling factors with $V_{i,j}^{(t)}$ to obtain $\tilde{L}_{i,j}^{(t)}$ for the next iteration, and then complete the operation of the proximity operator in the fifth line of Algorithm 4.3.1. The operation of the second line of Algorithm 4.3.1 must be performed before decoding the next symbol. This can be realized through CBU. That means $D_{j,j}$ is transferred to the diagonal PE, and the off-diagonal PE removes $\tilde{L}_{i,j}^{(t-1)}$ from their internal registers at the same time.

5.2.2 PE

The systolic array uses two types of PEs: the off-diagonal PE and the diagonal PE (as shown in Fig. 5.11), both supporting the following four operating modes.

1. Initialization of $\tilde{L}$: This operation mode is used for computation of the second line of Algorithm 4.3.1. All off-diagonal PEs need to be initialized to $\tilde{L}_{i,j}^{(t-1)} = 0$, and diagonal PEs need to be initialized to $D_{j,j}$ received from CBU.
2. Matrix multiplication: This operation mode is used for the computation of the fourth line 4 of Algorithm 4.3.1. The multiplier need make use of all inputs from the broadcast signal, subtract the multiplier output from $\tilde{L}_{i,j}^{(t-1)}$ in the first cycle of the matrix–matrix multiplication, then subtract the multiplier output from the accumulator in other cycles. Each PE stores its own $\tilde{L}_{i,j}^{(t-1)}$ value. At the k(th) cycle, all the PEs in Column k use the internally stored $\tilde{L}_{i,k}^{(t-1)}$ as input of the multiplier, rather than the signals from RBU.
3. Computation of square of two norm: This operation mode is used for the computation of the fifth line of Algorithm 4.3.1. The inputs of all multipliers are $V_{i,j}^{(t)}$. For the diagonal PEs, the result is transferred to the next PE in the same column. For off-diagonal PEs, the output of the multiplier adds $\sum_{n=j}^{i-1}\left(V_{n,j}^{(t)}\right)^2$ from the previous PE in the same column, and the result $\sum_{n=j}^{i}\left(V_{n,j}^{(t)}\right)^2$ is transferred to the next PE. If PE is in the last row, the result will be sent to the scale unit.
4. Scaling: This operation mode is used for the fifth line 5 of Algorithm 4.3.1. The input of the multiplier is the $V_{i,j}^{(t)}$ calculated by the scale unit and $D_{j,j}/\|v_j\|_2$ received from the CBU. The result $\tilde{L}_{i,j}^{(t)}$ is stored in each PE as $\tilde{L}_{i,j}^{(t-1)}$ of the next iteration.

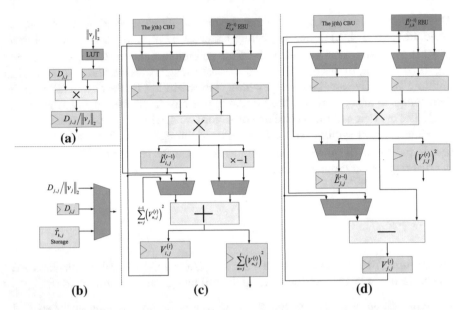

Fig. 5.11 Architecture details of the TASER algorithm. **a** the j(th) scale unit, **b** the j(th) CBU, **c** non diagonal PE, **d** diagonal PE. © [2018] IEEE. Reprinted, with permission, from Ref. [5]

5.2.3 Implementation Details

To prove the effectiveness of the TASER algorithm and the proposed systolic array, the FPGA and ASIC designs with different array sizes N are implemented in this section. All designs are optimized by using Verilog at the register transfer level. The implementation details are described as follows.

1. Fixed-point design parameters: To minimize hardware complexity and ensure near-optimal BER performance, here we adopt a 14-bit fixed-point design. All PEs except the last row of the triangular array use 8-bit decimal digits to represent $\tilde{L}_{i,j}^{(t-1)}$ and $V_{i,j}^{(t)}$, and PEs in the last row use 7-bit decimal digits. $\tilde{L}_{N,N}$ is stored in the register by using 5-bit decimal digits.

2. Computation of inverse SQRT: The computation of the inverse SQRT of the scale unit is realized by LUT and integrated by random logic. Each LUT contains 2^{11} items, and each word of each item contains 14 bits, of which 13 bits are decimal digits.

3. $\widehat{T}$ matrix memory: For FPGA, $\widehat{T}_{k,j}$ storage and LUT are implemented by using distributed random access memory (RAM), that is, without using block RAM. For ASIC, latch arrays built with standard units are used to reduce the circuit area [18].

4. RBU and CBU design: The implementation of RBU for PFGA design and for ASIC design is different. For FPGA, RBU in the i(th) row is an i-input multiplexer that receives data from all PEs in its row and sends approximate $\tilde{L}_{i,k}^{(t-1)}$ to these PEs. For ASIC, RBU consists of a bidirectional bus. Each PE in its row sends data

one by one with a tri-state buffer from which all PEs in the same row obtains data. CBU is also designed in a similar way: the multiplexer is used for FPGA design and the bus for ASIC design. In all target architectures, the i(th) RBU output is connected to the i(th) PE. For larger values of i, the path will have a larger fan out, and eventually become the critical path of the large-scale systolic array. The same will happen in CBU. To shorten the critical path, interstage registers will be placed at the input and output of each broadcast unit. Although that will add each TASER iteration by two extra cycles, the total data throughput increases because of the significant increase in clock frequency.

5.2.4 FPGA Implementation Result

Several FPGAs of different systolic array sizes N were designed and implemented on the Xilinx virtex-7 XC7VX690T FPGA, where $N = 9, 17, 33, 65$. The relevant implementation results are shown in Table 5.2. As expected, the resource utilization tends to increase with the square of array size N. For arrays of $N = 9$ and $N = 17$, the critical path is located in the MAC unit of PE, and for arrays of $N = 33$ and $N = 65$ arrays, the critical path is located in the row-broadcast multiplexer, thus limiting the data throughput when $N = 65$.

Table 5.3 compares the TASER algorithm with several existing massive MIMO data detectors, i.e., CGLS detector [6], NSA detector [1], OCD detector [7] and GAS detector [8]. These algorithms all use 128×8 massive MIMO system and are implemented on the same FPGA. The TASER algorithm can achieve data throughput comparable to that of CGLS and GAS, and has significantly lower latency than that of NSA and OCD. In terms of hardware efficiency (measured by data throughput per PFGA LUT), the hardware efficiency for the TASER algorithm is similar to that of CGLS, NSA, and GAS, while lower than that of OCD. For the 128×8

Table 5.2 FPGA implementation results of TASER with different array sizes

Matrix size	$N = 9$	$N = 17$	$N = 33$	$N = 65$
Number of BPSK Users/Time slot	8	16	32	64
Number of QPSK Users/Time slot	4	8	16	32
Resource quantity	1467	4350	13,787	60,737
LUT resource	4790	13,779	43,331	149,942
FF resource	2108	6857	24,429	91,829
DSP48	52	168	592	2208
Maximum clock frequency/MHz	232	225	208	111
Minimum delay/Clock cycle	16	24	40	72
Maximum throughput/(Mbit/s)	116	150	166	98
Power estimation[①]/W	0.6	1.3	3.6	7.3

① Power estimation at the maximum clock frequency and a supply voltage of 1.0 V

Table 5.3 Comparison of implementation results of different detectors for 128×8 massive MIMO systems

Detection Algorithm	TASER	TASER	CGLS [6]	NSA [1]	OCD [7]	GAS [8]
BER	Near-ML	Near-ML	Approximate MMSE	Approximate MMSE	Approximate MMSE	Approximate MMSE
Modulation mode	BPSK	QPSK	64 QAM	64 QAM	64 QAM	64 QAM
Preprocessing	Not include	Not include	Include	Include	Include	Include
Maximum number of iterations t_{max}	3	3	3	3	3	1
Resource quantity	1467(1.35%)	4350(4.02%)	1094(1%)	48,244(44.6%)	13,447(12.4%)	N.a.
LUT resource	4790(1.11%)	13,779(3.18%)	3324(0.76%)	148,797(34.3%)	23,955(5.53%)	18,976(4.3%)
FF resource	2108(0.24%)	6857(0.79%)	3878(0.44%)	161,934(18.7%)	61,335(7.08%)	15,864(1.8%)
DSP48	52(1.44%)	168(4.67%)	33(0.9%)	1016(28.3%)	771(21.5%)	232(6.3%)
BRAM18	0(0%)	0(0%)	1(0.03%)	32[1](1.08%)	1(0.03%)	12[1] (0.41%)
Clock frequency/MHz	232	225	412	317	263	309
Delay/Clock cycle	48	72	951	196	795	N.a.
Throughput/(Mbit/s)	38	50	20	621	379	48
Throughput/LUT	7933	3629	6017	4173	15,821	2530

①The BRAM36 used in these designs is equivalent to 2 BRAM18s

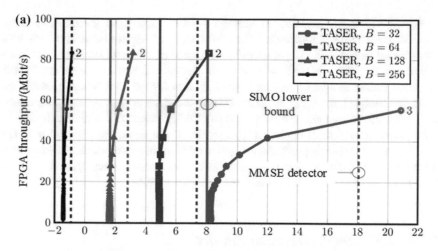

Minimum SNR/dB required to achieve 1% vector error rate.

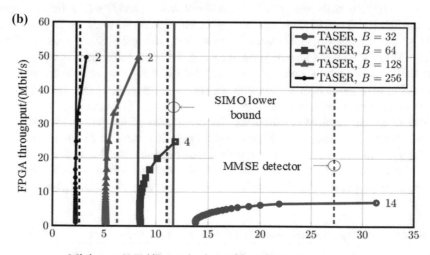

Minimum SNR/dB required to achieve 1% vector error rate.

Fig. 5.12 Trade-offs between throughput and performance in FPGA design. **a** BPSK, **b** QPSK © [2018] IEEE. Reprinted, with permission, from Ref. [5]

massive MIMO system, all detectors can achieve near-ML performance. Nevertheless, when considering the 32×32 massive MIMO system (Fig. 4.14a and b), the TASER algorithm has better BER performance than all other reference algorithms. However, CGLS, NSA, OCD, and GAS detectors can support 64 QAM modulation, while TASER can support only BPSK or QPSK, and the data throughput is linearly proportional to the number of bits per symbol. Thus, the TASER algorithm has no advantage in terms of data throughput and hardware efficiency.

Figure 5.12 shows the trade-off between the data throughput of the FPGA design based on the TASER algorithm and the minimum SNR required to achieve 1% VER for coherent data detection in massive MIMO systems. Meanwhile, the SIMO lower bound and the linear MMSE detection are used as references, where the MMSE detection serves as a basic performance limitation for the CGLS detection [6], the NSA detection [1], the OCD detection [7] and the GAS detection [8]. The trade-off between performance and complexity of the TASER algorithm can be achieved by the maximum number of iterations t_{max}, while the simulation proves that only few iterations can achieve theperformance beyond that of linear detection. As we can see from Fig. 5.12, TASER algorithm can achieve near-ML performance, and its FPGA design can achieve data throughput from 10Mbit/s to 80Mbit/s.

5.2.5   ASIC Implementation Results

Here, the ASIC with the systolic array size $N = 9$, $N = 17$ and $N = 33$ is implemented on the TSMC 40 nm CMOS, and the implementation result is shown in Table 5.4. The silicon area of the ASIC design increases proportionally with the square of the array size N, which can also be verified by Fig. 5.13 and Table 5.5. We can see that the unit area of each PE and scale unit remains basically the same, while the total area of PEs increases with N^2. The unit area of $\widehat{T}_{k,j}$ storage increases with N, where each storage contains one column of a $N \times N$ matrix. Different array sizes have different critical paths. When the array size $N = 9$, the critical path is located in the MAC unit of PE. When $N = 17$, the critical path is located in the inverse- SQRT LUT; and when $N = 33$, the critical path is located in the broadcast bus.

Table 5.6 shows the comparison of the ASIC implementation of TASER and the NSA detector in Ref. [9]. The NSA detector is the only ASIC design known for

Table 5.4 ASIC implementation results of TASER with different array sizes

Matrix size	$N = 9$	$N = 17$	$N = 33$
Number of BPSK users/Time slot	8	16	32
Number of QPSK users/Time slot	4	8	16
Kernel area/μm^2	149,738	482,677	1,382,318
Kernel density/%	69.86	68.89	72.89
Unit area/GE[①]	148,264	471,238	1,427,962
Maximum clock frequency/MHz	598	560	454
Minimum delay/Clock cycle	16	24	40
Maximum throughput/(Mbit/s)	298	374	363
Power estimation[②]/mW	41	87	216

① One gate equivalent (GE) refers to the area of a unit-sized NAND2 gate. ② Power estimation after the placement and routing at the maximum clock frequency and supply voltage of 1.1 V

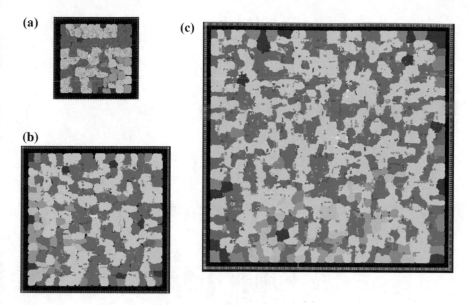

Fig. 5.13 ASIC implementation layout for TASER algorithm. **a** $N = 9$, **b** $N = 17$, **c** $N = 33$. © [2018] IEEE. Reprinted, with permission, from Ref. [5]

massive MU-MIMO systems. Although the data throughput of the ASIC design for TASER is significantly lower than that of the NSA detector, it owns better hardware efficiency (measured by throughput per unit area) and power efficiency (measured by energy per bit) because of its lower area and power consumption. Moreover, TASER can still realize the near-ML performance when the number of transmitted antennas is equal to the number of receiving antennas in a massive MU-MIMO system (Fig. 4.14a and b). In fact, the comparison in Table 5.6 is not completely fair. TASER design does not include the preprocessing circuit. However, the NSA algorithm [9] includes the preprocessing circuit and optimizes the broadband system with single-carrier frequency-division multiple access (SC-FDMA).

There have been a large number of ASIC designs for data detectors of traditional small-scale MIMO systems (see Refs. [11, 19]), and most of them can achieve the near-Ml performance and can even provide data throughput at the Gb/s level for small-scale MIMO systems. The efficiency of these data detectors for massive MIMO, however, has not been verified. The corresponding algorithm and hardware-level comparison is one of the future work directions.

Table 5.5 Area decomposition of the TASER algorithm with different ASIC array sizes

Matrix size	$N = 9$		$N = 17$		$N = 33$	
Area	Unit area	Total area	Unit area	Total area	Unit area	Total area
PE	2391(1.6%)	105,198(70.9%)	2404(0.5%)	365,352(77.5%)	2084(0.1%)	1,168,254(81.8%)
Scale unit	6485(4.4%)	25,941(17.5%)	6315(1.3%)	50,521(10.7%)	5945(0.4%)	95,125(6.6%)
$\widehat{T}_{k,j}$ storage	734(0.5%)	5873(4.0%)	1451(0.3%)	23,220(4.9%)	2888(0.2%)	92,426(6.5%)
Control unit	459(0.3%)	459(0.3%)	728(0.2%)	728(0.2%)	1259(0.1%)	1259(0.1%)
Other	#	10,793(7.3%)	#	31,417(6.7%)	#	70,898(5.0%)

Table 5.6 Comparison of ASIC implementation results of different algorithms

Detection algorithm	TASER	TASER	NSA [9]
BER	Near-ML	Near-ML	Approximate MMSE
Debug mode	BPSK	QPSK	64 QAM
Pre-process	Not include	Not include	Not include
Iterations number	3	3	3
CMOS process/nm	40	40	45
Voltage/V	1.1	1.1	0.81
Clock frequency/MHz	598	560	1000(1125[①])
Throughput/(Mbit/s)	99	125	1800(2025[①])
Kernel area/mm^2	0.150	0.483	11.1(8.77[①])
Kernel density/%	69.86	68.89	73.00
Unit area[②]/kGE	142.4	448.0	12,600
Power[③]/mW	41.25	87.10	8000(13,114[①])
Throughput/ Unit area[①]/[bit/(s GE)]	695	279	161
Energy/bit[①]/(pJ/b)	417	697	6476

① Suppose: $A \sim 1/\ell^2$, $t_{pd} \sim 1/\ell$ 和 $P_{dyn} \sim 1/\left(V_\ell^2 \ell\right)$, scaling the process to 40 nm and 1.1 V
② The number of the gates that do not contain storage
③ At the maximum clock frequency and given supply voltage

References

1. Wu M, Yin B, Wang G et al (2014) Large-scale MIMO detection for 3GPP LTE: algorithms and FPGA implementations. IEEE J Sel Top Sign Proces 8(5):916–929
2. Peng G, Liu L, Zhang P et al (2017) Low-computing-load, high-parallelism detection method based on Chebyshev Iteration for massive MIMO systems with VLSI architecture. IEEE Trans Sign Process 65(14):3775–3788
3. Peng G, Liu L, Zhou S et al (2017) A 1.58 Gbps/W 0.40 Gbps/mm^2 ASIC implementation of MMSE detection for \$128x8\$ 64-QAM massive MIMO in 65 nm CMOS. IEEE Trans Circuits & Syst I Regul Pap, PP(99):1–14
4. Peng G, Liu L, Zhou S et al (2018) Algorithm and architecture of a low-complexity and high-parallelism preprocessing-based K-Best detector for large-scale MIMO systems[J]. IEEE Trans Sign Process 66(7)
5. Castañeda O, Goldstein T, Studer C (2016) Data detection in large multi-antenna wireless systems via approximate semidefinite relaxation. IEEE Trans Circuits & Syst I Regul Pap, PP(99): 1–13
6. Yin B, Wu M, Cavallaro JR et al (2015) VLSI design of large-scale soft-output MIMO detection using conjugate gradients. In: IEEE International symposium on circuits and systems, pp 1498–1501
7. Wu M, Dick C, Cavallaro JR et al (2016) FPGA design of a coordinate descent data detector for large-scale MU-MIMO. In: IEEE International Symposium on Circuits and Systems, pp 1894–1897

 8. Wu Z, Zhang C, Xue Y et al (2016) Efficient architecture for soft-output massive MIMO detection with Gauss-Seidel method. In: IEEE International Symposium on Circuits and Systems, pp 1886–1889
 9. Yin B, Wu M, Wang G et al (2014) A 3.8 Gb/s large-scale MIMO detector for 3GPP LTE-advanced. In: IEEE international conference on acoustics, speech and signal processing, pp 3879–3883
10. Yan Z, He G, Ren Y et al (2015) Design and implementation of flexible dual-mode soft-output MIMO detector with channel preprocessing. IEEE Trans Circuits Syst I Regul Pap 62(11):2706–2717
11. Liao CF, Wang JY, Huang YH (2014) A 3.1 Gb/s 8 × 8 sorting reduced K-Best detector with lattice reduction and QR decomposition. IEEE Trans Very Large Scale Integr Syst 22(12):2675–2688
12. Zhang C, Liu L, Marković D et al (2015) A heterogeneous reconfigurable cell array for MIMO signal processing. IEEE Trans Circuits Syst I Regul Pap 62(3):733–742
13. Huang ZY, Tsai PY (2011) Efficient implementation of QR decomposition for gigabit MIMO-OFDM systems[J]. IEEE Trans Circuits Syst I Regul Pap 58(10):2531–2542
14. Mansour MM, Jalloul LMA (2015) Optimized configurable architectures for scalable soft-input soft-output MIMO detectors with 256-QAM. IEEE Trans Signal Process 63(18):4969–4984
15. Chiu PL, Huang LZ, Chai LW et al (2011) A 684 Mbps 57 mW joint QR decomposition and MIMO processor for 4 × 4 MIMO-OFDM systems. Solid State Circ Conf, 2011: 309–312
16. Wang JY, Lai RH, Chen CM et al (2010) A 2 × 2–8 × 8 sorted QR decomposition processor for MIMO detection 1–4
17. Zhang C, Prabhu H, Liu Y et al (2015) Energy efficient group-sort QRD processor with on-line update for MIMO channel pre-processing. IEEE Trans Circuits Syst I Regul Pap 62(5):1220–1229
18. Meinerzhagen P, Roth C, Burg A (2010) Towards generic low-power area-efficient standard cell based memory architectures. IEEE Int Midwest Symp Syst 129–132
19. Senning C, Bruderer L, Hunziker J et al (2014) A lattice reduction-aided MIMO channel equalizer in 90 nm CMOS achieving 720 Mb/s. IEEE Trans Circuits Syst I Regul Pap 61(6):1860–1871

Chapter 6
Dynamic Reconfigurable Chips for Massive MIMO Detection

The design of a dynamic reconfigurable chip for detecting massive multiple-input multiple-output (MIMO) signals mainly involves the signal detection algorithm, the model analysis and the architecture design for the reconfigurable signal detection processor. To study the reconfigurable signal detection processor, its implementation object, the signal detection algorithm, must be fully understood. As the design basis of the architecture for the reconfigurable signal detection processor, the design of the signal detection algorithm is the basis of the entire system [1]. The analysis of the signal detection algorithm mainly involves the behavior pattern analysis, parallelism of the mainstream signal detection algorithm, the operator extraction and the operator frequency statistics. The analysis of the signal detection algorithm directly determines the completeness of the signal detection function and many characteristics including frequency, power consumption and latency [2, 3]; it exerts a far-reaching influence on the algorithm development and prediction in the future.

6.1 Algorithm Analysis

How to accurately restore the signals transmitted by the user terminal on the base station (BS) is always a difficulty for the signal detection technology. According to whether the algorithm adopts the linear filter to perform signal detection, the signal detection algorithms are generally classified into linear signal detection algorithms and nonlinear signal detection algorithms. Compared with nonlinear signal detection algorithms, linear signal detection algorithms have a lower computation complexity; thus, they show great advantages in the case that the computation complexity of massive MIMO signal detection increases exponentially with the growth of the antenna array scale. However, with the development of radio communication technology, the channel complexity increases. When the channel condition is poor, linear

© Springer Nature Singapore Pte Ltd. and Science Press, Beijing, China 2019
L. Liu et al., *Massive MIMO Detection Algorithm and VLSI Architecture*,
https://doi.org/10.1007/978-981-13-6362-7_6

signal detection algorithms are inferior to nonlinear signal detection algorithms in terms of accuracy. Therefore, how to design a hardware system that supports both linear and nonlinear signal detection algorithms is an important research topic for the development of the next-generation radio communication technology.

6.1.1 Algorithm Analysis Method

Analysis of massive MIMO signal detection algorithms for reconfigurable computing involves behavior pattern analysis of massive MIMO signal detection, analysis of algorithm parallel strategy and extraction of core operators. First, to perform a systematic analysis on various massive MIMO signal detection algorithms, a behavior pattern analysis must be performed to identify the common features and special features of different algorithms. The algorithm features mainly include basic structure, operation type, operation frequency, data dependency between operations and data scheduling strategy. A set of representative algorithms with more common features is determined by performing feature analysis on each signal detection algorithm and extracting common features of multiple algorithms. Then, in order to fully exploit the performance advantages of the reconfigurable computation form for massive MIMO signal detection, parallel strategy analysis is required to be performed on massive MIMO signal detection algorithms. The results of the parallel strategy analysis provide the basis for parallelism and pipeline design in the algorithm mapping solution. The parallel strategy analysis takes a set of representative algorithms rather than a single algorithm as the study object. This helps inter-algorithm transfer of parallel features in a set of representative algorithms. With the development of massive MIMO technologies, new signal detection algorithms emerge one after another. If the parallel analysis based on a set of representative algorithms is adopted, after a new algorithm is included in a set of representative algorithms according to its features, an analysis can be performed on the algorithm by referring to the parallel strategy of a mapped algorithm in the set or even the mapping chart. This significantly saves the effort and time. After the analysis for the behavior pattern and parallel strategy, the core operators for the massive MIMO signal detection application are needed to be extracted; the core operators provide an important basis for the design of reconfigurable PEA, especially for reconfigurable PEs. The process of extracting core operators is actually quite difficult. You need to properly trade-off between the universality and complexity of operators, so as to meet the dual constraints of algorithms on performance and security.

Figure 6.1 shows the operation flow of the minimum mean square error (MMSE) detection algorithm. The main operation modules of the MMSE algorithm include the conjugate matrix multiplication, the matrix inversion, the matched filtering calculation and the channel equalization. Among them, the conjugate matrix multiplication,

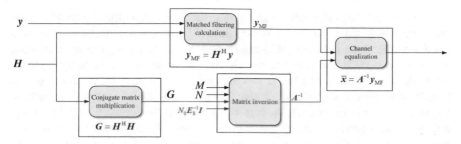

Fig. 6.1 Main operation flow of the massive MIMO MMSE detection algorithm

the matched filtering calculation and the channel equalization modules are constituted by complex matrix multiplications. The matrix multiplication features strong parallelism and simple data dependency. In addition, without data dependency between the conjugate matrix multiplication and the matched filtering calculation modules, the two modules support parallel processing. Therefore, the MMSE detection algorithm has high operation parallelism.

6.1.2 Common Features of Algorithms

Linear signal detection algorithms mainly include the zero-forcing (ZF) algorithm and the MMSE algorithm. The ZF algorithm does not consider the noise effect during the signal detection process. Therefore, the ZF algorithm is not applicable to complex channel conditions of massive MIMO. This section analyzes four detection algorithms in the set of MMSE algorithms: the Neumann series approximation (NSA) algorithm [4], the Chebyshev iteration algorithm [5], the Jacobi iteration algorithm [6] and the conjugate gradient algorithm [7]. The four algorithms are analyzed in the form of complex operation. Compared with the basic addition, subtraction, multiplication, and division operations, the multiply-accumulate and signed division operations account for a large proportion when performing operator abstraction at coarse-grained level. The multiply-accumulate mainly results from three operations, matrix multiplication, matrix–vector multiplication, and dot product of two vectors. According to the fixed-point analysis of algorithms, the 16-bit multiply-accumulate operation can meet the accuracy requirement. The signed division is mainly used in the initial value calculation and the iteration part of the Chebyshev iteration algorithm and the iteration part of the conjugate gradient algorithm. According to the fixed-point analysis results, the 16/8 divider can meet the accuracy requirement. Therefore, a set of 16-bit dividers with a parallelism of 2 can be designed in the arithmetic logic unit (ALU). In addition, for the sake of scheduling complexity and universality, the log-likelihood ratio (LLR) shall be extracted as a new operator to improve the performance of the signal detection processor.

The nonlinear signal detection algorithm mainly includes the K-best algorithm [8] and the standard deviation (SD) algorithm [9]. These two algorithms are developed from machine learning (ML) algorithms, which optimize the preprocessing and search parts of the original ML algorithms; the calculation complexity is greatly reduced. The K-best algorithm is developed from ML algorithms based on the breadth-first search, while the SD algorithm is the derivative of ML algorithms based on the depth-first search. According to the analysis results for nonlinear signal detection algorithms, the preprocessing is generally implemented by quadrature right-triangle (QR) decomposition and matrix inversion, which is similar to the preprocessing of linear signal detection algorithms. In addition, for both the preprocessing and the search parts (the number of search layers and search path are related to the antenna size), the size of data arrays to be processed increases with the antenna size and the calculation complexity increases exponentially with the antenna size. In this book, the CHOSLAR algorithm and TASER algorithm are discussed. The CHOSLAR algorithm is a simplified K-best algorithm. As a representative of ML algorithms, the CHOSLAR algorithm simplifies two aspects of the algorithm and hardware design in the original K-best algorithm: (1) The preprocessing part is simplified; (2) The number of search candidate nodes is reduced. Different from the mainstream two algorithms, the TASER algorithm [10] uses the matrix–vector operation and approximation nonlinear method to solve signal detection problems, which avoids the search for the original signal and significantly reduces the calculation complexity. For the sake of scheduling cost and performance of the signal detection processor, this book performs the coarse-grained operator abstraction for nonlinear signal detection algorithms, to reduce frequent configuration changes and scheduling among processing element (PE). The complex multiply-accumulate operation, signed division operation and LLR judgment operation for nonlinear signal detection algorithms are the same as those for linear signal detection algorithms. In addition, this book introduces the 16-bit real-number multiply-accumulate operator, interval judgment operator and 16-bit real root operator for solving 2-norm.

6.1.3 Computing Model

The computing model for reconfigurable massive MIMO signal detection mainly involves the processing modes and control rules of the massive MIMO signal detection application in the basic reconfigurable computing architecture. In fact, it can be interpreted as a modeling process of the massive MIMO signal detection application in the reconfigurable computing system.

In short, the massive MIMO signal detection algorithm can be considered as a transfer function from input to output, as shown in Fig. 6.2. Based on the idea of reconfigurable computing, the input can be further divided into data input and configuration input. The configuration input changes the relationship map between the data input and the output results to control the transfer function. The data input is commonly divided into the fixed input and the real-time input specific to the fea-

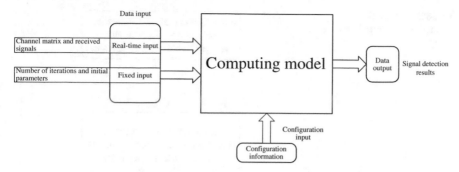

Fig. 6.2 Research on the computing model for reconfigurable massive MIMO signal detection

tures of the massive MIMO signal detection application. Generally, the fixed input involves the contents for which the input frequency is much higher than the computing frequency, including the initial parameters [e.g., channel matrix H obtained by channel estimation and signal-to-noise ratio (SNR)] in algorithm, the number of iterations, and the formula for generating iteration parameters, whereas the real-time input involves the contents for which the input frequency and computing frequency are comparable, including the received signals in algorithm and data-aided cyclic prefixes. Based on the analysis of massive MIMO signal detection algorithms, a reconfigurable computing model is created for each representative algorithm set, and the corresponding input, output, and matched transfer function are determined. This firmly connects the massive MIMO signal detection application with the reconfigurable computing, which clarifies the operating mechanism of the massive MIMO signal detection application on the reconfigurable architecture and provides a mathematical guidance for detailed design of the reconfigurable massive MIMO signal detection processor. Figure 6.3 shows the basic research methodology for the architecture design of the reconfigurable massive MIMO detection processor. The detailed analysis of the architecture design is provided in Sect. 6.2.

6.2 Data Path

The reconfigurable PE array (PEA) is the core computing part of the reconfigurable massive MIMO signal detection processor. The reconfigurable PEA and the corresponding data storage part constitute the data path of the reconfigurable massive MIMO signal detection processor. The architecture of the data path directly determines the flexibility, performance, and energy efficiency of the processor. The research on reconfigurable computing arrays mainly involves reconfigurable PEs, interconnection topology, heterogeneous modules, etc. As far as PEs are concerned, the granularity of basic operations in different massive MIMO signal detection algorithms varies greatly (from 1-bit basic logical operation to thousands-of-bits finite-

Fig. 6.3 Research
methodology for architecture
design

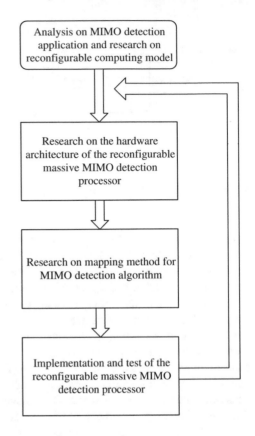

field operation). This book discusses the PE architecture with the mixed-granularity, which does not only involve the basic designs such as ALU, the data, and configuration interfaces and register but also involves the optimization for the proportion of PEs with different granularities in the array and their corresponding positions. In addition, the mixed-granularity also brings new challenges to the research on the interconnection topology. As the data processing granularity may be different for different PEs, the interconnections among PEs of different granularities may involve data mergence and data splitting. The interconnection cost and mapping property of the algorithm need to be considered for the heterogeneous interconnection architecture. The storage part of the data path provides data support for reconfigurable computing arrays. The compute-intensive and data-intensive reconfigurable massive MIMO signal detection processors need to perform lots of parallel computing; therefore, the data throughput of the memory is easy to become the performance bottleneck of the entire processor; this is denoted as the "memory wall" problem. Therefore, a cooperative design is required in aspects such as memory organization, memory capacity, memory access arbitration mechanism, and memory interface, so as to ensure that the performance of the reconfigurable computing arrays is not

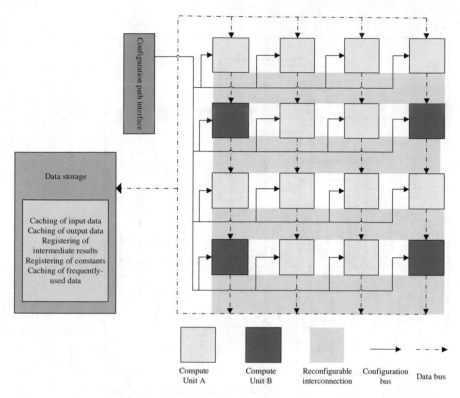

Fig. 6.4 Reconfigurable computing array and data storage

affected by the memory and reduce the additional area and power consumption incurred by the memory as much as possible. This book also makes corresponding research on the data access mode [11, 12]. Figure 6.4 provides a brief illustration of the reconfigurable PEA and data storage.

6.2.1 Structure of Reconfigurable PEA

As the main undertaker for processing the parallel part of computing tasks in the massive MIMO signal detection processor, the PEA module consists of master control interface, configuration controller, data controller, PEA controller, and PEA, as shown in Fig. 6.5.

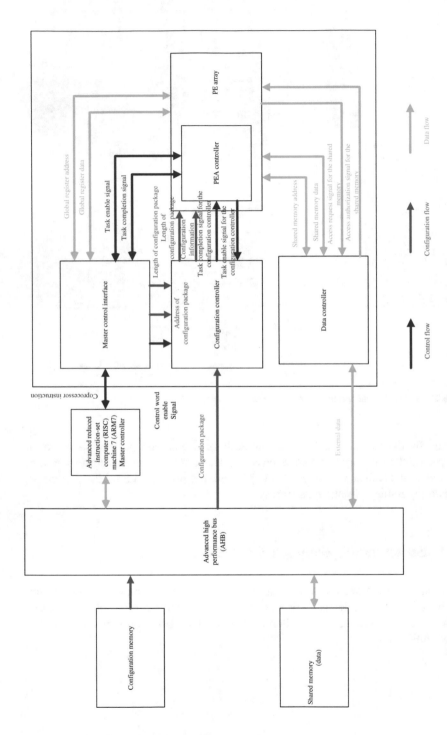

Fig. 6.5 Constituents of a PEA

6.2.1.1   PEA Submodules

The PEA may perform data exchange with the exterior using the master control interface, configuration controller, and data controller. The master control interface is a coprocessor or an AHB. As the main module of the master control interface, the advanced RISC machine (ARM) can write the configuration word to be executed and the dependent data into the interface. As the main module of the AHB, the configuration controller initiates a read request to the configuration memory and transfers the configuration package to the PEA. As another main module of the AHB, the data controller initiates read and write requests to the shared memory (on-chip shared memory that is mounted on the AHB and shared by the ARM7 and PEA; the data exchange between the shared memory and the primary memory is implemented by the data transfer of the direct memory access controller controlled by the ARM7) and completes the data transfer between the PEA and the shared memory. In the PEA, the most basic compute unit is PE; the most basic time unit is machine cycle (the machine cycle indicates the period from the time when a PE starts to execute a task in the configuration package to the time when execution of the task is ended. The memory access latency of each PE is uncertain. Therefore, scheduling on the basis of the clock cycle is very difficult. Scheduling on the basis of machine cycle instead of clock cycle simplifies the compiler design. The specific number of clock cycles occupied by each machine cycle is dynamically determined by the hardware in operation). Each PE has an ALU. In each machine cycle, the ALU performs one operation for four inputs (two 32-bit inputs and two 1-bit inputs) to obtain two outputs (one 32-bit output and one 1-bit output). After a PE completes the computation of one machine cycle, the PE waits for all the other PEs to complete the computation of the current machine cycle and then enters the next machine cycle together with all the other PEs. After completing the execution of the configuration package, PEs notify the PEA. After receiving the signal indicating that all PEs complete the execution, the PEA terminates the set of configuration information. PEs do not need to execute exactly equal number of machine cycles for a set of configuration package; one PE may terminate the set of configuration information ahead of time. In this programming model, complex external memory models such as data controller and shared memory are covered; the PEA external memory available to PEs is a continuous address space.

To support more complex control paradigms, two mechanisms are added to the preceding programming model: (1) A conditional execution control bit is added to each line of configuration for each PE. If conditional execution is enabled, the ALU of the PE determines whether to perform the computation of the machine cycle according to the 1-bit input information; (2) The configuration jump mechanism is added. If the 32-bit information output of the PE is written to the 16th register (R15) of the PE register file, the index of the configuration line to be executed by the PE in the next cycle is the index of the configuration line in this cycle plus the number in R15 (note that the number in R15 is a signed integer). To compress the configuration information and support more complex flow computing paradigms, three numbers of iterations are added to the preceding programming model: (1) The number of PEA top-layer iterations (**PEA top iter**). If the **PEA top iter** is not **0**, after executing the current configuration package once, the PEA executes the current configuration package another **PEA top iter** times. (2) The number of PE top-layer iterations (**PE top iter**). If **PE top iter** is not **0** for the current configuration package of a PE, after executing the configuration information about the PE in the current configuration package, the PE executes the configuration information another **PE top iter** times. Note, between repetitive executions, the PE does not need to wait for other PEs to complete the execution of the corresponding configuration information in the current configuration package. (3) The number of iterations for a PE configuration line (**PE line iter**). When a PE goes to a line of configuration information (corresponding to an operation of the ALU) in the current machine cycle, if the **PE line iter** is not **0** in the line of configuration information, the PE repetitively executes the line of configuration information for **PE line iter** times in the next **PE line iter** machine cycles. At each time of repetitive execution, the **in1/in2/out1_iter_step** field can be used to configure the address of **data source/destination** to be incremental (for details, see Sect. 6.3.4).

6.2.1.2   Behavior Description

The PEA workflow is as follows.

(1) The configuration controller uses the configuration arbiter to write the configuration package to the configuration cache of each PE.

(2) The PEA controller uses the configuration package enable signal to enable all PEs after receiving the task enable signal. The PE executes the configuration package in the cache; after executing a machine cycle, the PE sends a completion signal to the PEA controller and waits for the next PE enable signal. When all PEs provide the completion signals and the data controller provides the control completion signal of data, the PEA controller sends the PE enable signal of the next machine cycle. The PEA enters the next machine cycle. Thus, the synchronization of PEA machine cycles is realized.

(3) After the corresponding configuration packages of all PEs and the configuration package of the data controller are executed, the PEA controller sends a task completion signal to the master control interface and waits for the next instruction from the master control interface.

6.2.2   PE Structure

As the most basic compute unit in PEA, a PE consists of an ALU and a private register file. Figure 6.6 shows the basic structure of a PE. The most basic time unit of the PE is also machine cycle. A machine cycle corresponds to the time duration for a PE to complete one operation. In the same machine cycle, the global synchronization

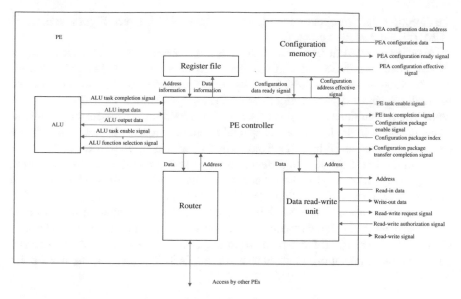

Fig. 6.6 Constituents of a PE

mechanism is adopted among PEs; that is, after a PE completes the computation of a machine cycle, the PE shall wait for the other PEs to complete their computations of the current machine cycle and then enter the next machine cycle together with the other PEs. Under the same set of configuration package, the PEA terminates the set of configuration information after receiving the feedback signal indicating that all PEs have completed the set of configuration package. However, different PEs do not need to execute exactly equal number of machine cycles for a set of configuration package.

6.2.2.1 ALU Design

The unit bit width of the parallel processed data in PE reconfigurable arrays is determined by its computation granularity. On one hand, if the computation granularity is too small, it cannot match with the signal detection algorithm that needs to be supported by the processor. If the bit truncation is selected mandatorily, the accuracy of the algorithm will be affected. If the collaboration of multiple PEs is adopted, the efficiency of interconnection resources, control resources, and configuration resources will be affected, thus eventually reducing the area efficiency and energy efficiency of the entire implementation. On the other hand, if the computation granularity is too large, only part of the bit width in the PE participates in the operation. This causes a redundancy of computing resources, thus affecting the overall performance such as area and latency. Therefore, the computation granularity shall match with the detection algorithm set that needs to be supported by the reconfigurable massive MIMO signal detection processor.

According to the brief summary and analysis on features of signal detection algorithms in Sect. 6.1, linear and nonlinear detection algorithms have their own features. The PE computation granularity is finally determined after fixed-point is performed for multiple signal detection algorithms. The analysis results show that the 32-bit word length is sufficient to support the accuracy requirements for current computations. In addition, the length of the special operators required by some algorithms can be controlled at 16-bit after fixed-point is performed. Therefore, in the ALU design of this book, the data concatenation and splitting operators and the operation for separately processing higher bits and lower bits are added; this will be introduced in detail in the design of ALU unit. The bit widths required by linear signal detection algorithms are basically around 32 bits. After fixed-point, simulation is performed for the two nonlinear signal detection algorithms described in this chapter. The simulation shows that 16-bit fixed-point word length is sufficient to meet the accuracy requirements of operations such as root. In the TASER algorithm, the fixed-point

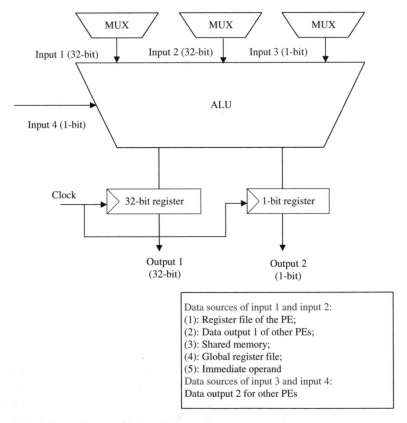

Fig. 6.7 Schematic diagram of the ALU data path

word length of the operator is 14-bit; however, 16-bit is recommended as far as the feasibility of hardware implementation is concerned. Generally speaking, the PE computation granularity of massive MIMO signal detection processors is recommended to be equal to or larger than 32-bit. Due to PE concatenation, the selected granularity shall be a power of two; therefore, the granularity can be selected as 32-bit. It is necessary to note that the PE processing granularity may be adjusted accordingly in actual architecture design to better satisfy application requirements if special algorithm sets are required.

ALU is the core compute unit in a PE. As a single-cycle combinational logic, ALU can perform binary and ternary operations for integers. The data paths of an ALU are classified as 1-bit data paths and 32-bit coarse-grained data paths. The 32-bit and 1-bit data paths can only interact with each other during ALU computation; they cannot interact with each other outside the ALU. In each machine cycle, ALU performs one operation for four input signals (two 32-bit and two 1-bit) and obtains two output signals (one 32-bit and one 1-bit). Figure 6.7 shows the data paths in detail. As shown in the figure, the ALU has four inputs, where input 1 and input 2 are 32-bit that participate in coarse-grained computation. The input 3 is a 1-bit input that is used for conditional selection and carry. The input 4 connected to the enabling-end is also a 1-bit signal that is used for condition control. The ALU performs normal computation if input 4 is 0, while the ALU is not enabled if input 4 is 1. The ALU has two outputs, where output 1 is 32-bit, and output 2 is 1-bit. The 1-bit output can be stored into a 1-bit shadow register so that it can be kept for a long time.

The operators in ALU mainly consists of common logics of signal detection algorithms, as shown in Table 6.1. In Table 6.1, in1, in2 and in3 indicate data input 1, 2 and 3, respectively; Out1 and Out2 indicate data output 1 and 2, respectively. According to the analysis on signal detection algorithms in the previous section, very similar operators with common features can be extracted from signal detection algorithms of the same type; these operators can be classified into the following categories.

Logic operators: AND, OR. XOR, etc.
Arithmetic operators: real signed addition, subtraction, multiplication, division, root and concatenation, complex addition, subtraction, multiplication and concatenation, unsigned addition, subtraction and absolute subtraction, etc.
Shift operators: logic shift, arithmetic shift, etc.
Detection-specific operators: 16-bit multiply-accumulate, 16-bit multiply-subtract, 16-bit chain addition, interval judgment, LLR judgment, etc.

ALU can implement many computing functions. A 5-bit function selection signal OpCode is used to select the specific computing function. ALU converts two 32-bit inputs and one 1-bit input into one 32-bit output and one 1-bit output according to the specific computing function; generally, the 1-bit output is used for conditional control or carry. Therefore, the computation of the condition bit (or carry) is integrated to the traditional 32-bit operators during operator design and the hardware implementability is considered in this book. According to the functions of in3 and Out2, the functions of

Table 6.1 Operators of the massive MIMO signal detection processor

No.	Function	Operation	Output 1	Output 2
0	Bitwise NOT	$T = {}'\mathrm{z}$	$\text{Out1} = {}'\mathrm{z}$	$\text{Out2} = {}'\mathrm{z}$
1	Signed addition	$T = \text{in1} + \text{in2} + \text{in3}$	$\text{Out1} = T[31:0]$	$\text{Out2} = T[31]$
2	Signed subtraction	$T = \text{in1} - \text{in2} - \text{in3}$	$\text{Out1} = T[31:0]$	$\text{Out2} = T[31]$
3	Signed multiplication	$T = \text{in1} \times \text{in2}$	$\text{Out1} = T[31:0]$	$\text{Out2} = T[0]$
4	Bitwise AND	$T[31:0] =$ $\text{in1\&in2\&}\{31\{1'b1\},$ $\text{in3}\}\}$ Enter 1s in the higher bits of in3 and then use in3 to perform bitwise AND operation with in1 and in2	$\text{Out1} = T[31:0]$	$\text{Out2} =$ $\&T[31:0]$
5	Bitwise OR	$T[31:0] = \text{in1} \mid \text{in2} \mid$ $\{31\{1'b0\}, \text{in3}\}$ Enter 0s in the higher bits of in3 and then use in3 to perform bitwise OR operation with in1 and in2	$\text{Out1} = T[31:0]$	$\text{Out2} = \mid T[31:0]$
6	Bitwise XOR	$T[31:0] =$ $\text{in1\^{}in2\^{}}\{31\{1'b0\}, \text{in3}\}$ Enter 0s in the higher bits of in3 and then use in3 to perform bitwise XOR operation with in1 and in2	$\text{Out1} = T[31:0]$	$\text{Out2} = {}^{\wedge}T[31:0]$
7	Absolute subtraction	$T = \mid \text{in1} - \text{in2} - \text{in3} \mid$	$\text{Out1} = T[31:0]$	$\text{Out2} = T[31]$
8	Select operation	$T[31:0] = \text{in3 ? in1 : in2}$	$\text{Out1} = T[31:0]$	$\text{Out2} = \text{in3}$
9	Logical left shift	$T[32:0] = \{\text{in1, in3}\} \ll$ in2	$\text{Out1} = T[31:0]$	$\text{Out2} = T[32]$
10	Logical right shift	$T[31:-1] = \{\text{in3, in1}\} \gg$ in2	$\text{Out1} = T[31:0]$	$\text{Out2} = T[-1]$
11	Arithmetic right shift	$T[31:0] = \{\text{in2}\{\text{in1}$ $[31]\}, \text{in1} \gg \text{in2}\}$ Compensate in2 in1 [31] s in the higher bits	$\text{Out1} = T[31:0]$	$\text{Out2} = \text{in1}[\text{in2}$ $-1]$
	Unsigned addition	$T = \text{in1} + \text{in2} + \text{in3}$	$\text{Out1} = T[31:0]$	$\text{Out2} = T[32]$
12	Unsigned subtraction	$T = \text{in1} - \text{in2} - \text{in3}$	$\text{Out1} = T[31:0]$	$\text{Out2} = T[32]$

(continued)

Table 6.1 (continued)

No.	Function	Operation	Output 1	Output 2
13	Leading zero detection	Calculate the number of leading zeros in the binary data corresponding to in1	Out1 = the number of leading zeros in in1	
14	Signed division	$T = \{\text{in1}/(\text{in2}[31{:}16]), \text{in1}/(\text{in2}[15{:}0])\}$	Out1 = $T[31{:}0]$	Out2 = 0
15	Root	$T = \{\sqrt{\text{in1}}[15{:}0], \sqrt{\text{in2}}[15{:}0]\}$	Out1 = $T[31{:}0]$	Out2 = 0
16	16-bit multiply-accumulate	$T = \text{in1}[31{:}16] \times \text{in2}[31{:}16] + \text{in1}[15{:}0] \times \text{in2}[15{:}0]$	Out1 = $T[31{:}0]$	Out2 = 0
17	16-bit multiply-subtract	$T = \text{in1}[31{:}16] \times \text{in2}[31{:}16] - \text{in1}[15{:}0] \times \text{in2}[15{:}0]$	Out1 = $T[31{:}0]$	Out2 = 0
18	16-bit chain addition	$T = \text{in1}[31{:}16] + \text{in2}[31{:}16] + \text{in1}[15{:}0] + \text{in2}[15{:}0]$	Out1 = $T[31{:}0]$	Out2 = 0
19	Data concatenation	$T = \text{in2}[15{:}0] <\!\!< 16 + \text{in2}[15{:}0]$	Out1 = $T[31{:}0]$	Out2 = 0
20	Complex addition	$T1 = \text{in1}[31{:}16] + \text{in2}[31{:}16]$ $T2 = \text{in1}[15{:}0] + \text{in2}[15{:}0]$	Out1 = $T1[15{:}0] \ll 16 + T2[15{:}0]$	Out2 = 0
21	Complex subtraction	$T1 = \text{in1}[31{:}16] - \text{in2}[31{:}16]$ $T2 = \text{in1}[15{:}0] - \text{in2}[15{:}0]$	Out1 = $T1[15{:}0] \ll 16 + T2[15{:}0]$	Out2 = 0
22	Complex multiplication	$T1 = \text{in1}[31{:}16 \times \text{in2}[31{:}16] \text{ in1}[15{:}0] \times \text{in2}[15{:}0]$ $T2 = \text{in1}[31{:}16] \times \text{in2}[15{:}0] + \text{in2}[31{:}16] \times \text{in1}[15{:}0]$	Out1 = $T1[15{:}0] \ll 16 + T2[15{:}0]$	Out2 = 0
23	Complex concatenation	$T1 = \text{in1}[23{:}16] \ll 8 - \text{in2}[23{:}16]$ $T2 = \text{in1}[7{:}0] \ll 8 - \text{in2}[7{:}0]$	Out1 = $T1[15{:}0] \ll 16 + T2[15{:}0]$	Out2 = 0
24	Interval judgment			
25	LLR judgment			

operators can be classified into four categories: (1) carry signal. For example, considering two 32-bit unsigned additions, the 33rd bit (T[32] bit) in the computation result is the overflow bit; the overflow bit of the lower 32 bits can be used as the carry signal for the higher 32 bits to implement a 64-bit addition. In hardware, the overflow bit is generated by outputting the 33rd bit of the computation result, which is very simple. Similarly, the shift of unsigned numbers can be implemented by logical left and right shifts. (2) Conditions for other PEs. For example, both the overflow bit of an unsigned subtraction (T[32]) and the sign bit of a signed subtraction (T[31]) can be used as the flag for comparing two numbers. If overflow occurs (i.e., bit borrowing occurs for T[32]) after an unsigned number a minus another unsigned number b, it indicates that a is smaller than b. Similarly, use the 1-bit outputs of AND, OR and XOR as three inputs to perform bitwise XOR in sequence to obtain T[31:0]; then perform AND, OR and XOR for each bit in T[31:0]; the obtained 1-bit output can be used as a condition of the logical expression. The 1-bit output of absolute subtraction can be used to determine whether the two inputs are equal, i.e., the output is 0 if the two inputs are equal; the output is 1 if the two inputs are not equal. (3) Conditional selection. Here it refers specifically to the conditional selection operations. Meanwhile, the operator can be used as the buffer of 1-bit data to transfer 1-bit output down in the pipeline. (4) No obvious use. For example, multiplication, arithmetic shift and leading zero detection.

The 32-bit input data of a PE may come from the shared memory (generally the input data of the whole computation process), internal register file of the PE (generally intra-PE computation intermediate data), 32-bit output data of the current PE and other neighbor PEs in the previous machine cycle (generally inter-PE computation intermediate data), immediate operand, and the global register file in the master control interface (generally the intermediate data obtained during the execution of the master controller). The 1-bit input of a PE may come from the 1-bit output of another PE in the previous machine cycle (i.e., short-term 1-bit intermediate result). The 32-bit output of a PE may be transferred to the shared memory, the register file in the PE and the global register file. Also, the computation results can be accessed by neighbor PEs in the next machine cycle regardless the format of the computation results of the PE in the current machine cycle (32-bit and 1-bit outputs).

6.2.2.2 Data Organization

Currently, a PE may access the following data: (1) input and output data during PEA computation in the shared memory shared by all Pes. The transfer of these data is controlled by ARM7. (2) short-term computational data in the private register file of the PE. (3) Data that needs to be exchanged during the operations of the ARM7 and the PEA in the global register file shared by all PEs in the PEA. As a data memory close to the PEA, the shared memory provides data input and output for the PEA under the control of the data arbiter. Table 6.2 provides the valid data sources and access costs, and Table 6.3 provides the valid data destination.

Table 6.2 Valid data sources and access costs

Input data	Data source	Use	Access cost
in1/in2	Directly access the shared memory, is used to read the external input data	Reading the external input data	$2 + m$
	Immediate operand	Reading the immediate operand of the compiler	0
	Access a register in the PE	Reading local long-term data computed by the PE	r
	Access output 1 of a PE in the previous machine cycle	Reading short-term data in the array	1
	Use a number in the register to access the shared memory indirectly	Achieve dynamic memory access during operation	$2 + m + r$
	Access the global register file in the master control interface	The data generated during the exchange operation with the ARM7	g
in3	Use the router to access output 2 of the neighbor PE in the previous machine cycle	Reading 1-bit computational data for conditional selection and carry	1
in4	Use the router to access output 2 of the neighbor PE in the previous machine cycle	Reading 1-bit control data for conditional execution	1

Table 6.3 Valid destination of data

Parameter	Meaning
ρ_v^K, ρ	Proportion of time that the vth output port of router K is occupied by its xth input port ($\rho_v^K = \sum_{x=1}^{U} \frac{\lambda_{x \to v}^K}{\mu_v^K}$, $\rho = \sum_{v=1}^{V} \rho_v^K$)
$\lambda_{x \to v}^K$	Average input rate of flit information (flit/cycle)
μ_v^K	Average service rate (cycle/flit)
U	Total number of input ports of a router
V	Total number of output ports of a router

6.2.3 Shared Memory

The shared memory is a multi-bank memory. Each shared memory has 16 banks, which is determined by the number of PEs in each PEA. So many banks can alleviate the memory access latency when memory access conflicts occur among PEs. In default, the address of a shared memory contains 10 bits, where the first two bits are label bits used to identify which bank the data is stored. The data is aligned word by word; each word has two bytes. Each bank is connected to an arbiter; meanwhile, each PE is connected to an arbiter. The priority of multiple PEs in accessing a bank

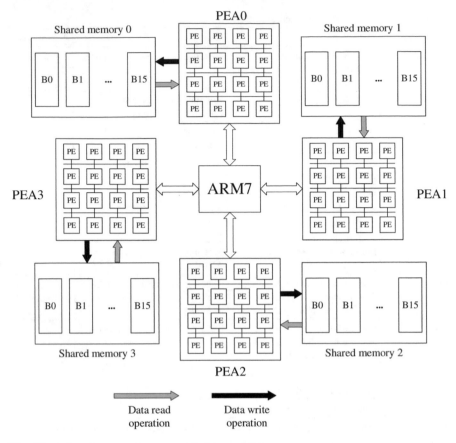

Fig. 6.8 A shared memory interacts with only one PEA

is determined by the arbiter. There is a dedicated interface between a shared memory and the PEA. The bit width of the address line for the dedicated interface is 4×8-bit, while the bit width of the data line is 4×32-bit. In each machine cycle, each bank can process one data access; a single-cycle shared memory can process a maximum of 16 data accesses (when all 16 banks initiate access requests).

In the beginning, each bank has 16 inputs for which a fixed priority is set in accordance with the order from 1 to 16. That is, if any conflict occurs during access (including read and write) of multiple inputs, corresponding memory access operations are performed in accordance with the input priority from 1 to 16. The arbiter supports broadcasting. If multiple PEs initiate data read requests to an address in a cycle, the arbiter can meet all requests during one cycle. The initial data in the shared memory is read from the external memory by the ARM7; the computation results are written to the external memory by the ARM7.

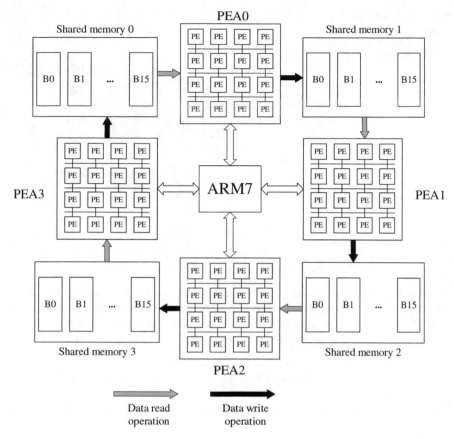

Fig. 6.9 A shared memory interacts with two neighbor PEAs

Access of a shared memory supports two modes: (1) interaction with only one PEA (the number of the PEA matches with that of the shared memory; for example, PEA0 only interacts with shared memory 0); (2) interaction with two neighbor PEAs. Figure 6.8 and 6.9 show the two modes.

6.2.4   Interconnection

Currently, in the reconfigurable system for massive MIMO detection, the communication among PEs is implemented via buses. However, compared with the traditional ASIC architecture, the size of the PEA can be limited to 4 × 4 because the adopted reconfiguration technology in the reconfigurable massive MIMO signal detection chip enables the PEA to be greatly downsized. On the basis of the bus architecture, this book provides the following solution specific to four groups of 4 × 4 PEAs: inter-

Fig. 6.10 Schematic
diagram of PE routing scope
(the PE in purple may access
each colored PE)

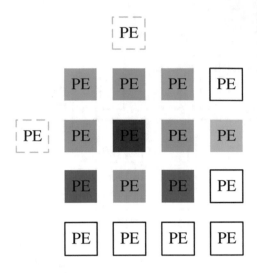

connection among PEs is implemented by means of routing. By means of routing, a PE may access the computation results of its neighbor PEs (the Euclidean distance of the PE is smaller than or equal to 2) in the previous machine cycle. Currently, the following four types of routing are provided for each PE, as shown in Fig. 6.10.

(1) Four neighbor PEs (top, bottom, left, and right);
(2) Four PEs with the distance of 2 to the PE (top jump, bottom jump, left jump, and right jump);
(3) Some PEs in the previous line and the next line (top left, bottom left, top right, and bottom right);
(4) Current PE.

Note that the modulo operation is used to map the target PE to a PE in the current array if the target PE is located beyond the current array. As shown in Fig. 6.10, the "top jump" target PE (the dotted box in the figure) is in (-1) line. There are four lines of PEs in total, and $(-1)\%4 = 3$. Therefore, the target PE is actually in the third line.

To meet the requirements of high data throughput and low latency for the next-generation mobile communication systems, many detection algorithms for the massive MIMO signal detection system (e.g., the NSA algorithm based on MMSE, the Chebyshev iteration algorithm, the conjugate gradient (CG) algorithm introduced previously, etc.) generally support very high parallelism in terms of hardware, to improve the detection efficiency of detection algorithms and thus improving the system performance. Moreover, in the reconfigurable system for massive MIMO signal detection, frequent data exchanges generally occur among PEs; this poses a challenge to the traditional bus structure in terms of communication latency and communication efficiency. In addition, communication technology is constantly evolving. The MIMO technology has experienced the development from common MIMO to massive MIMO since it emerged. The antenna array size becomes larger and larger;

the number of mobile terminals that the system can accommodate is also increased. With the development of the MIMO technology, new detection algorithms have been proposed. Therefore, the massive MIMO signal detection system in the future must support high scalability; the traditional bus structure cannot meet the requirement. Compared with the bus structure, the Network-on-Chip (NoC) has the following advantages [13]: (1) scalability. As its structure supports flexible changes, the number of resource nodes (RSNs) that can be integrated is not limited theoretically. (2) Concurrency. The NoC provides good parallel communication capability to improve data throughput and overall performance. The preceding advantages satisfy the demands of the massive MIMO signal detection system. (3) Multiple clock domains. Different from the single-clock synchronization of the bus structure, the NoC adopts the global asynchronization and local synchronization mechanism; each RSN has its own clock domain. Routing nodes are used to perform asynchronous communication among different nodes. Therefore, the area and power consumption problems caused by the huge clock tree in the bus structure are solved fundamentally.

The NoC consists of computing RSNs and communication network. The computing RSNs that usually compose of some intellectual property (IP) cores (e.g., digital signal processor (DSP), central processing unit (CPU), memory, input/output (I/O) unit, etc.) accomplish the generalized "computation" task. The communication network composed of the communication router (R), network interface (NI) and network topology link (NTL) implements high-speed communication of resources. The router mainly consists of computational logic, control logic, crossbar switch, and cache unit. The NI connecting the NoC interconnection network and the IP cores can implement the conversion of local bus protocols to on-chip network protocols, the packet disassembling and assembling and the separation of communication and computation tasks of the NoC. The NTL connects routers to form an on-chip communication network. Figure 6.11 shows the typical two-dimensional mesh-structure

Fig. 6.11 Two-dimensional mesh-structure NoC

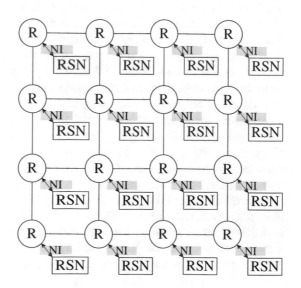

NoC. During the communication of the NoC, the router receives a packet from the source node and stores the packet into the cache unit of the input port first. Then, the computational logic and control logic of the router determine the transmission direction and arbitration channel to guide the direction of the packet flow. Finally, the packet is transmitted to the next router through the crossbar switch. This process is repeated until the packet reaches the destination and thus, implementing the data transmission between the source node and the destination node.

However, the network congestion is liable to occur to the traditional NoC; the data transmission and communication capability are limited [14]. Thus, the traditional NoC cannot meet the requirements of high throughput, high energy efficiency, low latency, and high reliability for evergrowing complex applications. With regard to the reconfigurable massive MIMO signal detection system, the dynamic reconfiguration occurs during signal detection, and thus the communication paths between different PEs may change accordingly. The topology and routing algorithm of the traditional NoC cannot be changed once they are determined. Due to failure to support dynamic behaviors of communication, the traditional NoC cannot meet the requirements of the reconfigurable system for dynamic reconfiguration. Therefore, the reconfigurable NoC with high flexibility, adaptivity, and configurability can improve the performance in complex communication mode. With high flexibility and configurability, the reconfigurable NoC may statically or dynamically reconfigure the topology, routing algorithm and router structure of the NoC for different applications or when congestion or a fault occurs on the network, to improve the communication capability of the NoC. Therefore, the reconfigurable massive MIMO signal detection system may use the reconfigurable NoC to interconnect different PEs.

Mapping optimization is one of the important research orientations for the NoC. The mapping mode affects the indexes of the NoC as well as the entire reconfigurable processor such as latency, power consumption, reliability, etc. The reconfigurable NoC may perform post-silicon customized configuration according to different communication features. The reconfigurable NoC may change its configurations such as topology and routing algorithm to meet the needs of different applications. This feature brings new challenges and requirements to the NoC mapping optimization. In brief, NoC mapping refers to the process of assigning the mapping objects in the application task graph to RSNs in the NoC topology based on certain rules to optimize the NoC target performance provided that the network topology, application task graph and design constraints are given. Figure 6.12 shows the mapping of a 3×3 two-dimensional mesh-structure NoC. Figure 6.12a shows the application characteristic graph (APCG), $G(C, A)$, which indicates the application. The APCG is a bidirectional graph, where each node $c_i \in C$ is an IP core in the application c_i (indicates the IP with the index of i; C indicates the set of all IPs in the APCG), each edge $a_{ij} \in A$ indicates the communication from c_i to c_j (a_{ij} indicates the edge from c_i to c_j; A indicates the set of all edges in the APCG), the weight of each edge V^{ij} indicates the traffic of edge a_{ij} unit: bit). Figure 6.12b shows the mapping result, where RSNi indicates the IP core with the index of i in the application task graph, Rk indicates the k routing node in the two-dimensional mesh topology. Routers communicate with each other via interconnection lines. The total number of the interconnection lines in

(a)   **(b)**

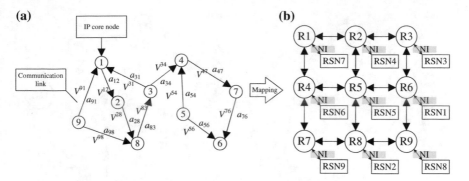

Fig. 6.12 **a** APCG, **b** Two-dimensional mesh-structure NoC after mapping

the NoC structure is defined as N. Obviously, $N = 24$ or the 3×3 two-dimensional mesh-structure NoC shown in Fig. 6.12b.

When the reconfigurable NoC is used as the interconnection mode of the reconfigurable system, the working status of the NoC directly affects the running status of the entire reconfigurable system. The reliability of the NoC significantly affects the reliability of the entire reconfigurable system because any fault in the NoC interconnection directly causes the failure of the entire reconfigurable system. In addition, the reconfigurable system requires a high reliability for its interconnection mode, e.g., a supper parallel computer requires its interconnection mode to work effectively for 10,000 h without packet loss [15]. As for the massive MIMO signal detection system, any fault in the NoC system may result in a great impact on the performance of the bit error rate (BER) for the detection algorithm. Therefore, the massive MIMO detection system has a high requirement for the reliability of the NoC system. However, the NoC is vulnerable to many factors such as crosstalk, electromagnetic interference (EMI), cosmic ray and power allocation disturbance; it is very challenging to maintain the reliability of the NoC [16–19]. Therefore, the research on the reliability of the NoC becomes particularly important. The NoC supports many different topologies and routing algorithms, most of which have been widely used in practice. Therefore, it is very important to develop a reliability-oriented mapping method that is applicable to different topologies and routing algorithms. However, to achieve this goal, many challenges need to be overcome; the most critical challenge is to develop a model that may quantitatively measure the reliability of the NoC architecture. Meanwhile, the model must be able to meet the requirement for high flexibility because the reconfigurable NoC may implement dynamic reconfiguration according to communication needs and when NoC congestion occurs or any interconnection line is faulty; the model is required to be able to well adapt to a new topology or routing algorithm. This book proposes a model for quantitatively computing the reliability of the reconfigurable NoC, the reliability cost model (RCM). In this model, all possible fault patterns of interconnection lines are taken into account and the reliability cost is deemed as a discrete random variable. Therefore, this model may be applied to

(a)   **(b)**

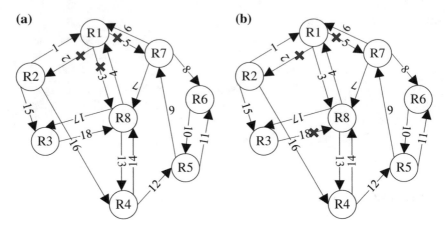

Fig. 6.13 Two fault patterns with three faulty interconnection lines. **a** Interconnection lines 2, 3 and 5 are faulty, **b** Interconnection lines 2, 5 and 18 are faulty [2018] IEEE. Reprinted, with permission, from Ref. [30]

more extensive NoC topologies and routing algorithms [20]. The RCM is introduced as follows.

To facilitate modeling, in the RCM, router faults are categorized as interconnection line faults; assume that the occurrence of faults in different interconnection lines are independent and the fault probability of the interconnection line with the index of j is p_j. When $n(n \leq N)$ interconnection lines are faulty, all $M = \binom{N}{n}$ fault patterns shall be considered. In the model, the reliability of NoCs with different topologies and routing algorithms is assessed by the reliability cost. The higher the reliability cost, the lower the reliability. The reliability cost for a source-destination pair is defined by a binary value, indicating whether a valid transmission path exists from the source to the destination. Figure 6.13 enumerates two different fault patterns. As shown in Fig. 6.13a, interconnection lines 2, 3, and 5 are faulty; thus, no valid communication path exists between R1 and R3 to transfer data from R1 to R3. In this case, the reliability cost is defined as 1 (expressed by $RC_{1,3}^{R_1,R_3} = 0$). In the second fault pattern shown in Fig. 6.13b, interconnection lines 2, 5 and 18 are faulty; however, data can be effectively transmitted through the path R1 $\rightarrow$ R8 $\rightarrow$ R3. Thus, the reliability cost is defined as 0 (expressed by $RC_{2,3}^{R_1,R_3} = 0$).

Based on the preceding definition, when n interconnection lines are faulty in a specific mapping mode, the reliability cost of fault pattern i in all M fault patterns can be expressed by Formula (6.1).

$$RC_{i,n} = \sum_{S,D} RC_{i,n}^{S,D} F^{S,D} \tag{6.1}$$

where $\mathrm{RC}_{i,n}^{S,D}$ indicates the reliability cost from the source S to the destination D for fault pattern i; its value can be obtained according to the preceding definition about reliability. $F^{S,D}$ indicates whether communication exists from the source S to the destination D; its value can be obtained by Formula (6.2).

$$F^{S,D} = \begin{cases} 1, & V^{S,D} > 0 \\ 0, & V^{S,D} = 0 \end{cases} \tag{6.2}$$

where $V^{S,D}$ indicates the traffic from the source S to the destination D. As the reliability cost varies dramatically according to fault patterns, all fault patterns are taken into account in the RCM and their expected values are used to assess the reliability costs of all fault patterns in a specific mapping mode. According to discussions in preceding sections, the fault probability varies with interconnection lines. As the fault probability of the interconnection line with the index of j is p_j, the probability that n interconnection lines are faulty can be expressed by Formula (6.3).

$$P_{I,n} = \prod_{j=0,j\in I}^{N} p_j \times \prod_{j=0,j\notin I}^{N} \left(1 - p_j\right) \tag{6.3}$$

where I indicates the set of indexes for interconnection lines that are faulty in the fault pattern i when n interconnection lines are faulty. Thus, in a specific mapping mode, the overall reliability cost is expressed by Formula (6.4).

$$\mathrm{RC} = \sum_{n=0}^{N} \sum_{i=1}^{M} \mathrm{RC}_{i,n} P_{I,n} \tag{6.4}$$

The RCM is applicable to various topologies and routing algorithms because it avoids the faulty links on the communication path shielded by the NoC architecture and improves the reliability. It is different from the other researches that use the standby core on the NoC side or the constraint that the source-destination pair to be close to a square (such constraint can only be implemented by the two-dimensional mesh topology [17, 21, 22]). In addition, the method of regarding the reliability cost as a discrete random variable not only improves the precision but also ensures a high flexibility of the model.

Except reliability, the power consumption of the NoC also has a great impact on the performance of the reconfigurable system. Researches show that the communication power consumption of the NoC accounts for more than 28% of the system's total power consumption [23, 24]; the proportion may even exceed 40% especially when multimedia applications are performed in the reconfigurable system [25]. Meanwhile, as has been introduced in Chap. 1, the next-generation mobile communication systems require low power consumption, which requires the massive MIMO signal detection system to be energy efficient. Therefore, the research on the NoC energy consumption is also crucial except reliability. Therefore, this book proposes an energy consumption quantitative model that is applicable to the reconfigurable NoC. Generally, energy consumption involves the static energy consumption and dynamic energy consumption. The static energy consumption is mainly affected by the operating temperature, technology level, and gate-source/drain-source voltage. For the same NoC in the same environment, in different mapping modes, the temperatures and technologies are the same, and the variation of the gate-source/drain-source voltage is very small; the difference in static energy consumption is relatively small. Therefore, the static energy consumption is not considered in the model. In contrast, the dynamic energy consumption (communication energy consumption) is considerably different in different mapping modes. Therefore, the communication energy consumption is a main indicator for assessing the energy consumption of a mapping mode. During energy consumption modeling, only the communication energy consumption is considered. In the research of this book, the indicator of the energy consumption of 1-bit message is used to assess the communication energy consumption. E_{Rbit} and E_{Lbit} indicate the energy consumptions for a 1-bit message through a router and an interconnection line, respectively. Based on these two indicators, when n interconnection lines are faulty, the communication energy consumption of all source-destination pairs in the case of i can be expressed by Formula (6.5).

$$E_{i,n} = \sum_{S,D} V^{S,D}\left[E_{\text{Lbit}}d_{i,n}^{S,D} + E_{\text{Rbit}}\left(d_{i,n}^{S,D} + 1\right)\right]F^{S,D} \tag{6.5}$$

where $d_{i,n}^{S,D}$ indicates the number of interconnection lines on the transmission path from the source S to the destination D in fault pattern i when n interconnection lines are faulty. $F^{S,D}$ has been defined by Formula (6.2). Likewise, the communication energy consumption also varies with the fault patterns; therefore, all fault patterns are taken into account and their expected values are used to express the overall communication energy consumption in a specific mapping mode, as shown in Formula (6.6).

$$E = \sum_{n=0}^{N}\sum_{i=1}^{M} E_{i,n}P_{I,n} \tag{6.6}$$

where $P_{I,n}$ has been defined by Formula (6.3); it indicates the probability of fault pattern i when n interconnection lines are faulty.

As presented in Chap. 1, currently, the demands of endlessly emerging application scenarios for data throughput and latency of communication systems become increasingly higher, which poses great challenges to massive MIMO signal detection. Therefore, except the abovementioned two indicators, reliability, and communication energy consumption, the performance (i.e., latency and throughput) also need to be considered as indicators for reconfigurable NoC. To this end, this book proposes a model specific to quantitative modeling for latency and a qualitative analysis model specific to throughput for the reconfigurable NoC [20, 26].

In this book, the wormhole switching technology of the NoC is considered when quantitatively modeling the latency. With the wormhole switching technology, the flit latency of the body flit and the tail flit are the same as the flit latency of the head flit. To facilitate analysis, the flit latency of the head flit is used to represent the flit latency of a transmission path during the modeling of latency in quantitative in this book. The flit latency of the head flit is defined as the interval from the time when the head flit is created in the source node to the time when the head flit is received by the destination node; it consists of three parts: (1) pure transmission latency. This latency indicates the time consumed from the transmission head flit in the source node to the receipt head flit in the destination node in the case of no faulty interconnection line or congestion. (2) waiting time caused by a faulty interconnection line. The research in this book takes into account that the head flit will wait for a clock cycle and try to transmit again until the transmission is successful when head flit transmission encounters a faulty interconnection line; thus, the latency is the interval from the time when the head flit encounters a faulty interconnection line to the time when transmission is successful. (3) waiting time caused by congestion. It is likely that transmission of multiple head flits needs to go through the same router or interconnection line at the same time; thus, congestion occurs. During modeling, the principle of first-come, first-served is adopted; the later head flit may perform transmission until the earlier head flit finishes processing; the waiting time is the latency of the third part. The following section analyzes the specific computing algorithms for the three parts of latencies. The computation of the pure transmission latency is simple. The transmission latency is defined as the interval from the time when a head flit is created in the source node to the time when the head flit is received by the destination node; it can be expressed by Formula (6.7).

$$LC_{i,n} = \sum_{S,D} \left[t_w d_{i,n}^{S,D} + t_r \left(d_{i,n}^{S,D} + 1 \right) \right] F^{S,D} \tag{6.7}$$

where t_w and t_r indicates the transmission time for a flit going through an interconnection line and a router, respectively. Both $d_{i,n}^{S,D}$ and $F^{S,D}$ have been defined in preceding sections. The research in this book takes into account that the head flit will wait for a clock cycle and try the transmission again until the transmission is successful when head flit transmission encounters a faulty interconnection line; the interval from the time when the head flit encounters a faulty interconnection line to the time when transmission is successful is defined as the wait latency for encountering an interconnection line fault. During modeling, due to the failure to accurately estimate the required number of clock cycles for each interconnection line to recover from a soft fault, the average waiting time is used to represent the waiting time caused by the faulty interconnection line j in the model; the average waiting time can be obtained by Formula (6.8).

$$LF_j = \lim_{T \to \infty} \left(p_j + 2p_j^2 + 3p_j^3 + \cdots + T p_j^T \right) = \frac{p_j}{\left(1 - p_j \right)^2} \tag{6.8}$$

where p_j indicates the fault probability of interconnection line j; T indicates the required number of clock cycles for the head flit to wait for. By referring to the existing research [27], the latency caused by congestion is processed with the first-come, first-served queue. In this queue, each router is regarded as a service desk. Under a deterministic routing algorithm, the transmission path of data is completely determined if the NoC structure and the source and destination nodes of data transmission are known. Therefore, the data waiting for being transmitted can only be in one queue; that is, there is one and only one service desk serving the data. However, with regard to an adaptive routing algorithm, the path and the next node of the data transmission are determined and changed according to the status of the current network. This means that the data may have multiple service desks. Therefore, the $G/G/m$-FIFO queue is used to estimate the waiting time caused by congestion in this book. As for this queue, both the arrival interval and the service time are regarded as an independent general random distribution. By referring to the Allen-Cunneen formula [28], the waiting time required for the transmission from the uth input port to the vth output port of router K can be expressed by Formulas (6.9), (6.10) and (6.11).

$$WT_{u \to v}^K = \frac{\overline{WT_0^K}}{\left(1 - \sum_{x=u}^{U} \rho_{x \to v}^K\right)\left(1 - \sum_{x=u+1}^{U} \rho_{x \to v}^K\right)} \tag{6.9}$$

$$\overline{WT_0^K} = \frac{P_m}{2mp} \times \frac{C_{A_{u \to v}^K}^2 + C_{S_v^K}^2}{\mu_v^K} \times \rho_v^K \tag{6.10}$$

$$P_m = \begin{cases} \frac{\rho^m + \rho}{2}, & \rho \geq 0.7 \\ \rho^{\frac{m+1}{2}}, & \rho < 0.7 \end{cases} \tag{6.11}$$

In Formula (6.10), $C_{A_{u \to v}^K}^2$ indicates the variable coefficient for the queue of router K. In the research of this book, the queue to each router on the network is assumed to be the same as the queue to the network. Therefore, $C_{A_{u \to v}^K}^2$ is assumed to be equal to the variable coefficient C_A^K of the queue to the NoC; its value is determined by the application. Similarly, $C_{S_v^K}^2$ indicates the variable coefficient for the service queue of router K; its value is determined by the distribution of the service time. As shown in Fig. 6.14a, the service time of the ith output port of R4 consists of the following three parts: (1) The transmission time for the data going through R5 without congestion; (2) The waiting time for the data distribution from the input port j to the output port k; (3) The time required for waiting for the output port k to be idle; that is, the service time of the output port k on R5. As each output port of R5 has a great impact on the service time of the output port i on R4, this problem is solved by creating a correlation tree in this research. During the creation of a correlation tree, if the router connecting the output port k of R5 communicates with R5, the router is added to the tree; otherwise, the router is not added to the tree. As shown in Fig. 6.14b, the real line indicates that the router is added to the tree; the dotted line indicates that the router is not added to the tree. The process of creating the correlation tree continues

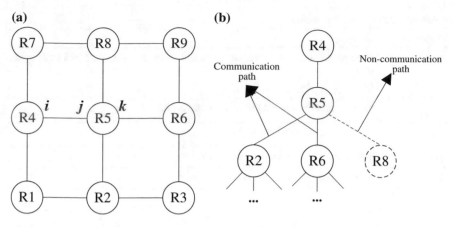

Fig. 6.14 **a** An example for a topology, **b** The correlation tree for the topology. © [2018] IEEE. Reprinted, with permission, from Ref. [30]

until the router only communicates with PEs and does not communicate with the other routers. After a correlation tree is created, the service time of leaf nodes is first computed; then the service time of parent nodes is computed using inverse recursion; see Formula (6.12). In Formula (6.12), $\overline{S_v^K}$ indicates the average service time of the output port v on router K; $\overline{(S_v^K)^2}$ indicates the second-order moment of the service time distribution of output port v of router K.

$$\overline{S_v^K} = \sum_{x=1}^{U} \frac{\lambda_{u \to x}^K}{\lambda_x^K} \times \left(t_w + t_r + WT_{u \to x}^{K+1} + \overline{S_v^{K+1}} \right)$$

$$\overline{(S_v^K)^2} = \sum_{x=1}^{U} \frac{\lambda_{u \to x}^K}{\lambda_x^K} \times \left(t_w + t_r + WT_{u \to x}^{K+1} + \overline{S_v^{K+1}} \right)^2$$

$$C_{S_v^K}^2 = \frac{\overline{(S_v^K)^2}}{(\overline{S_v^K})^2} - 1 \tag{6.12}$$

The parameters used in Formulas (6.9)–(6.11) are determined by the application and the structure of the router; Table 6.4 specifies their meanings. After the three parts of latencies are computed, the latency in fault pattern i can be expressed by Formula (6.13) when n interconnection lines are faulty.

$$L_{i,n} = LC_{i,n} + \sum_{S,D} \left(\sum_{K=1}^{d_{i,n}^{S,D}+1} WT_{U(K) \to V(K)}^{R(K)} + \sum_{j=1}^{d_{i,n}^{S,D}} LF_{L(j)} \right) F^{S,D} \tag{6.13}$$

where $R(K)$ indicates the function for computing the index of the Kth router on the communication path from the source S to the destination D. $U(K)$ and $V(K)$ indicate

Table 6.4 Definitions of parameters

Parameter	Meaning
ρ_v^K, ρ	Time percentage that the vth output port of router K is occupied by its xth input port. $(\rho_v^K = \sum_{x=1}^{U} \frac{\lambda_{x \to v}^K}{\mu_v^K}, \; \rho = \sum_{v=1}^{V} \rho_v^K)$
$\lambda_{x \to v}^K$	Average input rate of flit information (flit/cycle)
μ_v^K	Average service rate (cycle/flit)
U	Total number of the input ports in a router
V	Total number of the output ports in a router

the function for computing the indexes of the input port and the output port on the router, respectively. $L(j)$ indicates the function for computing the index of the jth interconnection line. Similar to the previous analysis, all faults shall be taken into account; thus, the total latency can be computed by Formula (6.14).

$$L = \sum_{n=0}^{N} \sum_{i=1}^{M} L_{i,n} P_{I,n} \tag{6.14}$$

Qualitative analysis is performed for throughput by means of bandwidth limitation. The traffic of each node is balanced by Formula (6.15), which helps avoid congestion. Thus, the performance is guaranteed.

$$\sum_{a_{ij}} \left[f\left(P_{\mathrm{map}(c_i),\mathrm{map}(c_j)}, b_{ij} \right) \times V^{ij} \right] \le B(b_{ij}) \tag{6.15}$$

where $B(b_{ij})$ is the bandwidth of link b_{ij} (the link connecting the ith and the jth nodes of the NoC). c_i, c_j and V^{ij} have been defined as the parameters of the APCG. The binary function $f\left(P_{\mathrm{map}(c_i),\mathrm{map}(c_j)}, b_{ij} \right)$ indicates whether link b_{ij} is used by path $P_{\mathrm{map}(c_i),\mathrm{map}(c_j)}$; it is defined by Formula (6.16).

$$f\left(P_{\mathrm{map}(c_i),\mathrm{map}(c_j)}, b_{ij} \right) = \begin{cases} 0, & b_{ij} \notin P_{\mathrm{map}(c_i),\mathrm{map}(c_j)} \\ 1, & b_{ij} \in P_{\mathrm{map}(c_i),\mathrm{map}(c_j)} \end{cases} \tag{6.16}$$

As for numerous general function processors at present and in the future (such as the future load data processors for processing scientific, earth and communication tasks that are defined by European and American space agencies), the requirements for their communication devices are high reliability, low power consumption and high performance [21], which are also the requirements for the massive MIMO signal detection system. Therefore, how to find a mapping mode in which the reliability, communication energy consumption, and performance can be optimized simultaneously and properly balanced when mapping an application to the reconfigurable NoC for operation, is a very important research topic. Nowadays, many research results are provided for modeling and optimizing the three indicators at home and abroad. Although certain progress has been achieved in reliability, communication energy consumption, and performance, these methods have the following common problems on the whole: (1) Basically, only the energy consumption and reliability are computed by means of quantitative modeling. However, as an indicator that is as important as reliability and communication energy consumption, performance is only considered qualitatively at most; the impact of performance is not considered in most researches. The way for considering performance has substantial defects in the applications of the reconfigurable NoC at present and in the future. (2) Most of these methods are only applicable to specific topologies and routing algorithms. For example, in some researches, mapping is performed on a rectangular NoC or the mapped graph on the NoC is required to be rectangle and as close as possible to square; this limits their methods to two-dimensional mesh topologies [21, 22]. Although researches on mapping methods independent of topologies have been implemented in recent two years [29], the proposed models are too simple that only consider the energy consumption or latency quantitatively. As for the other indicators, only qualitative analysis is performed at most; such methods greatly reduce the accuracy for finding the optimal mapping mode and the difficulty in improving the flexibility. (3) After a multi-objective is modeled by the multi-objective joint optimization model, the computation load of the proposed mapping methods is relatively large. In addition, the mapping methods cannot meet the requirement of dynamic reconfiguration. In recent two years, the dynamic remapping method is proposed to implement multi-objective joint optimization specific to the dynamic reconfiguration requirement in some literature [29]. In these researches, the modeling contents are too simple, which reduces the accuracy for finding the optimal mapping mode. In addition, according to their experimental results, the remapping also requires a long time that cannot well meet the requirement of the dynamic reconfigurable NoC. This book carries out some researches on the multi-objective joint optimal mapping algorithm for reliability, communication energy consumption, and performance when mapping an application to the reconfigurable NoC; some results have been achieved. Specific to the reconfigurable NoC and based on the quantitative models for reliability cost, energy consumption and latency and the qualitative analysis model for throughput, this book proposes two multi-objective joint optimal mapping solutions. The following sections introduce the two solutions, respectively.

In the first solution [30], this book proposes an efficient mapping method for co-optimization of reliability, communication energy, and performance (CoREP). Specific to CoREP, a total cost model is created for the mapping, and the energy-latency product (ELP) is introduced into the model for the first time to assess both the communication energy consumption and the latency. Meanwhile, the requirement variance of reliability and ELP for different application scenarios is considered. For example, the main requirement of most mobile devices is low energy consumption, whereas the main requirement of most space systems is high reliability. Therefore, the CoREP model uses the weight parameter $\alpha \in [0, 1]$ to distinguish the optimization weights of reliability and ELP. In the reconfigurable NoC, a higher reliability and a smaller ELP are always demanded. Based on these considerations, the total cost for any mapping mode can be expressed by Formula (6.17).

$$\text{Cost} = \alpha\text{NRC} + (1 - \alpha)\text{NELP} \tag{6.17}$$

where NRC indicates normalized reliability cost and NELP indicates normalized ELP. In addition, specific to the CoREP model, this book proposes a mapping method based on the priority and ratio oriented branch and bound (PRBB).

Formula (6.18) can be used to perform normalized processing for reliability cost, where RC is the reliability cost defined by Formula (6.4). N_{RC} indicates the normalized parameter for reliability cost. The normalized parameter is obtained by considering the worst case, i.e., the reliability cost when the number of faulty interconnection lines on the entire NoC reaches the tolerable limit value.

$$\text{NRC} = \text{RC}/N_{\text{RC}} \tag{6.18}$$

The ELP is defined as the product of energy consumption and latency, as shown in Formula (6.19).

$$\text{ELP} = \text{E} \times \text{L} \tag{6.19}$$

where E and L indicate the energy consumption and the latency defined in Formulas (6.6) and (6.14), respectively. The normalized ELP can be computed by using Formula (6.20), where N_{ELP} indicates a normalized parameter that is obtained by considering the worst case (i.e., the data to be transmitted goes through the longest and most congested communication path).

$$\text{NELP} = \text{ELP}/N_{\text{ELP}} \tag{6.20}$$

The CoREP computation model proposed in this book is applicable to various frequently-used topologies and routing algorithms today. The main reasons are as follows: (1) The CoREP model only takes into account the number of faulty interconnection lines on all communication paths and selects the mapping mode with the least number of faulty interconnection lines. This process is not restricted by topologies and routing algorithms. Therefore, the model is applicable to various topologies and routing algorithms. However, the other similar algorithms take into account that the graph of the mapped source-destination pair shall be as close as possible to square to improve the reliability [22]. This requires that the topology that is mapped to the NoC can only in the mesh form, which cannot meet the requirement of the reconfigurable NoC for high flexibility. (2) The CoREP model dynamically computes the communication energy consumption; the computation of the communication energy consumption updates with each change of the communication path. This computation is also independent of topologies and routing algorithms. In previous researches of the same type, the communication energy consumption can only be computed when the communication path is known in advance, which limits the computation mode to mesh topologies. (3) The latency quantitative model uses the $G/G/m$-FIFO queue to assess the latency. The computation mode is independent of various topologies and can be applied to various deterministic and adaptive routing algorithms; this model is much more flexible than the model proposed in Ref. [27] that is only applicable to deterministic routing algorithms. To sum up, the CoREP model proposed in this book has a high flexibility and is applicable to various frequently-used topologies and routing algorithms for computing the reliability, communication energy consumption and latency. This is a great improvement compared with the previous work.

After performing the quantitative modeling analysis on target parameters using the CoREP model, the issue of looking for the joint optimal mapping mode can be defined as follows: For a given NoC consisting of routers and PEs based on any topology and routing algorithm in practical application, look for a mapping function and use the mapping function to find the node that maps the actual PE to the NoC so that to satisfy the condition of achieving the minimum total cost computed by Formula (6.17). Specific to this issue, this book proposes the PRBB mapping algorithm with low computation complexity.

First, the branch and bound (BB) method is introduced. The BB method is a common method to obtain the computation complexity of non-deterministic polynomial problems (NP-problems) [31]. With the BB method, a search tree is created to look for the optimal value of any optimal target function. This book provides an example of a simple search tree, as shown in Fig. 6.15. The search tree shows the process of mapping an application with three IPs to an NoC with three routing nodes. With regard to the NoC mapping topic, each node in the search tree indicates a possible mapping mode. As shown in Fig. 6.10, each node is represented by a box; each number in a box indicates the index of an IP in practical application. The position of each number in a box indicates the index of the routing node in NoC to which the IP with the index represented by the number is mapped; a space indicates that the NoC node has not been mapped by any IP. For example, "3 1 _" indicates that IP3 and IP1 are mapped to the first and second routing nodes of the NoC, respectively; no IP is mapped to the third node of the NoC. According to the above definitions, the root node of the search tree indicates empty mapping, which means that no IP is mapped to the NoC; it is the beginning of the entire mapping process. With the deepening of the mapping process, IPs start to be mapped to the first and latter nodes on the NoC; this is partial mapping. In the search tree, partial mapping is represented by an intermediate node, e.g., the intermediate node "3 1 _" is a partial mapping mode. The entire mapping process continues until all the leaf nodes that indicate all IPs have been mapped are created. During the creation of a search tree, only the nodes that are mostly likely to be optimal mapping modes are created, while the nodes that cannot be optimal mapping modes are deleted to reduce the computation for the search. For any intermediate node, if its minimum cost is greater than the maximum cost of the found optimal node, the intermediate node is impossible to become the optimal mapping mode; thus, the intermediate node and all its child nodes cannot be created. This significantly reduces the computation for searching the optimal mapping mode. In

Fig. 6.15 Search tree for mapping an application with three IPs to the NoC. © [2018] IEEE. Reprinted, with permission, from Ref. [30]

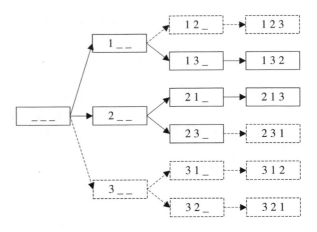

the BB algorithm, the most important step is the methods for computing the deletion conditions and the upper and lower bounds of cost for the intermediate nodes, which can be well solved by the PRBB algorithm.

Referring to the BB algorithm, this book proposes the PRBB algorithm to look for the optimal mapping mode for mapping an application to the NoC. The PRBB algorithm takes advantage of the branch node priority recognition technique and the usage of the partial cost ratio technique to improve the efficiency in searching for the optimal mapping mode. With the branch node priority recognition technique, the PRBB algorithm performs computation according to the priority assigned to a node. According to the description of a search tree in previous sections, an intermediate node closer to the root node has more child nodes, i.e., an intermediate node closer to the root node incurs a larger computation cost. Therefore, if nonoptimal mapping modes on such nodes can be identified and deleted first, the complexity of the entire search algorithm may be greatly reduced. In this research, intermediate nodes close to the root node are prioritized; the closer to the root node the intermediate node is, the higher its priority is. During the search for the optimal mapping mode, the intermediate node of higher priority is computed and determined whether to be created or deleted at the earliest possible time. In this way, nodes of higher priorities with nonoptimal mapping modes can be found and deleted as early as possible, which can improve the search efficiency of the algorithm. The partial cost ratio is used for the deletion conditions. The accuracy may be reduced if intermediate nodes are deleted too quickly; therefore, the PRBB algorithm uses the partial cost ratio to trade-off between the speed and accuracy. The cost of any mapping mode can be looked as the sum of the partial costs incurred by different numbers of faulty interconnection lines. The partial cost ratio is defined as the ratio between two adjacent partial costs, as shown in Formula (6.21).

$$\text{ratio}_{n+1,n} = \frac{\text{Cost}^{n+1}}{\text{Cost}^n} \qquad (6.21)$$

where $\text{Cost}^n = \sum_{i=1}^{M} \left[\frac{\alpha R_{i,n}}{N_R} + \frac{(1-\alpha)E_{i,n}L_{i,n}}{N_{\text{ELP}}} \right] P_{I,n}$ indicates the cost when n interconnection lines are faulty, i.e., the nth partial cost.

It has been discussed in previous sections that the CoREP modeled under the condition that the fault probability varies with interconnection lines. In this research, two different values p_h and p_l are used to simply verify that the non-unitary form of fault probability is true. The fault probabilities of different interconnection lines are distinguished by the traffic on interconnection lines. Taking into consideration that more traffic would result in larger energy consumption, which then would result in the accumulation of temperature; interconnection lines are more likely to be faulty in higher temperature. Therefore, in this research, the fault probability of interconnection lines in the area of high traffic is assumed to be p_h, and the fault probability of interconnection lines in the area of low traffic is assumed to be p_l. This is a simplified example for verifying that the fault probability varies with interconnection lines. The model proposed in this book can work with more complex and comprehensive models with different fault probabilities; this will be implemented in future researches. With the simplified model in this book, $P_{I,n}$ can be obtained by Formula (6.3), where p_j is substituted by p_h or p_l.

When n changes to $n + 1$, only the change of a small among of paths can change $RC_{i,n}$. In addition, due to the small change in the number of interconnection lines on the transmission path, the change in latency caused by $E_{i,n}$ and pure transmission is also very small. During the computation, the latency caused by faulty interconnection lines is computed using the average latency. Therefore, the change of $L_{i,n}$ is relatively small. Thus, Formula (6.21) can be simplified to Formula (6.22).

$$\text{ratio}_{n+1,n} < \frac{N - n}{n + 1} \times \frac{1}{4(1 - p_l)^2} \tag{6.22}$$

Once the fault probability of an interconnection line exceeds 0.5, the entire network can hardly work properly. Therefore, the fault probability of each interconnection line is assumed to be smaller than or equal to 0.5. According to Formula (6.21), the partial cost ratio decreases rapidly with the increase of n. Therefore, for a very large n, $\text{ratio}_{n+1,n}$ is close to 0 and can be ignored. In addition, the nth partial cost can be expressed by Formula (6.23).

$$\text{Cost}^n = \text{Cost}^0 \times \prod_{k=1}^{n} \text{ratio}_{k,k-1} \tag{6.23}$$

Therefore, the total cost can be expressed by the sum of the above partial costs and ignoring the cost when n is very large. In this way, the process for computing partial costs can be simplified, thus reducing the computation load.

In PRBB, the simplification in cost computation may cause decrease in accuracy; thus, Formula (6.24) is used to express deletion conditions to relatively improve the accuracy for finding the optimal mapping mode.

$$\text{LBC} > \min\{\text{UBC}\} \times \left(1 + \text{ratio}_{1,0}\right) \tag{6.24}$$

where LBC indicates the lower bound of the cost in partial mapping mode; it is computed based on the following three parts: (1) The partial reliability cost incurred by the communication between mapped IPs $\text{LBC}_{m,m}$, which is obtained by Formula (6.4); (2) The partial reliability cost incurred by the communication between non-mapped IPs $\text{LBC}_{u,u}$, which is computed on the basis of the communication between the closest possible cores; (3) The partial reliability cost incurred by communication between a mapped IP and a non-mapped IP $\text{LBC}_{m,u}$, which is computed based on the optimal mapping mode. Therefore, LBC can be obtained by using Formula (6.25). UBC indicates the upper bound of reliability cost in partial mapping mode; it is the reliability cost obtained by temporarily using the greedy algorithm to map non-mapped IPs to the NoC.

$$\text{LBC} = \text{LBC}_{m,m} + \text{LBC}_{u,u} + \text{LBC}_{m,u} \tag{6.25}$$

To reduce the time for finding the optimal mapping mode, the PRBB mapping algorithm adopts two methods to reduce the computation complexity. The first method is to use Formula (6.24) to reduce the number of intermediate nodes to be searched for. When the deletion condition expressed by Formula (6.24) is met, the intermediate node and all its child nodes are deleted; otherwise, the node and its child nodes are inserted to the search tree for further comparison. When the deletion condition is used to delete nonoptimal mapping modes, the algorithm would retain the obtained cost values for the mapping modes estimated to be close to the optimal mapping mode and performs further comparison. Thus, this ensures that the accuracy of optimal mapping mode is not sacrificed by the acceleration algorithm. The second method is to use the average value of different mapping modes to simplify $\text{ratio}_{2,1}$.

The methods proposed in this book are specially designed for NoC-based reconfigurable systems. Therefore, this book provides the corresponding workflow. Figure 6.16 shows the workflow of mapping an application to an NoC-based reconfigurable system using the CoREP model and the PRBB algorithm. For a certain NoC and a given application, the optimal mapping solution is first provided by the PRBB algorithm on the basis of the total cost computed by the CoREP model. During the execution of the application on the NoC, the topology and routing algorithm of the NoC are required to be able to be reconfigured accordingly to cope with special cases such as sudden permanent faults and application demands. After the topology is reconfigured, the system would use the PRBB algorithm to find the corresponding optimal mapping mode according to the reconfigured topology and routing algorithm.

Next, the computation load of the PRBB algorithm is compared with that of the BB algorithm. As it is quite difficult to accurately compute the number of deleted intermediate nodes, this book compares the differences of the two algorithms in computation load in three cases, optimal case, general case and worst case. The computation load is the least in the optimal case. Therefore, this book assumes that most intermediate nodes can be deleted and only one intermediate node is left on

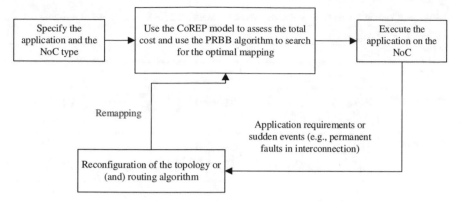

Fig. 6.16 Workflow of mapping an application onto a NoC-based interconnected reconfigurable system by using the CoREP model and the PRBB method. © [2018] IEEE. Reprinted, with permission, from Ref. [30]

each branch for further comparison. In the worst case, the other extreme, only one intermediate node can be deleted on each branch and all the other nodes need to be retained for further comparison, which results in the largest computation load. In the general case, assume that only k nodes are retained on each branch. Compared with the BB algorithm, the PRBB algorithm is mainly used to delete the intermediate nodes close to the root node as soon as possible; this means that the BB algorithm consumes a larger computation cost than the PRBB algorithm. Therefore, assume that one more node is retained on each branch for further comparison using the BB algorithm compared with that using the PRBB algorithm. During the computation, similar basic operations are performed in each cycle; therefore, the number of cycles is defined as the time complexity of an algorithm and is used to assess the operation complexity of an algorithm. Table 6.5 provides the operation complexity for the three cases, where N_{NoC} indicates the number of nodes on the NoC. Figure 6.17 shows the reduced percentage of the computation load for PRBB compared with that of BB. In this figure, **optimistic** indicates the reduced percentage in complexity in the optimal case; **pessimistic** indicates the reduced percentage in complexity in the worst case; $k = 3, 4, \ldots, 9$, indicates the reduced percentage in complexity in the general case. As can be seen from the figure, the reduced percentages of computation load for PRBB in all cases are positive numbers, which means that the PRBB algorithm performs less computations in searching for the optimal mapping mode. In addition, as can be seen from Fig. 6.17, when the NoC size gets larger, the proportion of reduction in computation increases. This means that the PRBB mapping method proposed in this book is more applicable to larger reconfigurable NoCs.

To make the research results be more persuasive, an experimental analysis is firstly performed in this book for the optimization results of reliability, communication energy consumption, and performance to verify that CoREP and PRBB are effective in co-optimization of the three indicators. Then, an experiment is conducted on joint optimization of two indicators among reliability, communication energy consump-

Table 6.5 Complexity analysis and comparison between the PRBB and the BB algorithms

Different cases	Algorithm	
	BB	PRBB
Optimal case	$O\left(2^{N_{NoC}+1} \times N_{NoC}^3\right)$	$O\left(N_{NoC}^{4.5}\right)$
General case	$O\left(N_{NoC}^3\right) \times$ $\left[\frac{k^{N_{NoC}+1}-k}{(k-1)^2} - \frac{N_{NoC}}{k-1}\right]$	$O\left(N_{NoC}^{2.5}\right) \times \left[\frac{(k-1)^{N_{NoC}+1}-k+1}{(k-2)^2} - \frac{N_{NoC}}{k-2}\right]$
Worst case	$O\left(N_{NoC}^3\right) \times$ $\sum_{j=0}^{N_{NoC}-1} \prod_{i=0}^{j} (N_{NoC} - i)$	$O\left(N_{NoC}^{2.5}\right) \times$ $\left\{\sum_{j=0}^{N_{NoC}-1}\left[\prod_{i=0}^{j} (N_{NoC} - 1)\right](n-j)^2 + n\right\}$

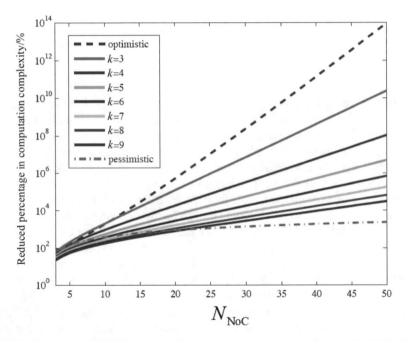

Fig. 6.17 Percentage of PRBB's reduction in computation complexity compared with BB. © [2018] IEEE. Reprinted, with permission, from Ref. [26]

tion, and performance to illustrate that CoREP and PRBB can realize multi-objective joint optimization.

In the experiment, a software platform is designed with the C++ algorithm to compute the optimal mapping mode and the time required for finding the optimal mapping mode. The obtained optimal mapping mode is emulated on the NoC emulator with accurate cycle created by the SoCDesigner to obtain the figures such as reliability, communication energy consumption, latency and throughput. In the emulator, each node of the NoC includes one router and one PE; the experimental environments of routers are identical, as shown in Table 6.6. As the PRBB method is independent of

Table 6.6 Routing emulation environment

Switching technology	Wormhole
Arbitration strategy	Time slice cycle
Whether the virtual channel technology is used	Yes
E_{Rbit}	4.171 nJ/bit
E_{Lbit}	0.449 nJ/bit

the applications of the switching technology, arbitration strategy and virtual channels, these parameters are the same as those for BB for a fair comparison. E_{Rbit} and E_{Lbit} indicate the energy consumptions required for transmitting a single-bit message through a router and an interconnection line, respectively; they are obtained by using an open-source NoC emulator [32]. Table 6.6 provides the values obtained by emulating the energy consumption model with the Power Complier of the Synopsys company. The following two steps are used to compute the communication energy consumption. (1) Compute the numbers of routers and interconnection lines on the communication path and multiply the obtained numbers by E_{Rbit} and E_{Lbit}, respectively, to obtain the total communication energy consumption. (2) The total energy consumption is used to divide the total number of transmitted flit messages to obtain the communication energy consumption required for transmitting a unit flit message [22]. Although this method cannot obtain the exact value of energy consumption, it provides an effective method for comparing the energy consumption of different mapping modes. During the emulation, faults with a probabilities of p_h and p_l are randomly injected to the interconnection lines on the NoC according to the different locations of interconnection lines. The reliability is assessed by the probability of properly transmitting a flit message from the source node to the destination node [32]. As an estimation of probability, the reliability unit is set to "1" in this emulation. In the experiment, assume that each packet consists of eight flit messages (one head flit message, one tail flit message and six body flit messages). Finally, the throughput is obtained by computing the average number of flit messages that can be transmitted by each node.

This book firstly analyzes the optimization results of reliability, communication energy consumption, and performance. First, the flexibility is verified. In the experiment, NoCs with four different topology and algorithm combinations are selected; these topologies and routing algorithms are selected on the basis of a survey report on the NoC. By referring to 60 latest articles on NoC, the report classifies the 66 topologies and 67 routing algorithms used in the articles and provides the following information: 56.1% of NoCs use mesh/Torus topologies; 12.1% of NoCs use customized topologies; 7.6% of NoCs use ring topologies; 62.7% of NoCs use deterministic routing algorithm; the remaining 37.3% of NoCs use adaptive routing algorithms. Accordingly, Torus, Spidergon, deBruijnGraph and mesh NoC topologies are selected in the experiment; the corresponding routing algorithms are OddEven, CrossFirst, Deflection and Full-adaptive. For details, see Table 6.7. In addition, eight

Table 6.7 Topology and routing algorithm combinations

NoC No.	Topology		Routing algorithm	
	Name	Category	Name	Category
1	Torus	Mesh/Torus	OddEven	Adaptive routing
2	Spidergon	Ring	CrossFirst	Deterministic routing
3	deBruijnGraph	Customized	Deflection	Deterministic routing
4	Mesh	Mesh/Torus	Full-adaptive	Adaptive routing

Table 6.8 Application information used to verify the flexibility when the reliability, communication energy consumption, and performance are jointly optimized

Application name	Number of IPs	Meaning of application
DVOPD	32	Dual-video object plane decoder
VOPD	16	Video object plane decoder
PIP	8	Picture in picture
MWD	12	Multi-window display
H264	14	H.264 decoder
HEVC	16	High-efficiency video codec
Freqmine	12	Data mining application
Swaption	15	Computer combination using Monte Carlo simulation

different applications are adopted to conduct the experiment; the details of these applications are shown in Table 6.8. The first four applications are common multimedia applications in reality; the latter four applications feature high traffic to verify that the reconfigurable NoC system can meet the high complexity requirement. H264 [33] and HEVC [34] are two complex and latest video coding standards, while Freqmine and Swaption are obtained by referring to princeton application repository for shared memory computers (PARSEC). The same numbers of IPs in the first four applications are selected as the Ref. [21]; the number of IPs in the latter four applications is obtained according to the rule of the balanced inter-IP traffic. In the experiment, the NoC size shall meet both the minimum size requirement of the application and the special requirement of the topology.

The weight comparison between the reliability and ELP is not required for verifying the flexibility of the CoREP model and the PRBB algorithm. Therefore, weight set to 0.5 ($\alpha = 0.5$) to indicate that the two indicators are equally important. As the entire NoC system cannot run properly in most cases when the fault probability of an interconnection line is greater than 0.5, the fault probability of an interconnection line shall be smaller than or equal to 0.5 ($p_l = 0.5, 0.1, 0.01, 0.001, 0.0001$, the corresponding $p_h = 0.5, 0.5, 0.1, 0.01, 0.001$). Due to lack of a mapping method that is applicable to different topologies and routing algorithms at the present and the failure to use the enumeration method that takes too long to search for the optimal solution, the proposed mapping method is compared with a classical simulated annealing (SA) algorithm in the experiment. As a probability algorithm, the SA algorithm is used to search for the local optimal solution of a target function [35]. However, it lacks the model for assessing the reliability, communication energy consumption, and performance. As the computation complexity for assessing different models is different, for a fair comparison, the SA algorithm is compared with the PRBB algorithm using the CoREP model proposed in this book.

In the experiment, the eight applications listed in Table 6.8 are mapped to the four different NoC combinations listed in Table 6.7 using the PRBB algorithm and the SA algorithm. The results show that all the eight applications can be successfully mapped to the NoC with the PRBB algorithm. This demonstrates that the mapping method is efficient under various topologies and routing algorithms. In addition, the average results of the optimal mapping mode in the case of eight applications and four NoC combinations can be shown by Fig. 6.18. Figure 6.18a shows the increment of average reliability. When p_l is very small, any mapping mode is highly reliable; thus, it is quite difficult to improve the reliability in this case. However, when p_l increases, the improvement of reliability also increases; it reflects that the PRBB algorithm can find a mapping mode with a better reliability than SA algorithm. Figure 6.18b shows the average runtime ratio of different fault probabilities (SA/PRBB). As can be seen from the figure, the runtime of the SA algorithm is at least 500 times that of the PRBB algorithm. This advantage is due to the two technologies adopted by the PRBB algorithm for reducing computation. In addition, when the fault probability of an interconnection line decreases, the runtime ratio increases. This is because the smaller the fault probability of an interconnection line, the less the number of faulty interconnection lines; thus, the PRBB algorithm takes less time to find the optimal

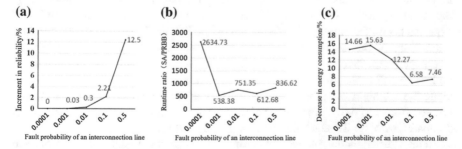

Fig. 6.18 **a** Increment in average reliability, **b** Average runtime ratio (SA/PRBB), **c** Change of reduction in average communication energy consumption with the interconnection line fault probability. © [2018] IEEE. Reprinted, with permission, from Ref. [30]

mapping mode. For the SA algorithm, it computes most mapping modes; thereby its computation time is not greatly reduced. Figure 6.18c describes the change of the reduction in average communication energy consumption with the fault probability. As can be seen from the figure, the optimal mapping mode found by the PRBB algorithm consumes less communication energy than that found by the SA algorithm in all cases. A greater fault probability would lead to more faulty interconnection lines. Therefore, when the fault probability of an interconnection line increases, the communication energy consumption increases due to the processing of more faulty interconnection lines. However, the reduction in communication energy consumption is not directly related to the fault probability because both algorithms use the same model to assess the communication energy consumption. As latency is directly related to throughput, a detailed comparison between the latency and the throughput are provided in the latter sections.

According to the experimental results, the multi-objective joint optimization model CoREP can optimize the reliability, communication energy consumption, and performance for assessing mapping modes specific to various NoC topologies and routing algorithms, thus ensuring the flexibility of the method. Meanwhile, compared with the SA algorithm, the PRBB algorithm takes less time to find the optimal mapping mode. This shows a great advantage of the PRBB algorithm.

The following experiment illustrates the advantage of the PRBB algorithm. In the experiment, the PRBB algorithm is compared with a latest algorithm of the same type in reliability, communication energy consumption, latency, throughput and computation time. It has been discussed in previous sections that Refs. [17, 22] also consider the co-optimization of two indicators; however, they are restricted to specific topologies and routing algorithms. In addition, the method proposed in Ref. [17] considers different cases from the method proposed in this book; thus, it is difficult to make a fair comparison. Thereby, this book compares the PRBB algorithm with the BB algorithm proposed in Ref. [22]. For a fair comparison, the two-dimensional mesh topology and XY routing algorithm used by the BB algorithm are selected. Table 6.9 shows the selected eight applications and the corresponding NoC sizes. The first four items are the same as those of the BB algorithm for a fair comparison; the reasons of selecting the latter four items are the same as those for the latter four items in Table 6.8. To compare the optimal mapping modes obtained using the PRBB algorithm with the BB algorithm, this book conducts a large number of experiments on the emulator implemented by the SoCDesigner. Table 6.10 provides a summary of the experimental results on the injection of 50,000 sets of faults. The experiment involves the maximum value, minimum value and average value under two different optimization weights. In addition, all experiments are conducted using the SA algorithm; the experimental results are shown in Table 6.10. As can be seen from Table 6.10, the PRBB algorithm has great advantages in every aspect compared with the BB algorithm and the SA algorithm.

In the CoREP model, the weight parameter α is used to weigh the reliability and ELP. Therefore, the value selection for α is quite important. In previous experiments, the selected value for α is the same as that in the BB algorithm for a fair comparison. The experiment in this book takes an application as an example to study the

Table 6.9 Features, data size and corresponding NoC size of the practical applications

Application	Number of IPs	Minimum/maximum communications	NoC size
MPEG4	9	1/942	9
Telecom	16	11/71	16
Ami25	25	1/4	25
Ami49	49	1/14	49
H264	14	3280/124,417,508	16
HEVC	16	697/1,087,166	16
Freqmine	12	12/6174	16
Swaption	15	145/747,726,417	16

Table 6.10 Comparison results between the PRBB algorithm and the BB algorithm, and the PRBB algorithm and the SA algorithms when the reliability, communication energy consumption, and performance are optimized simultaneously

Parameter	Compared with BB			Compared with SA		
	Maximum	Minimum	Average	Maximum	Minimum	Average
α = 0.2						
Increment in reliability (%)	106	0.01	10	208	−4	12
Reduction in communication energy consumption (%)	40	4	24	59	−13	28
Reduction in latency (%)	49	4	17	40	0.8	20
Throughput optimization (%)	22	5	9	22	4	9
Time (compared with PRBB)	20×	1×	3×	4477×	111×	1041×
α = 0.6						
Increment in reliability (%)	106	0.01	10	208	−4	12
Reduction in communication energy consumption (%)	40	4	24	59	−13	28
Reduction in latency (%)	49	4	17	40	0.8	20
Throughput optimization (%)	22	5	9	22	4	9
Time (compared with PRBB)	20×	1×	3×	4477×	111×	1041×

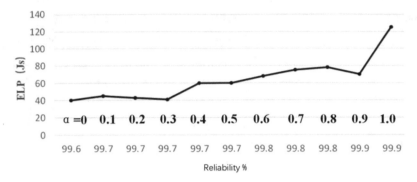

Fig. 6.19 The change of reliability and ELP with the change of α on the condition of $p_l = 0.01$ and $p_h = 0.1$. © [2018] IEEE. Reprinted, with permission, from Ref. [30]

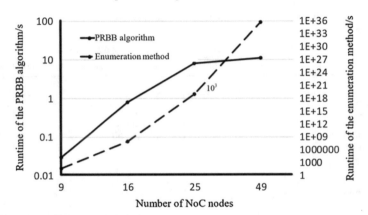

Fig. 6.20 Comparison between the computation time of the PRBB algorithm and that of the enumeration method as the NoC size changes. © [2018] IEEE. Reprinted, with permission, from Ref. [30]

changes of experimental results when α changes on the condition of $p_l = 0.01$ and $p_h = 0.1$. Figure 6.19 shows the experimental results. According to the experimental results, when α increases, the reliability of the optimal mapping mode found by the PRBB algorithm increases accordingly (or remains unchanged). In the meantime, as a sacrifice, the ELP also increases. To meet the requirements of different application scenarios, α shall be properly selected so as to effectively trade off between the reliability and the ELP.

It has been analyzed in previous sections that the computation load of the PRBB algorithm is highly dependent on the NoC size, i.e., the PRBB computation load is highly correlated with the number of nodes in the NoC. Therefore, this book also studies the change of the PRBB computation load with the change of the number of NoC nodes. Figure 6.20 shows the research results. For comparison, Fig. 6.20 also shows the results of using the enumeration method to search for the optimal mapping mode. The computation load is still expressed by the computation time.

Like the theoretical analysis results, the computation time increases as the NoC size increases. However, the increase rate of the computation time for the PRBB algorithm is far smaller than that for the enumeration algorithm. This is because the PRBB algorithm adopts the two technologies as has been mentioned in previous sections.

As mentioned in the previous sections, the method proposed in this book is not only applicable to co-optimization of the three indicators but also can implement joint optimization for any two of the three indicators. The following section verifies the optimization results for any two of the three indicators, reliability, communication energy consumption, and performance. The CoREP model is designed to assess the reliability, communication energy consumption, and performance at the same time. Of course, CoREP can assess only two of the three indicators by certain means. As the work for the co-optimization of reliability and performance is not much, this book does not provide experiments on this aspect. This book provides the experiments on the co-optimization of reliability and communication energy consumption and the co-optimization of communication energy consumption and performance to illustrate that the CoREP model is also applicable to co-optimization of two indicators.

To use the CoREP algorithm to perform co-optimization of reliability and communication energy consumption, you only need to remove the latency from Formulas (6.19) and (6.20). Then, the PRBB algorithm is used to search for the optimal mapping mode. For the co-optimization in reliability and communication energy consumption, the experiments regarding these two indicators are used to verify the effectiveness of the PRBB algorithm.

Firstly, the flexibility of the method is verified. Similarly, multiple applications are mapped to the four NoC combinations described in Table 6.7. In the experiment in this section, the 14 different applications provided in Table 6.11 are emulated and tested. The first eight applications are extracted from practical multimedia applications, while the latter four applications are selected for the same reason as the latter four applications in Table 6.8. The last two applications are random applications computed by TGFF [36]. The other experimental condition settings are the same as those in the previous experiments.

Figure 6.21a shows that the increment in the reliability of the mapping mode decreases as the fault probability decreases. This is because the reliability of any mapping mode is high when the fault probability decreases. As can be seen from Fig. 6.21b, the reduction in energy consumption decreases as the fault probability of an interconnection line decreases. As for the runtime, when the fault probability decreases, the PRBB algorithm takes less time to process faulty interconnection lines; however, the runtime of SA has nothing to do with the fault probability of interconnection lines. Therefore, the runtime ratio (SA/PRBB) increases as the fault probability decreases, as shown in Fig. 6.21c. As can be seen from the experiments in this section, the CoREP model and the PRBB algorithm have very a high flexibility in co-optimization of reliability and communication energy consumption; they are applicable to most different topologies and routing algorithms. Although the PRBB algorithm cannot always find a more optimal mapping mode than the SA algorithm,

Table 6.11 Actual information used in the flexibility verification

Application name	Number of IPs	Meaning of application
DVOPD	32	Dual-video object plane decoder
VOPD	16	Video object plane decoder
MPEG4	9	MPEG4 decoder
PIP	8	Picture in picture
MWD	12	Multi-window display
mp3enc mp3dec	13	Mp3 encoder and mp3 decoder
263enc mp3dec	12	H.263 encoder and mp3 decoder
263dec mp3dec	14	H.263 decoder and mp3 encoder
H264	14	H.263 decoder
HEVC	16	High efficiency video codec
Freqmine	12	Data mining application
Swaption	15	Computer combination using Monte Carlo simulation
random1	16	Generated by TGFF
random2	16	Generated by TGFF

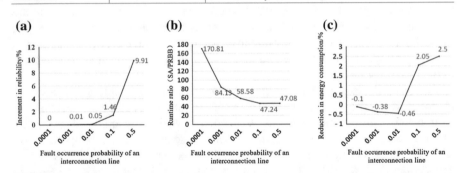

Fig. 6.21 **a** Average increment in reliability, **b** Average runtime ratio (SA/PRBB), **c** Changes of the average reduction in communication energy consumption as the fault probability of an interconnection line changes. © [2018] IEEE. Reprinted, with permission, from Ref. [26]

the PRBB algorithm has a significant advantage in runtime, thus better meeting the requirements of reconfigurable NoCs.

The previous experiments have illustrated that the PRBB method also has a high flexibility in co-optimization of reliability and communication energy consumption. The following section conducts an experimental verification over the accuracy of the optimal mapping mode obtained on the basis of the PRBB algorithm. According to the description in previous sections, researches on co-optimization of reliability and communication energy consumption are also provided in Refs. [17, 22, 29]; however, these two researches are not flexible and are only applicable to two-dimensional mesh topologies and deterministic routing algorithms. Reference [21] provides the

Table 6.12 Average communication energy consumption, performance and runtime cost of the PRBB algorithm, Refs. [8, 16] compared with those of the enumeration method

	PRBB	Reference [8]	Reference [16]
Runtime acceleration (%)	99.34	99.12	93.46
Cost of communication energy consumption (%)	4.12	23.70	13.18
Performance cost (%)	7.08	24.37	13.18

experimental results on the cost of finding the global optimal mapping mode using the enumeration method in Refs. [17, 21]. The experiments in Ref. [21] use the following applications: MPEG4, VOPD, 263 decoder and 263 encoder and mp3 encoder. The number of IPs required by these applications is smaller than or equal to 9. Therefore, these applications can be mapped to a 3×3 NoC. In this way, the time consumed by the enumeration method to find the optimal mapping mode is acceptable. Therefore, similar experiments to the ones in Ref. [25] are done in the research of this book and the obtained experimental results are compared with those of the reference. Table 6.12 shows the comparison results. As can be seen from Table 6.12, compared with the enumeration method, the PRBB algorithm features the maximum speed improvement, the minimum performance cost and the minimum increment in communication energy consumption. In other words, the optimal mapping mode found by the PRBB algorithm is closest to the global optimal mapping mode.

To compare the PRBB algorithm with the BB algorithm in Ref. [22], this book conducts comprehensive experiments with the same experimental conditions as those used in previous sections for verifying the reliability and energy consumption. The experimental results in Table 6.13 show that the PRBB algorithm superior to the BB algorithm in every aspect. The specific data shows that the reliability of 70% of optimal mapping modes found by the PRBB algorithm is 43% higher than that of the optimal mapping modes found by the SA algorithm. The communication energy consumption of only about 45% of optimal mapping modes found by the PRBB algorithm is inferior to that of the optimal mapping modes found by the SA algorithm. Meanwhile, the runtime of SA is much longer than that of the PRBB (about 1668 times of the runtime of the PRBB). To sum up, the PRBB algorithm achieves a better trade-off between reliability and communication energy consumption compared with the BB algorithm and the SA algorithm.

In the first scheme, the experiments are analyzed mainly in two aspects. For the first aspect, the experiments are conducted for the co-optimization of reliability, communication energy consumption, and performance. The experiments show that that the CoREP model and the PRBB algorithm proposed in this book have a very high flexibility and are applicable to the current frequently-used topologies and routing algorithms. Meanwhile, the comparison experiments with the BB algorithm and the SA algorithm also show that the CoREP model and the PRBB algorithm have

Table 6.13 Comparison results between the PRBB algorithm and the BB algorithm, and between the PRBB algorithm and the SA algorithm when the reliability, communication energy consumption, and performance are optimized simultaneously

Parameter	Compared with BB			Compared with SA		
	Maximum	Minimum	Average	Maximum	Minimum	Average
$\alpha = 0.2$						
Increment in reliability (%)	113.3	0	8.7	43.7	−4.4	2.9
Reduction in communication energy consumption (%)	46.3	1.8	23.6	32.6	−8.9	5.82
Time (compared with PRBB)	11.8×	1×	3.5×	1168×	17.2×	326×
$\alpha = 0.6$						
Increment in reliability (%)	32.1	−0.1	3.9	43.7	4.1	2.1
Reduction in communication energy consumption (%)	55.5	0	19.7	32.6	−17.4	3.3
Time (compared with PRBB)	11.8×	1×	2.8×	1656×	17.7×	293×

higher advantages in aspects such as reliability, communication energy consumption, latency, throughput, and runtime. For the second aspect, the experiments are conducted for the co-optimization of any two indicators of reliability, communication energy consumption, and performance. Likewise, the experiments show that the CoREP model and the PRBB algorithm have a very high flexibility and higher advantages compared with the BB algorithm and the SA algorithm. In addition, the experiments indicate that the CoREP model and the PRBB algorithm have a wide range of application scenarios; they are not only applicable to the co-optimization of three indicators but also applicable to the co-optimization of two indicators.

The following section introduces the second mapping scheme proposed in this book for implementing multi-objective joint optimization [37]. In scheme 2, the reliability efficiency model (REM) is proposed on the basis of the transformations of the preceding quantitative models for reliability, energy consumption and latency, and the qualitative model for throughput. Using the REM, this book proposes the priority and compensation factor oriented branch and bound (PCBB) mapping method. The PCBB mapping method assigns priorities to tasks to improve the running speed and uses the compensation factor to trade off between the accuracy and the computing speed. After the system is reconfigured, the PCBB mapping method supports dynamic remapping to find the optimal mapping mode.

During the process of mapping an application to the NoC and searching for the multi-objective joint optimal mapping mode, how to trade-off among multiple objectives is very important. In fact, the reliability is usually improved at the cost of the communication energy consumption and performance. In the second scheme proposed in this book, the concept of reliability efficiency is used to trade off among the reliability, communication energy consumption, and performance. The reliability efficiency is defined by the reliability gain of unit ELP (RC/ELP). The requirements for the reliability, communication energy consumption, and performance vary according to application scenarios. Therefore, this research uses weight parameter γ to distinguish their importance. The reliability efficiency defined in this book can be expressed by Formula (6.26).

$$RC_{eff}^{S,D} = \frac{\left(RC^{S,D} - minre\right)^{\gamma}}{1 + E^{S,D} \times L^{S,D}} \tag{6.26}$$

where minre indicates the minimum reliability requirements of the system for a communication path; γ indicates the weight parameter; these two parameters can be specified by users according to application requirements. The overall reliability efficiency for a specific mapping mode can be expressed by Formula (6.27).

$$RC_{eff} = \sum_{S,D} RC_{eff}^{S,D} F^{S,D} \tag{6.27}$$

where $F^{S,D}$ has been defined by Formula (6.2). The following section introduces the computation of the reliability, communication energy consumption and latency in Formula (6.26).

The reliability of a source-destination pair can be computed using Formula (6.28).

$$RC^{S,D} = \sum_{n=0}^{N} \sum_{i=1}^{M} RC_{i,n}^{S,D} P_{I,n} \tag{6.28}$$

where $RC_{i,n}^{S,D}$ has been defined in the previous sections; it indicates the reliability cost from the source S to the destination D in fault pattern i when n interconnection lines are faulty; its value can be obtained according to the preceding definition about reliability. $P_{I,n}$ has been defined in Formula (6.4). Similar to Formula (6.5), the communication energy consumption of a source-destination pair can be expressed by Formula (6.29).

$$E_{i,n}^{S,D} = V^{S,D}\left[E_{\text{Lbit}}d_{i,n}^{S,D} + E_{\text{Rbit}}\left(d_{i,n}^{S,D} + 1\right)\right] \tag{6.29}$$

All parameters in Formula (6.29) have been defined in the previous sections. Considering all fault patterns, the communication energy consumption of a source-destination node pair can be obtained by Formula (6.30).

$$E^{S,D} = \sum_{n=0}^{N}\sum_{i=1}^{M} E_{i,n}^{S,D} P_{I,n} \tag{6.30}$$

When it comes to performance, the latency and reliability must be considered. In the second scheme proposed in this book, the latency is still modeled quantitatively, while the throughput is analyzed qualitatively using bandwidth limitation, as shown in Formula (6.15). Like the latency modeling in the previous sections, the computation of the latency in scheme 2 is also divided into three parts. As the first part, the pure transmission latency can be computed by Formula (6.31).

$$LC_{i,n}^{S,D} = t_{\text{w}}d_{i,n}^{S,D} + t_{\text{r}}\left(d_{i,n}^{S,D} + 1\right) \tag{6.31}$$

The waiting latency caused by faulty interconnection lines and congestion can be computed by Formulas (6.8)–(6.12). When n interconnection lines are faulty, the latency for the faulty pattern i can be expressed by Formula (6.32).

$$L_{i,n}^{S,D} = LC_{i,n}^{S,D} + \sum_{K=1}^{d_{i,n}^{S,D}+1} WT_{U(K)\to V(K)}^{R(K)} + \sum_{j=1}^{d_{i,n}^{S,D}} LF_{L(j)} \tag{6.32}$$

Taking into account all fault patterns, the latency of a source-destination pair can be expressed by Formula (6.33).

$$L^{S,D} = \sum_{n=0}^{N}\sum_{i=1}^{M} L_{i,n}^{S,D} P_{I,n} \tag{6.33}$$

The mapping method used in scheme 2 is still a mapping algorithm based on BB. The proposed PCBB method is divided into two parts: (1) mapping process. Search for the joint optimal mapping mode for reliability, communication energy consumption, and performance when an application is mapped to the NoC for the first time. (2) Remapping process. When emergencies such as hard faults and topology change request of the application occur in the reconfigurable NoC system, dynamic remapping is conducted to ensure that the application can execute better on the current NoC. During the mapping process, the PCBB method also uses two technologies to improve the computing efficiency and accuracy. The first technology is the priority-based assignment technology; it first sorts the IPs to be mapped in descending order according to the total traffic and then assign the highest priority to the IP with the highest traffic. During the mapping process, the IP with the highest priority is first mapped to ensure that the intermediate nodes closer to the root node can be mapped to the IPs with higher traffic. The intermediate nodes that are impossible to be the optimal mapping mode are deleted as early as possible to improve the computing efficiency. In the BB algorithm, if the upper bound for the gain of an intermediate node is smaller than the maximum gain so far, the node is impossible to be the optimal mapping mode; it needs to be deleted. Of course, an accurate estimation of the upper bound for the gain of each intermediate node would enable the found mapping algorithm to be more accurate, but it also incurs more computation load. Therefore, the two factors should be traded off effectively. The PCBB algorithm introduces the compensation factor β in the deletion conditions to trade off between the computation load and the computation accuracy, as shown in Formula (6.34).

$$\text{UB} < \max\{R_{\text{eff}}\}/(1 + \beta) \tag{6.34}$$

where UB indicates the upper bound for the reliability gain of an intermediate node. The computation of UB consists of three parts: (1) the upper bound for the reliability gain between mapped IPs, $\text{UB}_{m,m}$; (2) the upper bound for the reliability gain between a mapped IP and a non-mapped IP, $\text{UB}_{m,u}$; (3) the upper bound for the reliability gain between non-mapped IPs, $\text{UB}_{u,u}$. If the condition in Formula (6.34) is met, the intermediate node is deleted; otherwise, the intermediate node is retained for further comparison.

During the execution of an application on the NoC, the NoC is required to be dynamically reconfigured for some special cases such as hard faults in interconnection lines, routers and Pes, and the topology change request of the application. To run the application better on the NoC, remapping is also required to search for the current optimal mapping mode. As the NoC reconfiguration is implemented in real time, the remapping is also required to be implemented in real time. Taking this into account, this book stores the mapping algorithm running before the reconfiguration into the memory and defines the mapping algorithm as the current optimal mapping mode. The current optimal reliability gain is obtained by the current optimal mapping mode. Then, the PCBB mapping algorithm proposed in the previous sections is used to update the mapping mode. The comparison in the search tree starts from the last optimal mapping mode; this can significantly reduce the computation load to ensure that the remapping can be implemented in real time. The overall mapping method involves the mapping pseudocode and the remapping pseudocode, as shown in the Algorithm 6.1

Algorithm 6.1 Mapping pseudocode

```
if(need reconfiguration){

    Read(LastMap);

    MaxGain = LastMap->gain;

    MaxUpperBound = LastMap->upperbound;

}

else{

    MaxGain = -1;

    MaxUpperBound = -1;

}

initialization;

while(Q is not empty){

    Establish(child);

    if(child->gain < Maxgain or child->upperbound < MaxUpperBound){

        delete child;

    }

    else{

        insert child in Q;

        if(all IPs are mapped)

            BestMap = child;


            if(child->gain > MaxGain)

                MaxGain = child->gain;

            if(child->upperbound > MaxUpperBound)

                MaxUpperBound = child->upperbound;

        }

    }
```

Use the method for analyzing the computation complexity of the PRBB algorithm to analyze the computation complexity of the PCBB algorithm; the analysis results show that the computation complexity of the PCBB algorithm is the same as that of the PRBB algorithm. Table 6.14 shows the analysis results.

The above is the second solution proposed in this book. The solution introduces the concept of reliability efficiency and describes the computation process for the reliability efficiency. The multi-objective joint optimization model of reliability efficiency is used to introduce the mapping and the remapping of the PCBB algorithm and interpret the process of mapping an application to the NoC.

The research in this book uses three groups of comparative experiments to illustrate that the proposed REM and PCBB mapping method have the advantages of high efficiency, reconfigurability, and accuracy. First, the PCBB algorithm and the SA algorithm are, respectively, used to map an application to the NoC with the two-dimensional mesh topology and the XY routing algorithm. The PCBB method is then compared with the SA method to verify its high efficiency. Then, on the condition of the reconfiguration of the NoC, the application is dynamically remapped to illustrate the reconfigurability of the PCBB method. Finally, the optimal mapping mode found by the PCBB algorithm is compared with those found by the BB algorithm and the SA algorithm to illustrate the accuracy of the PCBB method proposed in this book. The background used by the three groups of experiments is the same as that used by the experiments for verifying the PRBB algorithm.

First, the high efficiency of the PCBB algorithm is verified. The research in this book maps the applications introduced in Table 6.9 to the NoC with the two-dimensional mesh topology and the XY routing algorithm using the PCBB algorithm and the SA algorithm, respectively, and compares the PCBB runtime with the SA runtime. In the experiments, $p_l = 0.5, 0.1, 0.01, 0.001, 0.0001$, and the corresponding $p_h = 0.5, 0.5, 0.1, 0.01, 0.001$. The compensation factor β is set to 0 to ensure that the PCBB algorithm can find the optimal mapping mode as soon as possible. According to the comparison results shown in Table 6.14, in any case, the PCBB algorithm uses less time than the SA algorithm to find the optimal mapping mode. To be specific, the runtime of the SA algorithm is 189–1871 times that of the PCBB

Table 6.14 Runtime comparison between the PCBB algorithm and the SA algorithm

Application	PCBB/s	SA/s	Ratio (SA/PCBB)
MPEG4	0.02	37.42	1871
Telecom	0.65	156.66	241
Ami25	7.38	1397.29	189
Ami49	18.14	13,064.38	720
H264	0.50	486.78	973
HEVC	0.64	340.27	531
Freqmine	0.87	335.50	385
Swaption	0.45	385.46	856

Table 6.15 Remapping time comparison after topologies and routing algorithms are changed

Application	Torus/s	Spidergon/s	deBruijnGraph/s	Mesh/s
MPEG4	0.02	0.01	0.01	0.03
Telecom	0.57	0.56	0.57	0.56
Ami25	3.67	3.7	3.81	3.66
Ami49	8.06	6.26	6.96	6.81
H264	0.56	0.56	0.57	0.57
HEVC	0.56	0.57	0.58	0.56
Freqmine	0.57	0.57	0.57	0.57
Swaption	0.57	0.56	0.57	0.57

runtime. This advantage mainly benefits from the two technologies for reducing the computation load in the PCBB algorithm.

After the NoC is reconfigured, the mapping mode also needs to be updated to enable the applications to execute better on the NoC. The research in this book uses a large number of experiments to verify the reconfigurability of the proposed method. The practical applications are first mapped to the NoC with the two-dimensional mesh topology and the XY routing algorithm. Then, the topologies and routing algorithms are changed to the combinations shown in Table 6.7 to adapt the application requirements. Therefore, the eight applications in Table 6.9 need to be remapped to the four NoC combinations in Table 6.7, and the remapping time is recorded. According to the experimental results shown in Table 6.15, the reconfiguration time increases as the increase of the NoC size. This is consistent with the theoretical analysis results.

The hard faults in interconnection lines, routers, and PEs can be rectified by introducing redundancies. However, this book does not consider this reconfiguration process and only discusses the time required by remapping after the NoC is reconfigured. Like discussed in previous, router faults are categorized as interconnection line faults. Table 6.16 shows the time required for remapping after the NoC is reconfigured due to interconnection line faults. As can be seen from Table 6.16, the PCBB algorithm can quickly perform remapping and find the optimal mapping mode after NoC reconfiguration.

As the latest Ref. [38] studies the issue of remapping after a faulty PE is replaced by a redundant element, it is selected as the contrast object. To be fair, the applications used for remapping time comparison are the same as those used in Ref. [38]. Table 6.17 shows the comparison results. As can be seen from Table 6.17, the PCBB algorithm can perform dynamic remapping. Although PCBB takes a slightly longer time than LICF for remapping, the PCBB algorithm is much more efficient than the MIQP. As the search space of the PCBB algorithm is much larger than those of LICF and MIQP, the results in Table 6.17 are sufficient to state that the PCBB algorithm can implement dynamic remapping.

The preceding experiments show that the REM and the PCBB mapping method feature high efficiency and reconfigurability; the following experiments verify the

Table 6.16 Remapping time comparison due to faulty interconnection lines (uint: ms)

Application	Number of faulty interconnection lines			
	3	6	9	12
MPEG4	10	12	13	12
Telecom	25	24	23	25
Ami25	50	54	53	54
Ami49	83	83	81	80
H264	24	23	25	23
HEVC	26	27	24	27
Freqmine	24	25	25	24
Swaption	25	24	22	22

Table 6.17 Remapping time comparison due to faulty PEs (unit: s)

Application	NoC size	Number of faulty PEs	LICF/s	MIQP [38] /s	PCBB [38] /s
Auto-Indust (9IPs)	4 × 4	2	0.01	0.2	0.03
		4	0.02	2.51	0.04
		6	0.04	51.62	0.06
		7	0.04	177.72	0.08
TGFF-1 (12IPs)		2	0.01	0.44	0.02
		3	0.02	1.34	0.05
		4	0.03	4.3	0.06

accuracy of the PCBB mapping method. With the same experiment conditions as used in preceding experiments, the optimal mapping modes found by the PCBB algorithm are compared with those found by the BB algorithm and the SA algorithm. Table 6.18 shows the comparison results. As can be seen from Table 6.18, on average, the optimal mapping mode found by the PCBB algorithm is better than those found by the BB algorithm and the SA algorithm. From the minimum point of view, the optimal mapping mode found by the PCBB algorithm is slightly superior to that found by the BB algorithm but is inferior to that found by the SA algorithm. However, compared with the SA algorithm, the PCBB algorithm can find the optimal mapping mode more quickly. In other words, the SA algorithm obtains small advantages in reliability, communication energy consumption, latency, and throughput at the cost of plenty of runtime, which is hardly accepted to the reconfigurable system with a high requirement for the runtime. Although the PCBB algorithm sacrifices a small amount of reliability, communication energy consumption, latency and throughput, it obtains a significant advantage in runtime. Above all, from the average point of view, the PCBB algorithm has great advantages in reliability, communication energy consumption, latency and throughput compared with the SA algorithm. Compared with the SA algorithm, the PCBB algorithm achieves about 5.3% increment in reli-

Table 6.18 Comparison of the optimal mapping modes found by the PCBB, BB and SA algorithms

Parameter	Compared with BB			Compared with SA		
	Maximum	Minimum	Average	Maximum	Minimum	Average
Increment in reliability (%)	106.8	−0.96	13	111.4	−1.95	5.3
Reduction in communication energy consumption (%)	46.5	−1.1	22.4	39.4	−22.3	7.9
Reduction in latency (%)	37.1	2.4	15.5	25.3	−3.5	8.9
Throughput optimization (%)	22.2	0.7	9.3	22.2	3.5	8.5

ability, 7.9% reduction in communication energy consumption, 8.9% reduction in latency and 8.5% improvement in maximum throughput on average.

In the experiments for verifying the PRBB and PCBB schemes, all applications are mapped to NoCs with the same router structure. The research in this book implements a simple router structure using the Verilog language and performs the hardware cost verification for the router structure. The main purpose of this book is to design an accurate multi-objective joint optimization model and an efficient mapping algorithm rather than the hardware implementation. Therefore, only a simple router is designed. The resistor–transistor logic (RTL) design is implemented by the Verilog language and its functional verification is performed on the development board of the Altera DE2-115 field programmable gate array (FPGA). Then, the layout area obtained by reverse extraction and post-layout simulation is used to indicate the hardware cost. Generally, a router consists of the I/O unit, crossbar switch module, virtual channel module and route allocation and arbitration module. Table 6.19 lists the estimated required number of logic gates to implement such a simple router. In addition, the post-layout simulation analysis is conducted on the RTL router structure that is implemented with the Verilog language using the 65 nm technology of Taiwan Semiconductor Manufacturing Company (TSMC). Figure 6.22 shows the automatically extracted layout. As can be seen from Table 6.19 and Fig. 6.22, about 207,000 gates are required to implement such a router; the hardware area is about 0.2 mm^2. In the 80-core NoC structure [39] proposed by Intel, the area of a single router structure is about 0.34 mm^2. Although the router structure for the experiment in this book is simpler than the one proposed by Intel, the area of the single router structure is 0.2 mm^2. The hardware cost of the proposed design is not high, which is totally acceptable.

This book uses three groups of experiments to illustrate the advantages of the REM model and the PCBB algorithm. First, the PCBB algorithm and the SA algorithm are used, respectively, to map an application to the NoC with the two-dimensional mesh topology and the XY routing algorithm. The comparison results show that the

Table 6.19 Estimation on the number of logic gates in a simple router structure

Logic unit	Size/10 thousand gates
I/O unit	4.6
Crossbar switch module	0.9
Virtual channel module	14.5
Route allocation and arbitration module	0.7
Total	20.7

Fig. 6.22 Layout of a single router structure. © [2018] IEEE. Reprinted, with permission, from Ref. [37]

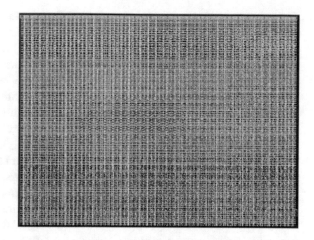

PCBB algorithm has a high efficiency. Then, the experiments of remapping based on dynamic reconfiguration of the NoC are used to illustrate the reconfigurability of the REM and the PCBB algorithm. Finally, the experiments that compare the PCBB algorithm with the BB and the SA algorithms in many aspects are used to illustrate the high accuracy of the PCBB algorithm in searching for the optimal mapping mode. In addition, the hardware cost analysis is performed on a router in the simulated NoC structure in the research of this book. The implementation and verification of the RTL for a simple router illustrates that the hardware cost of the router is within the acceptable range.

The experimental results show that the CoREP and the REM models proposed in this book can implement the quantitative modeling for the reliability, communication energy consumption and latency and qualitative analysis on the throughput. Compared with the researches of the same type, the models proposed in this book can obtain multi-objective joint optimization, which is the first innovation. Secondly, the CoREP-based PRBB algorithm and the REM-based PCBB algorithm proposed in this book have very high flexibility. Moreover, the PCBB mapping method is capable of performing remapping based on the NoC reconfiguration, which results in a high flexibility and reconfigurability. Most of current researches of the same type are limited to specific topologies and routing algorithms and few of them can implement reconfigurability, which is the second bright spot of the method proposed in this book. Furthermore, all the mapping methods proposed in this book can find the optimal mapping mode in a very short time. The efficient mapping modes are vital to reconfigurable NoCs, thus high efficiency is the third innovation of the proposed algorithms.

6.3   Configuration Path

The research on the configuration method of the reconfigurable massive MIMO signal detection processor mainly involves the research on the organization mode of configuration information, the configuration mechanism and the design method for the configuration hardware circuit. The research on the organization mode of configuration information mainly involves the definition of configuration bits, the structure organization, and the compression of the configuration information [40–42]. As the massive MIMO signal detection algorithm has a high computation complexity and involves some operations requiring more configuration information (e.g., large lookup tables), the required configuration information is usually massive. As a result, the organization and compression of the configuration information become the key for the massive MIMO signal detection algorithm to execute efficiently on the reconfigurable processor. The research on the configuration mechanism mainly solves the problem of how to schedule the configuration information corresponding to the computing resources. Generally, the massive MIMO signal detection algorithm needs to frequently switch among multiple subgraphs; therefore, a corresponding configuration mechanism needs to be established to minimize the impact of configuration switch on execution performance. Eventually, both the organization mode of the configuration information and the configuration mechanism must be supported by the

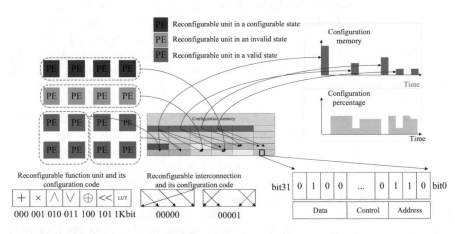

Fig. 6.23 Research on the configuration information organization mode and research on the configuration mechanism

configuration hardware circuit. Design of the configuration hardware circuit mainly involves designs of the configuration memory, the configuration interface and the configuration control. Figure 6.23 provides a brief description of the research on the organization mode of the configuration information and the configuration mechanism.

6.3.1 Control Design

The workflow of the reconfigurable PEA (collaborated with ARM) is as follows.

(1) The master control ARM writes the PEA configuration word (the medium for the ARM to control the PEA) and the data (a maximum of 15) from the master control ARM required by some PEA computation to the global register file of the master control interface via the ARM7 coprocessor instruction or AHB communication protocol. The size of the PEA configuration word is one word, including the address (20-bit) and the length (12-bit) of the PEA configuration information; the PEA configuration word is placed at the global register file address 0. The other data (32-bit) is placed at the global register file address 1–15. Once receiving the write configuration word of the ARM7 host, the master control interface enables the configuration controller via the configuration word enable signal (the signal is valid at high level for a single cycle) and delivers the configuration word to the configuration controller. Then, the ARM7 writes all intermediate data to the global register file and enables the PEA controller using the task enable signal (the signal is also valid at high level for a single cycle).

(2) The configuration controller carries the configuration package (a set of configuration information for PEA) from the primary memory and parses the index of the PE corresponding to each line in the configuration package. The configuration package includes the sequential control of the entire PEA, the sequential control of each PE, the functions of PEs in each machine cycle and the operations to be completed by the data controller in each machine cycle. The design principles for the configuration package are as follows. ① Various computation modes shall be taken into account; ② The redundant information shall be compressed as much as possible. Then, the configuration controller sends the configuration package and the corresponding PE index (signal line) of each line to the PEA to distribute the configuration information to each PE. After the configuration information is delivered, the configuration controller uses the execution completion signal (the signal is valid at low level) to notify the PEA.

(3) After the configuration controller and master control enable signal are in place, the PEA controller enables the PEA and the data controller to execute the computation according to the configuration package. In the configuration package, the sequence is defined in units of machine cycle. PEs execute computations according to the sequential information and current machine cycle enable defined in the configuration package. After the completion of the configuration information in

a line, a PE sends a task completion signal to notify the PEA. After receiving the task completion signals of all PEs, the PEA enters the next machine cycle. After a configuration package is completely executed, a PE sends a configuration package completion signal to notify the PEA controller.

(4)   The data controller is in charge of the data interaction between the PEA and the shared memory. The data controller can automatically detect the broadcast behavior during memory access and broadcast the corresponding data to each PE.

6.3.2   Master Control Interface

6.3.2.1   Function Description

The master control interface is the slave interface of the AHB interface or the ARM7 coprocessor interface. The ARM7 can write configuration words to the master control interface. Also, the master control interface can exchange data with the ARM7 via a register file. The register file of the master control interface is the medium for direct data exchange between the PEA and the ARM7. According to the read and write addresses of the global register file, the master control interface can be divided into the following three subfunctions (see Fig. 6.24).

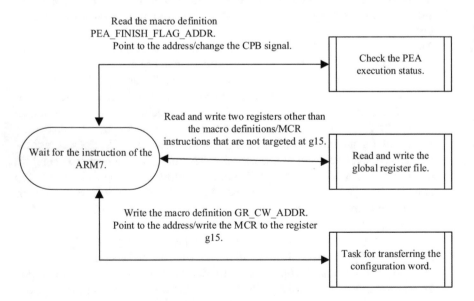

Fig. 6.24   Workflow chart of the master control interface

(1) Cache the configuration words written by the ARM7. Forward the configuration words to the configuration controller to transfer the configuration information.
(2) Notify the ARM7 of the PEA execution status.
(3) Provide a global register file for a fast data exchange between the PEA and the ARM7 in operation.

The ARM core of the System-on-Chip (SoC) platform does not support coprocessor instructions and only supports the AHB protocol. However, the performance of the coprocessor instructions is better on the RTL platform. Therefore, the following two schemes are provided to implement the master control interface: the AHB protocol-based scheme and the coprocessor instruction-based scheme.

6.3.2.2 Behavior Description

(1) The ARM7 packages the header address and the length of the configuration package corresponding to the task that is to be assigned to the PEA to a configuration word and writes the configuration word to the 16th register (g15) of the PEA's global register file with the microcontroller (MC) coprocessor instruction or the AHB protocol. Once detecting that the configuration word is written into g15, the global register file immediately sends a configuration word valid signal to notify the configuration controller. The configuration controller shake hands with the master control interface via the control word in-place signal, reads the data of g15 via the control word data signal, and start carrying configuration information.
(2) The ARM7 continues to dynamically exchange data with the first 15 registers (g0–g14) of the global register file during the PEA computation using master control register (MCR)/memory request controller (MRC) instructions or the AHB protocol. Each PE can read and write the data in the global register file during execution. The data in the global register file can be used as the number of iterations in the configuration package or 32-bit input signals. The data in the global register file except g15 can be used as the 32-bit output signals of PEs.
(3) If the AHB protocol is used, the PEA enters the execution state once data is written to g15. Whereas, if the coprocessor instructions are used, the master control ARM7 needs to execute an additional CDP instruction (the opcode is 4'b1111) to enable the PEA to enter the execution state after the relevant global registers are written. After receiving the write g15 instruction (AHB protocol) or CDP instruction (coprocessor instruction protocol), the master control interface sends a task to enable signal to notify the PEA controller. Then, the master control interface hangs any operation of the ARM7 specific to the PEA until the PEA controller sets the task completion signal to low. Under the AHB protocol,

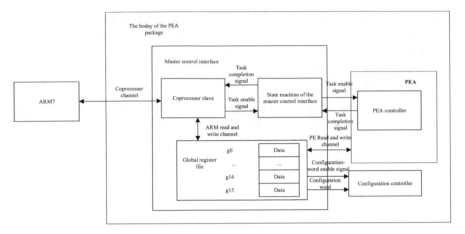

Fig. 6.25 The modular structure of the master control interface

the ARM7 reads the g14 register to implement a hang, i.e., **1** indicates a hang by the driver; **0** indicates the task is completed. Under the coprocessor instruction protocol, the PEA controller sets the control program assist (CPA) signal to low to notify the ARM7 that the task has not been completed. Figure 6.25 shows the modular structure of the master control interface.

6.3.3   Configuration Controller

The configuration controller module is in charge of the parsing, reading, and distribution of the configuration information. This section introduces the specific functions of the configuration controller according to the workflow chart of the configuration controller shown in Fig. 6.26.

(1) Parse the configuration information-related configuration word received by the master control interface.

(2) Read the configuration information from the last level cache (LLC) according to the size of the configuration information in the top-layer PEA configuration field in the configuration information.

(3) Judge whether the configuration information has been saved in the PE cache according to the first 32 bits of the read-in configuration information. If the configuration information has been saved in the cache, directly enable the PEA to execute the configuration information. Otherwise, continue to read-in configuration information.

(4) Distribute the read-in configuration information to each PE.

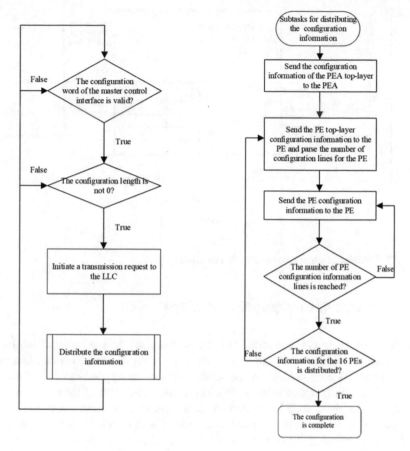

Fig. 6.26 Workflow chart of the configuration controller

As shown in Fig. 6.27, the principal part of the configuration controller module is the configuration controller submodule, which exchanges information with the PEA and the master control interface. Meanwhile, the configuration controller module connects to the AHB bus with an AHB master device to access the LLC for configuration information. The configuration controller mainly supports two functions: (1) Initiate an access request to the LLC according to the configuration word received by the master control interface; (2)Transfer the configuration information read from the LLC to the configuration arbiter of the PEA.

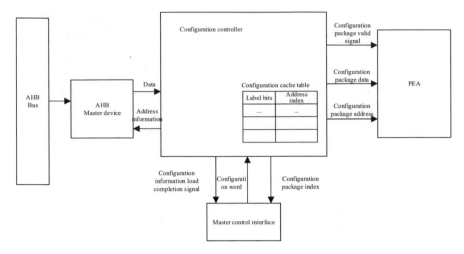

Fig. 6.27 Modular graph of the configuration controller

6.3.4   Design of Configuration Package

The PEA hardware configuration package is compiled on the basis of the Excel scheduling list. Each row in the scheduling list corresponds to each configuration line of each PE; each column in the scheduling list corresponds to each function section of the configuration information. The abstraction levels of the scheduling list are improved and the spatial dimension of the scheduling list is increased (the spatial dimension is increased from the single time dimension) for the software configuration package so as to enhance the programming readability and the convenience for each PE to execute sequential control.

6.3.4.1   Basic Signs

{ }   The content in braces indicates an independent morpheme. A morpheme can be a parameter, a command, the configuration information of a PE and the entire configuration package.

[]   The content in square brackets indicates an optional parameter. For simplicity in writing, this parameter can be omitted and directly use the default value.

%   The content behind % indicates comments. Comments can be added anywhere.

6.3.4.2   Top-Layer Commands

\PeaTop{number of iterations} {contents of the configuration package} is the configuration package command at the very beginning.

(1) Number of iterations: The number of iterations for the PEA top-layer, which indicates the number of additional executions for the set of configuration package; its value is a positive integer ranging from 0 to 63 or a global register (see "Data Source" in this section).
(2) Configuration package: The content is the configuration information of each PE. Currently, each PEA has 16 PEs; therefore, each configuration package includes the configuration information for a maximum of 16 PEs.
    \PeTop{number of iterations}{number of iteration start line}{configuration line} is the PE configuration information command at the very beginning.
(3) Number of iterations: The number of iterations for the PE top-layer, which indicates the number of additional PE executions for the set of configuration information; its value is a positive integer ranging from 0 to 63 or a global register.
(4) Number of iteration start line: The value N indicates that the repetitive execution starts from the Nth configuration line. The first configuration line is regarded as line 0. For details about how to compute the number of iteration start line, see "Computing the Configuration Line Number" in this Sect.
(5) Configuration line: It is the specific configuration information. The configuration information of a PE can include a maximum of 32 configuration lines; each configuration line may be executed for one or multiple machine cycles.

6.3.4.3 Basic Format of Configuration Lines

A configuration line can be in any of the following formats. The following introduces the meanings of various statements and how to write valid statements.

\Wait {number of machine cycles}
ALU instruction \Wait {number of machine cycles}
\If {condition} {ALU instruction}
\If {condition} {ALU instruction} \Wait {number of machine cycles}
\For {number of iterations} {ALU instruction}
\For {number of iterations} {ALU instruction\Break {condition}}
\For {number of iterations} {ALU instruction \Wait {number of machine cycles}}
\For {number of iterations} {ALU instruction \Wait {number of machine cycles} \Break {condition}}
\For {number of iterations} {\If {condition} {ALU instruction}}
\For {number of iterations} {\If {condition} {ALU instruction} \Wait {number of machine cycles}}
\While {ALU instruction}
\While {ALU instruction}
\While {ALU instruction\Break {condition}}

\While {ALU instruction\Wait {number of machine cycles}}
\While {ALU instruction\Wait {number of machine cycles} \Break {condition}}
\While {\If {condition} {ALU instruction}}
\While {\If {condition} {ALU instruction} \Wait {number of machine cycles}}

6.3.4.4   Data Source and Destination

The data source/destination is used as a command parameter. As the PEA is mix-grained, the sources/destinations of the coarse-grained data (32-bit) and the fine-grained data (1-bit) need to be specified, respectively. This type of commands are in three forms: (1) \target{address}; it indicates an address of the access target. (2) Integer immediate operand; it indicates that the source is an immediate operand. (3) \target[increment]{address}; the increment is a decimal integer ranging from -15 to $+15$. It indicates that the data source/destination address is target $+ n \times$ increment when the configuration is accessed during its nth execution if the configuration is executed for multiple times. All **increment**s are optional parameters, i.e., it can be omitted; its value is **0** by default. In LaTex, the coarse-grained data is shown in bold, and fine-grained data is shown in normal font. In practical application, as the fine-grained outputs are always accessed by other PEs in the next machine cycle, their destinations do not need to be specified separately. Tables 6.20, 6.21 and 6.22 show the coarse-grained input data, the fine-grained input data and the coarse-grained output data, respectively. In the tables, SM is short for the shared memory, and RF is short for the register file.

6.3.4.5   Instruction Field (ALU Command)

An instruction field command expresses the computation behavior of the ALU for the PE of a configuration line. The instruction field command is the principal part of a configuration line. A configuration line must have one and only one instruction field command.

Form: \instruction name {coarse-grained output} {input1} {input2} {input3}; Refer to the table for the names; input1 and input2 are coarse-grained input data; input3 is fine-grained input data; Out1 is coarse-grained output data; the fine-grained output Out2 does not require special execution. Refer to "Data Source and Destination."

Contents: The contents express the computation behavior of the ALU in the configuration line.

Computation of line number: The "number of iteration start line" of \Jump and PeTop for output destination involves the computation of the configuration line number. The following points need to be noted for the computation of the configuration line number:

Table 6.20 Coarse-grained input data

Form	Syntax	Meaning
\SM[increment]{target}	The **increment** is a decimal integer ranging from −15 to +15. The **target** can be written in two forms: (1) **BankNum-BankAddr**, where the **BankNum** is a decimal positive integer ranging from 0 to 3; it indicates the four banks of the shared memory; the **BankAddr** is a decimal positive integer ranging from 0 to 1023; it indicates the address in each bank. \SM{1-513} indicates the access of the data at address 513 in bank1 of the shared memory. (2) **0xAddr**, where the **Addr** is a hexadecimal positive integer ranging from 000 to FFF; there are 4096 addresses in total. \SM{0xBFF} indicates the access of the data at address 3071 in the shared memory (i.e., the address 1023 in bank 2, \SM[2-1023])	Directly access the shared memory to read the external input data
Immediate operand	Integer, decimal or hexadecimal starting with 0x. The length of an immediate operand is 32-bit; whether an immediate operand is a signed integer or an unsigned integer needs to be determined according to the type of the ALU instruction	Access a register in the PE. Read the local long-term data computed by the PE
\RF[increment]{target}	The **increment** is a decimal integer ranging from −15 to +15; the **target** is an unsigned integer ranging from 0 to 15	Access a register in the PE. Read the local long-term data computed by the PE
\PE{offset of x-axis x}{offset of y-axis y}	In a 4 × 4 array, the location of the current PE is the origin; the x-axis is positive in the right direction; the y-axis is positive in the upward direction. For example, \PE{1}{−1} indicates the PE with the coordinates of $(1, -1)$ relative to the current PE. If the index of the current PE is 0, \PE{1}{−1} refers to PE5. A PE can access all PEs with the target distance of smaller than or equal to 2, i.e., x + |y| ≤ 2	Access the coarse-grained output of a PE in the previous machine cycle. Read the short-term data in the array
\SMbyRF[increment]{target}	The **increment** is a decimal integer ranging from −15 to +15. The **target** is an unsigned integer ranging from 0 to 15. Note that the **increment** here indicates the increment of the value in the register rather than the index of the register; that is, if \SMbyRF[N]{M}, RF[M] is first accessed to obtain the base address; then, the address of accessing to the shared memory is RF[M] + i×N in the ith iteration cycle	Use the value of a register to indirectly access the shared memory to achieve dynamic memory access during operation

(continued)

Table 6.20 (continued)

Form	Syntax	Meaning
\GR{target}	The **target** is an unsigned integer ranging from 0 to 15	Data generated during the exchange operation between the global register file of the master control interface and the ARM7
\SMbyPE[increment] {offset of x-axis x} {offset of y-axis y}	The **increment** is a decimal integer ranging from −15 to +15. The coordinate range is the same as that for \PE. Note that **increment** here indicates the increment of PE output, i.e., the increment for the address of the shared memory to be accessed that is obtained by a PE	Use OUT1 of a PE in the previous machine cycle as the index to access the shared memory to achieve the dynamic memory access during the operation

Table 6.21 Fine-grained input data

Form	Syntax	Meaning
\PE{offset of x-axis x}{offset of y-axis y}	Similar to the coarse-grained \PE	Use the router to access the fine-grained output of the neighbor PE in the previous machine cycle to read the 1-bit computation data
Immediate operand	The value is 0 or 1	1-bit immediate operand 0 or 1. Note: For the operations (except for the AND operation) that do not involve fine-grained inputs, fill in 0 for all the fine-grained inputs, while fill 1 for the AND operation

Note The coarse-grained output data are roughly the same as the coarse-grained input data. The coarse-grained output data do not involve the immediate operand and the \PE; the "0" and the \Jump forms are added. For the items that are the same as those for the fine-grained input data, only entries are provided; the detailed interpretations are not provided again

Table 6.22 Coarse-grained output

Form	Syntax	Meaning
	0	The output of the PE in this machine cycle is not specially specified. Generally, the output is an intermediate data lives a short life and is only accessed by another PE in the next machine cycle
\Jump	\Jump	The computation result of the PE in this machine cycle is an offset. For example, if the number of the configuration line in the current machine cycle is M, the computation result is N, and the output destination is \Jump; then, the configuration line to be executed by the PE is no longer the usual $M + 1$ but $M + N$ (N can be negative) after the line of configuration information is executed. For the details about how to compute the offset of \Jump, see "Computing the Number of the Configuration Line"

(1) The line number of the configuration line with the first ALU command after p is 0.

(2) After that, the line number is increased by 1 each time an ALU command appears.

(3) If the \Wait command does not modify other statements, the line number is increased by 1. If the \Wait command modifies the instruction field or PeTop, it is not used as a line of separate configuration and does not occupy a line number. The \Wait command appears separately as a line of configuration in either of the following two cases: (1) If N successive \Wait commands appear after a condition field command, each of the $N - 1$ \Wait commands except the first \Wait command occupies a line of configuration information; the line number is increased by 1 for each of them. (2) If the previous configuration includes an iteration field, the \Wait command next to the braces { } of the iteration field occupies a line of configuration information; the line number is increased by 1.

6.3.5 Mapping Method

To fully take advantages of the reconfigurable architecture, the rational and smooth configuration of the massive MIMO signal detection algorithm on the reconfigurable architecture is crucial. The reconfigurable architecture is a novel computing architecture different from the traditional Von Neumann architecture. Besides the traditional instruction flow and data flow, the configuration flow is introduced, which makes the mapping of the massive MIMO signal detection application onto the reconfigurable platform more complex, as mentioned in Refs. [43, 44]. As shown in Fig. 6.28, the key links of mapping include generating the data flow diagram of the massive MIMO signal detection algorithm, dividing the data flow diagram into different subdiagrams, mapping subdiagrams to the reconfigurable massive MIMO signal detection PEA and generating the corresponding configuration information. The process of generating a data flow diagram mainly involves the expansion of the core loop, the scalar replacement and the distribution of intermediate data. In the process of partitioning the data flow diagram, the complete data flow diagram is divided into multiple subdiagrams with data dependencies in the time domain based mainly on the computing resources of the reconfigurable PEA. The process of mapping subdiagrams to the reconfigurable massive MIMO signal detection PEA mainly involves the mapping of the subdiagrams with specific PEs and interconnections in the PEA hardware and generates valid configuration information eventually.

The following section takes a relatively complex matrix traversal for example to show the collaborative operation between the master control ARM core and the PEA array. As a type of matrix decomposition, the LDL decomposition has important applications in the MMSE detection algorithm for massive MIMO signal detection. The traversal of a lower triangular matrix involved in the LDL decomposition is very interesting. This section gives a detailed introduction on how to map the traversal access process onto the PEA. By using the spatial mapping mode, all the computations

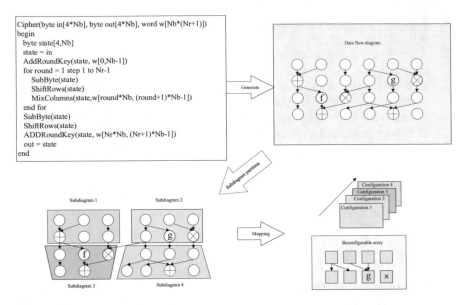

```
Cipher(byte in[4*Nb], byte out[4*Nb], word w[Nb*(Nr+1)])
begin
   byte state[4,Nb]
   state = in
   AddRoundKey(state, w[0,Nb-1])
   for round = 1 step 1 to Nr-1
       SubByte(state)
       ShiftRows(state)
       MixColumns(state,w[round*Nb, (round+1)*Nb-1])
   end for
   SubByte(state)
   ShiftRows(state)
   ADDRoundKey(state, w[Nr*Nb, (Nr+1)*Nb-1])
   out = state
end
```

Fig. 6.28 Mapping of the signal detection algorithm onto the reconfigurable architecture

after the data fetching can be conveniently converted to a data flow diagram, thus being mapped to the PEA. Therefore, this process will not be covered here.

Figure 6.29 shows the traversal of a lower triangular matrix involved in the LDL decomposition. The square in the figure represents a matrix that is a conjugate symmetric matrix; therefore, only its lower triangular element needs to be saved. Before the data fetching, the elements in the matrix are arranged tightly in the memory in column priority. That is, data is placed in the memory in the following order: the elements (m elements) in the first column of the lower triangular matrix, the elements ($m-1$ elements) in the second column of the lower triangular matrix, ..., one element in the mth column. During the access, the number of accesses for an $m \times m$ matrix is m. In each access, all the elements in an inscribed rectangle of the lower triangular matrix are traversed in the column priority. That is, in the first access, all the elements (m elements) in the first column are accessed. In the second access, the latter $m-1$ elements in the first column are first accessed and the latter $m-1$ elements in the second column are then accessed. In the ith access, the access order is as follows: the latter i elements in the first column, the latter i elements in the second column, ..., the latter i elements in the ith column, as shown in Fig. 6.29.

The following section discusses how to map the data fetching process onto the PEA. The regularity of memory access addresses is poor. Therefore, the research considers using a PE to compute the next address to be accessed and using another PE to read data via an indirect access. The PEA involves many numbers of iterations, which can be mapped to the cyclic traversal. As shown in Fig. 6.29, the memory access involves three layers of cycles: (1) The 1st access, 2nd access, ..., mth access are initiated to the lower triangular matrix. (2) In the ith access, the 1st column, the

Fig. 6.29 Traversal of a lower triangular matrix involved in the LDL decomposition

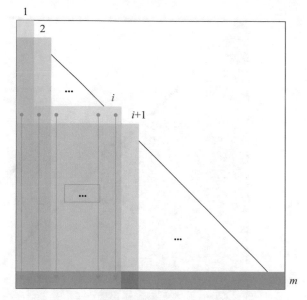

The number of accesses for an matrix is . In each access, all the elements in an inscribed rectangle of the lower triangular matrix are traversed in the column priority.

2nd column, ..., the ith column are accessed in sequence. (3) During the access of the ith column, the ith element, the $i + 1$th element, ..., the mth element are accessed in sequence. The PEA configuration information-related to iteration also involves three layers: the number of iterations for the top-layer of the PEA configuration information **PEA_TOP**, the number of iterations for the top-layer of the PE configuration information **PE_TOP** and the number of iterations for PE configuration lines **PE_CONF**. In the master control ARM, the coprocessor can be called for multiple times to implement the iterations. This book makes the three types of iterations involved in the abovementioned accesses, respectively, correspond to the number of times that the coprocessor is called by the master control ARM, the number of iterations for the top-layer of the PE configuration information and the number of iterations for PE configuration lines, as shown in Fig. 6.30. The number of columns and rows m is determined at runtime. Therefore, m can be written to the global register file of the PEA by the ARM. The ith access is also initiated by the ARM by calling the coprocessor, and the memory access behavior is related to i. Therefore, the number i also needs to be written to the global register file by the ARM. Likewise, $m - i - 1$ in Fig. 6.30 also needs to be written to the global register file by the ARM. Then, the number of iterations for the top-layer of the PE configuration information and the number of iterations for PE configuration lines can be determined according to the values in the global register file.

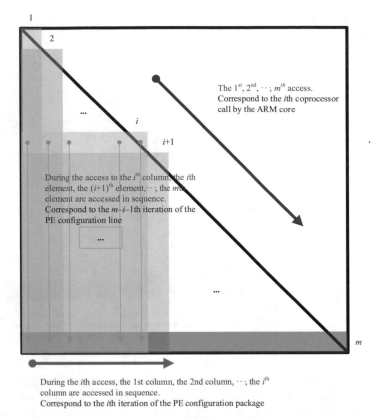

Fig. 6.30 Correspondence of the three numbers of iterations in the traversal process of a lower triangular matrix

Figure 6.31 shows the data flow diagram of the PEA during the traversal of a lower triangular matrix. PE2 is assigned to compute the addresses. The coarse-grained address signal of PE2 corresponds to the address of another PE for indirect memory access; other PEs can read the data by indirectly accessing the memory via the router. PE1 is used to compute the accumulated value of the address signal each time. All the numbers of iterations are configured to be obtained from the global register file (refer to "Access of the global register file to obtain the number of iterations" in Sect. 6.3.2). The actual number of iterations is written to the global register file by the ARM according to the value of i before the PEA is enabled. At the initial time, **Address = 0**. At the first execution, the **Address** needs to jump i steps forward to reach the position of the first data that needs to be accessed when the coprocessor is called the ith time. Here, the i needs to be stored into the private register RF0 of PE1 from the global register GF0 in advance. Then, the **Address** is increased

triangular matrix

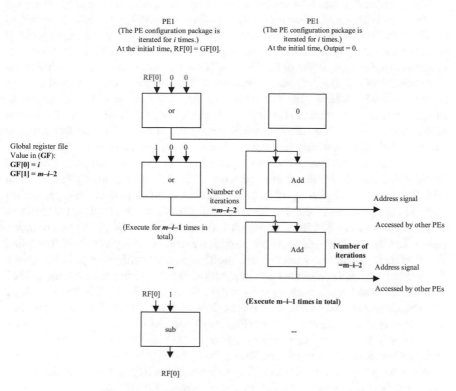

Fig. 6.31 Data flow diagram of the PEA during the traversal of a lower triangular matrix

by 1 for each iteration. If the number of iterations for a PE configuration line is $m - i - 2$, the **Address** is increased by 1 for $m - i - 1$ times; thus, the data in the first column is traversed during the ith access. The **Address** is increased by 1 each time of the first $m - i - 1$ accesses; after $m - i - 1$ data in each column is accessed, the configuration information reaches the end. At this time, all the addresses of the data in the first column during the ith access are generated. Before ending the configuration execution, PE1 needs to subtract the data in register RF0 by 1 to get ready for the next execution. As the number of iterations for **PE_TOP** is $m - i$, the preceding process needs to be executed for another i times. Take the second execution as an example. At the end of the first execution, the **Address** has reached the last element of the first column in the matrix. At the beginning of the second execution, the **Address** needs to be increased by $i - 1$ to point to the next element to be accessed [the shared memory saves the results in the lower triangular matrix in sequence; i.e., the element behind $(1, m)$ is $(2,2)$]. At this time, the PE1 has subtracted 1 from the data in RF0 at the end of the first execution; therefore, the data obtained from RF0 is $i - 1$ at the beginning of the second execution. Therefore, the PE2 can then continue to accumulate the

PE1 output to the **Address** after idling for a machine cycle. So far, most of the problems about the data flow diagram have been resolved, and the scheduling list can be directly obtained. However, there are still some problems need to be resolved for the data initialization. For the 0th iteration of PE_TOP, the private register RF0 of the PE1 needs to be initialized to the value of the global register GF0; the coarse-grained output address signal of the PE2 **Address** needs to be initialized to **0**. For the first iteration of PE_TOP, the private register RF0 of the PE2 and the output address signal of the PE2 **Address** need to retain the results at the end of the 0th iteration. In the configuration shown in the following algorithm, the requirement that values of parameters in the first iteration retain the results at the end of the 0th iteration can be satisfied because the results of the last iteration are not reset by the complete PE configuration information during the iteration. However, the data initialization at the beginning of the 0th iteration cannot be satisfied. If a machine cycle is used to initialize data at the beginning of the 0th iteration, the data initialization is performed again at the beginning of the first iteration. As a result, the requirement that values of parameters in the first iteration retain the results at the end of the 0th iteration cannot be satisfied. Two methods are provided to resolve this problem in this book. The first method illustrates how to call multiple sets of configuration information. The second method is recommended in practical use. In the first method, the ARM7 calls an additional configuration package to perform the initialization before calling the coprocessor to read an inscribed rectangle each time. The configuration package runs for only one machine cycle; the PE1 reads the global register GF0 and writes the results to the private register RF0 (refer to "Access of the global register file to obtain coarse-grained inputs" in Sect. 6.3.2); the PE2 performs the operation of adding zeros and writes the results to the output of its routing unit. The Algorithm 6.2 shows the final C program for calling the coprocessor for i times.

Algorithm 6.2 The final C program for calling the coprocessor for i times

```
#include<stdlib.h>

#include<stdio.h>

Int main()

{

Int index, matrix_size;
```

Char name[20], index_str[5];

//prepare data

//copy in Shared Memory()

for(index=0; index<matrix_size; index++)   {

   _callCoprocessor("Intitializing", "Init-ConfigPack.txt", "InitProfileResult", 1, index);

   itoa(index, index_str, 10);

   name = strcat ("LULT", index_str);

   _callCoprocessor(name, "LULT-ConfigPack.txt", strcat(name, "ProfileResult"), 2, index, matrix_size-index-2);

                    }

//copyoutSharedMemory()

//printf results

}

There are two points that need to be paid attention: (1) The ARM7 serial program includes a loop. In the loop, the initialization configuration package is first called to initialize the PE and then the configuration package in the LDL part is called to read an inscribed rectangle of the lower triangular matrix to perform data processing. During the call, additional parameters are used to transfer data to the global register file (refer to the description of how the coprocessor interface writes the global register file in Sect. 6.3.2). (2) To distinguish the profile document generated at each time the LDL is called, the strcat function is used to perform string concatenation and the concatenation result is used as the document name; the itoa function is used to convert integers to strings that are used as distinguishing marks. The PEA needs to read the configuration package from the outside to the PEA each time a set of configuration package is called; logically speaking, the cost of such a call is not small. In fact, to cope with the cyclic call of multiple sets of configuration packages, there is a configuration cache inside the PEA, which can store four sets of configuration packs in the PEA. The replacement of the configuration cache follows the simple order. That is, if the number of frequently called configuration packages is not larger than 4, only the cost of the first call (requiring 50–500 clock cycles) is high. Because the configuration information needs to be moved from the external memory in the first call; the time cost for subsequent calls is only the cost for the ARM to enable the PEA (only requiring 5–10 clock cycles). Therefore, the cost is small when the configuration packages are called in the order of 1-2-1-2-1-2-.... However, if the configuration packages are called in the order of 1-2-3-1-2-3-4-5-1..., the first set of configuration package is kicked out from the PEA cache when the fifth set of configuration package is called. Then, the data needs to be moved from the external memory again if the first set of configuration package is called again. That is, the number of configuration packages that are frequently called in a sub-application within a period of time should not be larger than four. If eight sets of configuration information are called in an application in the following order: 1-2-3-4-5-6-7-8-1-2-3-4-5-6-7-8-1-2-3-4-5-6-7-8-…, you should try to divide it into the calls with a order of 1-2-3-4-1-2-3-4-1-2-3-4…. During the process, save the intermediate data of the fourth set of configuration information, and call the remaining four sets of configuration information in the order of 5-6-7-8-5-6-7-8-5-6-7-8…, where the fifth set of configuration information reads the intermediate data to reduce the cost of the configuration information switch. In the second method, the number of the start line for PE top-layer iteration (see "Computing the Iteration Line Number" in Sect. 6.3.4) is used to perform configuration. The number of the start line for PE top-layer iteration indicates the number of the line each of n iterations starts with after the PE executes all its configuration lines for the first time in the case that the number of iterations for the PE top-layer is n. Therefore, you only need to initialize the private register RF0 of the PE1 and the coarse-grained output **Address** of the PE2 in the line 0 of the PE1 and PE2. Then, set the number of the start line for PE top-layer iteration **PE_TOP[ITER_LINE]** to **1**, and set the number of iterations for the PE top-layer to be from **GF0**. In this way, after executing all the configuration information for the first time, the PE directly starts the execution from line 1, i.e., the PE skips the initialized line. Thus, the PE does not need to call a set of initialization configuration package before calling the set of configuration package. As a result, the preceding C program can be simplified. That is, the program for calling the initialization configuration package can be deleted to simplify the execution process.

References

1. Tessier R, Pocek K, Dehon A (2015) Reconfigurable computing architectures. Proc IEEE 103(3):332–354
2. Yu Z, Yu Z, Yu X et al (2014) Low-power multicore processor design with reconfigurable same-instruction multiple process. IEEE Trans Circuits Syst II Express Briefs 61(6):423–427
3. Zhu J, Liu L, Yin S et al (2013) Low-power reconfigurable processor utilizing variable dual VDD. IEEE Trans Circuits Syst II Express Briefs 60(4):217–221
4. Wu M, Yin B, Wang G et al (2014) Large-scale MIMO detection for 3GPP LTE: algorithms and FPGA implementations. IEEE J Sel Top Signal Process 8(5):916–929
5. Peng G, Liu L, Zhang P et al (2017) Low-computing-load, high-parallelism detection method based on Chebyshev iteration for massive MIMO systems with VLSI architecture. IEEE Trans Signal Process 65(14):3775–3788
6. Peng G, Liu L, Zhou S, et al (2017) A 1.58 Gbps/W 0.40 Gbps/mm^2 ASIC implementation of MMSE detection for $128x8$ 64-QAM massive MIMO in 65 nm CMOS. IEEE Trans Circuits Syst I Regul Pap (99):1–14
7. Jin J, Xue Y, Ueng Y L, et al (2017) A split pre-conditioned conjugate gradient method for massive MIMO detection. IEEE Int Workshop Signal Process Syst 1–6

8. Peng G, Liu L, Zhou S, et al (2018) Algorithm and architecture of a low-complexity and high-parallelism preprocessing-based K-best detector for large-scale MIMO systems[J]. IEEE Trans Signal Process 66(7)

9. Winter M, Kunze S, Adeva EP et al (2012) A 335 Mb/s 3.9mm2 65 nm CMOS flexible MIMO detection-decoding engine achieving 4G wireless data rates 13B(4):216–218

10. Castañeda O, Goldstein T, Studer C (2016) Data detection in large multi-antenna wireless systems via approximate semidefinite relaxation. IEEE Trans Circuits Syst I Regul Pap 99:1–13

11. Liu L, Chen Y, Yin S et al (2017) CDPM: Context-directed pattern matching prefetching to improve coarse-grained reconfigurable array performance. IEEE Trans Comput Aided Des Integr Circuits Syst 99:1

12. Yang C, Liu L, Luo K et al (2017) CIACP: A correlation- and iteration- aware cache partitioning mechanism to improve performance of multiple coarse-grained reconfigurable arrays. IEEE Trans Parallel Distrib Syst 28(1):29–43

13. 周阳. 面向多种拓扑结构的可重构片上网络建模与仿真[D]. 南京航空航天大学硕士论文 (2012)

14. Achballah AB, Othman SB, Saoud SB (2017) Problems and challenges of emerging technology networks-on-chip: a review. Microprocess Microsyst 53

15. Dally WJ, Towles BP (2004) Principles and practices of interconnection network 299(6):707–721

16. Hu J, Marculescu R (2003) Exploiting the routing flexibility for energy/performance-aware mapping of regular NoC architectures 688–693

17. Chou CL, Marculescu R (2011) FARM: Fault-aware resource management in NoC-based multiprocessor platforms. Des Autom Test Eur Conf Exhib 1–6

18. Kohler A, Schley G, Radetzki M (2010) Fault tolerant network on chip switching with graceful performance degradation. Comput-Aided Des Integr Circ Syst IEEE Trans on 29(6):883–896

19. Chang YC, Chiu CT, Lin SY, et al (2011) On the design and analysis of fault tolerant NoC architecture using spare routers. Design automation conference, pp 431–436

20. Chen WU, Deng CC, Liu LB et al (2015) Reliability-aware mapping for various NoC topologies and routing algorithms under performance constraints. Sci China 58(8):82401

21. Khalili F, Zarandi HR (2013) A reliability-aware multi-application mapping technique in networks-on-chip. Euromicro international conference on parallel, distributed, and network-based processing, pp 478–485

22. Ababei C, Kia HS, Hu J, et al (2011) Energy and reliability oriented mapping for regular networks-on-chip. In: ACM/IEEE international symposium on networks-on-chip, pp 121–128

23. Kim JS, Taylor MB, Miller J, et al (2003) Energy characterization of a tiled architecture processor with on-chip networks. In Proceedings international symposium on low power electronics and design (ISLPED), pp 424–427

24. Kahng A B, Li B, Peh L S, et al. ORION 2.0: a fast and accurate NoC power and area model for early-stage design space exploration. In Design, automation & test in Europe conference & exhibition, pp 423–428

25. Das A, Kumar A, Veeravalli B (2013) Energy-aware communication and remapping of tasks for reliable multimedia multiprocessor systems. In: IEEE international conference on parallel and distributed systems, pp 564–571

26. Liu L, Wu C, Deng C et al (2015) A flexible energy and reliability-aware application mapping for NoC-based reconfigurable architectures. IEEE Trans Very Large Scale Integr Syst 23(11):2566–2580

27. Kiasari AE, Lu Z, Jantsch A (2013) An analytical latency model for networks-on-chip. IEEE Trans Very Large Scale Integr Syst 21(1):113–123

28. Bolch G, Greiner S, de Meer H, et al (1998) Queueing networks and Markov chains. Wiley-Interscience, pp 904–4507

29. Khalili F, Zarandi HR (2013) A fault-aware low-energy spare core allocation in networks-on-chip. Norchip. IEEE, pp 1–4

30. Wu C, Deng C, Liu L et al (2015) An efficient application mapping approach for the Co-Optimization of reliability, energy, and performance in reconfigurable NoC architectures. IEEE Trans Comput Aided Des Integr Circuits Syst 34(8):1264–1277

31. Gerez SH (1999) Algorithms for VLSI design automation. Wiley, Hoboken, United States, pp 5–9
32. Ye TT, Benini L, De Micheli G (2002) Analysis of power consumption on switch fabrics in network routers. In: Proceedings on the design automation conference. IEEE, pp 524–529
33. Wiegand T, Sullivan G J, Bjøntegaard G, et al (2003) Overview of the H.264/AVC video coding standard. IEEE Trans Circ Syst Video Technol 13(7):560–576
34. Sullivan GJ, Ohm J, Han WJ et al (2012) Overview of the high efficiency video coding (HEVC) standard. IEEE Trans Circuits Syst Video Technol 22(12):1649–1668
35. Bertsimas D, Tsitsiklis J (1993) Simulated annealing. Stat Sci 8(1):10–15
36. Dick RP, Rhodes DL, Wolf W (1998) TGFF: task graphs for free. In: Proceedings of the sixth international workshop on hardware/software codesign (CODES/CASHE '98), pp 97–101
37. Wu C, Deng C, Liu L et al (2017) A multi-objective model oriented mapping approach for NoC-based computing systems. IEEE Trans Parallel Distrib Syst 99:1
38. Li Z, Li S, Hua X, et al (2013) Run-time reconfiguration to tolerate core failures for real-time embedded applications on NoC manycore platforms. In: IEEE international conference on high PERFORMANCE computing and communications & 2013 IEEE international conference on embedded and ubiquitous computing, pp 1990–1997
39. Hoskote Y, Vangal S, Singh A et al (2007) A 5-GHz mesh interconnect for a teraflops processor. IEEE Micro 27(5):51–61
40. Atak O, Atalar A (2013) BilRC: An execution triggered coarse grained reconfigurable architecture. IEEE Trans Very Large Scale Integr Syst 21(7):1285–1298
41. Wei S, Wei S, Wei S, et al (2017) Minimizing pipeline stalls in distributed-controlled coarse-grained reconfigurable arrays with triggered instruction issue and execution. In: Design automation conference, p 71
42. Liu L, Wang J, Zhu J et al (2016) TLIA: Efficient reconfigurable architecture for control-intensive kernels with triggered-long-instructions. IEEE Trans Parallel Distrib Syst 27(7):2143–2154
43. Yin S, Liu D, Sun L, et al (2017) DFGNet: mapping dataflow graph onto CGRA by a deep learning approach. In: IEEE international symposium on circuits and systems, pp 1–4
44. Lu T, Yin S, Yao X, et al (2017) Memory fartitioning-based modulo scheduling for high-level synthesis. In: IEEE international symposium on circuits and systems, pp 1–4

Chapter 7
Prospect of the VLSI Architecture for Massive MIMO Detection

5G is a more advanced mobile communications network deployed in 2018 and later, which mainly includes the following technologies: the millimeter wave technology [1] (26, 28, 38, and 60 GHz) that is able to provide a transmission rate as high as 20 Gbit/s; the massive MIMO technology that can provide "a performance that is 10 times that of the 4G network" for the 5G communications network. As another important technology for 5G, "the low- and medium-frequency band 5G" (5G New Radio) that leverages the frequencies ranging from 600 MHz to 6 GHz, especially 3.5 to 4.2 GHz. Extended and evolved from 4G communications, 5G that represents the development tendency of new generation information communications is going to penetrate every field in the future society; thus, it will construct an omnidirectional user-oriented information ecosystem. This chapter prospects the future application scenarios and hardware development from three aspects: server, mobile terminal, and edge computing, which correspond to the subsequent sections.

7.1 Prospect of Server-Side Applications

7.1.1 Outline of 5G Communications Characteristics

The differentiation of application scenarios for 5G communications proposes the engineering requirements on communications services mainly from the perspectives of equipment quantity, communications bandwidth, and performance, i.e., deep coverage, ultra-low power consumption, ultra-low complexity, ultra-high density, ultimate capacity, ultimate bandwidth, deep ecological consciousness, strong security, ultra-high reliability, ultra-low latency, and perfect mobility, etc., as shown in Fig. 7.1.

On the premise of ensuring even improving the quality of service (QoS) for communications, high data rate, low latency and low power consumption are the most essential requirements. In terms of solutions for the establishment of 5G new radio, the key technologies arise such as massive MIMO [2], millimeter wave bands/visible

© Springer Nature Singapore Pte Ltd. and Science Press, Beijing, China 2019
L. Liu et al., *Massive MIMO Detection Algorithm and VLSI Architecture*,
https://doi.org/10.1007/978-981-13-6362-7_7

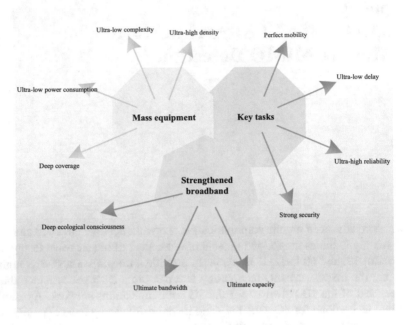

Fig. 7.1 Requirements and characteristics of 5G communications

light transmission, filter-bank-based multicarrier (FBMC) modem, dense networking and heterogeneous network, device-to-device (D2D) and in-vehicle network [3] and onboard network, software-defined networking (SDN) [4], cognitive radio networks, and green communications [5].

According to the report of GSMA, until 2025, 5G network will be commercially used in 111 countries and regions throughout the world. Before the 5G technology is laid in a large scale and provided for consumers, two transitions must be accomplished. First, the mobile operators must upgrade their network infrastructures into 5G equipment. Currently, the primary 5G equipment suppliers are Huawei and Zhongxing Telecommunications Equipment (ZTE) from China, Ericsson from Sweden, and Nokia from Finland. Second, the mobile phone manufacturers need to keep up with the pace to embed 5G wireless signal receivers into mobile phones, making full preparation for the 5G network.

At the early stage of the commercialization of 5G, operators will initiate extensive network construction. Revenues of equipment manufacturers from the investment on the 5G network equipment will become the primary source of direct economic output of 5G [6]. According to the *White Paper of the Impacts of 5G on Economy and Society*, it is estimated that the network equipment and terminal devices will bring the manufacturers a total revenue of approximately RMB 450 billion yuan in 2020, accounting for 94% of the direct economic output. In 2025, the middle stage of the commercialization of 5G, the expenditures from users, other industrial terminal devices and telecom services will grow constantly, which are expected to rise by RMB 1400 billion yuan and RMB 700 billion yuan, respectively, accounting for 64% of the direct economic output. In 2030, the middle and later stage of the commercialization of 5G, internet enterprises and information service industries related to 5G will become the backbone of the direct economic output, which will increase the economic output to RMB 2600 billion yuan, occupying 42% of the direct economic output.

In light of this, we can conclude that in the near future, the commercialization of 5G will result in a great revolution in the basic manufacturing industry and product substitution in equipment manufacturing industry, shining with extremely high commercial value and investment space. Thus, multiple national equipment manufacturers have devoted substantial human and material resources to industries related to 5G.

7.1.2 Outline of the Server-Side Characteristics

Server is a common name for the type of equipment working based on the network environment, which is usually undertaken by various kinds of computers. Unlike a terminal, a server acts as the control and service center of the network, which serves various terminal devices (usually undertaken by various kinds of computation equipment) that are connected to it; it has a high requirement for the computing performance. The three common server architectures include the cluster architecture, the load balancing architecture, and the distributed server architecture. The cluster architecture refers to integrating multiple servers to handle the same service, and it seems that there is only one server from the perspective of client. One advantage of the cluster architecture is that it can use multiple computers to conduct parallel computations to achieve a higher computing speed. The other advantage of the cluster architecture is that it can use multiple computers to backup, which ensures the proper operation of the entire system even if any machine is broken. Established upon the existing network structure, the load balancing architecture can offer a low-cost, effective, and transparent method to extend the bandwidth of network equipment and server, increase the throughput, strengthen the processing capability for network data, and enhance the network flexibility and availability. The distributed resource sharing server is a theoretical computing model server form that studies the geographic information distributed on the network and the database operations affected; it can distribute data and programs to multiple servers. The distributed architecture contributes to the distribution and optimization of tasks in the entire computer system, overcomes the defects in traditional centralized system where strained resources of central hosts and response bottlenecks occur, and addresses issues such as data heterogeneity, data sharing, and computing complexity in the geographic information system (GIS) of network, which is a significant progress in GIS. To ensure the security of important data, the cluster server architecture is mainly used in communications industry. The load balancing that aims at sharing the access loads and avoids temporary network traffic jam, is mainly applied in electronic business websites. The distributed servers are born to achieve the cross-sector high-speed access of multiple single nodes. At present, the distributed server is the first choice for the purpose like content delivery network (CDN).

As an exclusive communications system that a user sends files or accesses remote system or network through remote links, communications server can simultaneously provide communications channels for one or more users as per the software and hardware capabilities. Generally, communications servers are featured with the following functions: the gateway function that provides connections between the user and the host by converting data formats, communications protocols, and cable signals; the service access function that allows remote users to dial-in; the modem function that offers internal users with a group of asynchronous modems for dial-in access to remote systems, information services, or other resources; the bridge and router function that maintains the dedicated or dial-in (intermittent) links to remote local area networks (LAN) and automatically transmits data groupings among LANs; and the e-mail server function that automatically connects other LANs or electronic post offices to collect and transmit e-mails.

Since the performance of a server is crucial to that of the network, 5G and even beyond 5G communications raise the following requirements on the server: strong data processing capability for handling the access of the data in large flow, high stability, and reliability, have a full-functional system with ensured data security, and etc. As is mentioned in Chap. 1, from the perspective of hardware implementation, the main superiority of the ASIC method for implementing a data processing module is that it is able to obtain the optimum overall merit of performance and power consumption, which can satisfy the sharply rising computation capability required by massive MIMO detection chips, and achieve high throughput, high energy efficiency, and low latency. Nowadays, with the rapid development of mobile communications, the deficiency of flexibility prevents it from being further extensively applied. However, in the processing of compute-intensive data, reconfigurable processors cannot only achieve high throughput, low energy consumption, and low latency, but also exhibit unique advantages in flexibility and scalability. Additionally, benefiting from the reconfigurability of hardware, this architecture is possible to execute system update and error correction during the operation of the system, which poses dominant privileges in extending the service life and guaranteeing the release time of products. Thus, the reconfigurable processor becomes a significant and promising research subject in the development of communications in the future.

7.1.3 Server-Side Application

As the latest standard of the global communications, 5G does not confine its significance to a higher speed or improved mobile broadband experience, instead, its mission is especially to connect new sectors and encourage new services, e.g., advocating industrial automation, large-scale IoT, smart home, and autonomous driving,

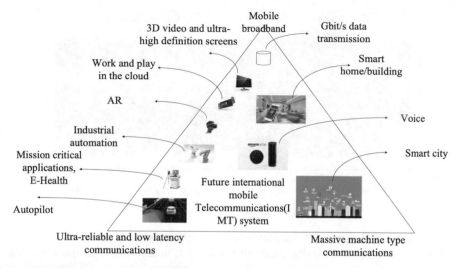

Mobile broadband

3D video and ultra-high definition screens

Gbit/s data transmission

Work and play in the cloud

Smart home/building

AR

Voice

Industrial automation

Mission critical applications, E-Health

Smart city

Autopilot

Future international mobile Telecommunications(IMT) system

Ultra-reliable and low latency communications

Massive machine type communications

Fig. 7.2 Main usage scenarios of 5G communications

etc. Correspondingly, these sectors and services have higher requirements for the networks, which are higher reliability, lower latency, wider coverage, and higher security. Therefore, a flexible, effective, and scalable network is in urgent demand to meet different requirements from all walks of life.

In June 22, 2015, the conference of ITU-RWP5D held by International Telecommunications Union (ITU) defined the three main usage scenarios of the future 5G: the enhanced mobile broadband, the ultra-reliable and low latency communications, and the massive machine type communications. The specific scenarios cover Gbit/s mobile broadband data access, smart home, smart building, voice, smart city, three-dimensional video, ultra-high definition screens, work and play in the cloud, AR, industrial automation, mission-critical application, self-driving car, etc., as shown in Fig. 7.2.

From October 26 to 30, 2015, when World Radiocommunication Conference 2015 (WRC-15) was held in Geneva, Switzerland, ITU-R officially approved three resolutions that were beneficial to the promotion of future research process of 5G and nominated the official name of 5G as "IMT-2020". Out of the main usage scenarios, business requirements, and challenges of mobile Internet and IoT, "IMT-2020" recategorized the main usage scenarios of 5G into four based on the specific network function requirements: continuous wide-area coverage, high traffic capacity hotspot, low power consumption and a large number of connections, and low latency and high reliability, which are basically consistent with the three major usage scenarios of ITU. "IMT-2020" only further subdivides the mobile broadband into continuous wide area coverage and high traffic capacity hotspot, as shown in Fig. 7.3.

Continuous wide-area coverage and high traffic capacity hotspot scenarios are mainly designed to meet the mobile internet business requirements in 2020 and later, which are also primary traditional 4G scenarios. Continuous wide-area coverage is

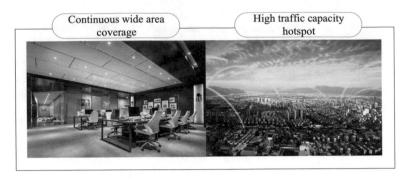

Fig. 7.3 Continuous wide-area coverage and high traffic capacity hotspot scenarios

the fundamental coverage method of mobile communications targeting the assurance of users' mobility and service continuity to offer seamless and high-speed service experience. Its primary challenge comes from the needs to ensure a 100 Mbit/s higher data rate for users anytime and anywhere, which is more obvious in harsh environments such as base station coverage edge and high-speed moving. The scenarios requiring high traffic capacity hotspot are mainly oriented at local hotspot areas to provide users with ultra-high data rate to satisfy the extremely high traffic density demands on the network, which need to be supported by multiple technologies. For instance, super intensive networking can effectively multiplex spectral resources and significantly promote frequency multiplexing efficiency in the unit area; full spectrum access can make the full use of low-frequency and high-frequency spectral resources to achieve a higher data rate.

The scenarios with low power consumption and a large number of connections, low delay, and high reliability (Fig. 7.4) mainly aim at IoT services, which are the scenarios newly extended in 5G dedicated to solving the problem that the conventional mobile communications cannot well support the IoT and vertical industrial applications. The scenarios with low power consumption and a large number of connections are generally for the circumstances where sensing and data collection are targeted and featuring with small data packets, low power consumption, and vast connections such as smart city, environmental monitoring, intelligent agriculture, forest fire prevention, etc. In these usage scenarios, a large number of terminals are widely distributed, which not only require the network to support over 100 billion connections to meet the connection density demand of 1 million/km^2 but also guarantee an ultra-low power consumption and cost. The low latency and high-reliability usage scenarios are primarily leveraged for special application requirements of vertical industries such as internet of vehicles (IoV) and industrial control. These usage scenarios have extremely high requirements on latency and reliability; they need to provide users with end-to-end latency at millisecond level and close to 100% service reliability guarantee. Table 7.1 lists the main usage scenarios and key challenges of performance for 5G.

The specific usage scenarios are introduced as follows.

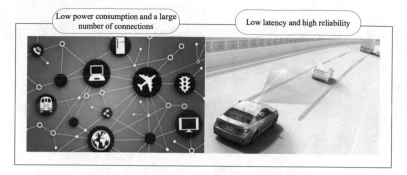

Fig. 7.4 Scenarios with low power consumption and a large number of connections, and low latency and high reliability

Table 7.1 Main usage scenarios and key challenges of performance for 5G

Scenarios	Key challenges
Continuous wide-area coverage	100 Mbit/s user experienced data rate
High traffic capacity hotspot	User experienced data rate: 1 Gbit/s Peak data rate: tens of Gbit/s Traffic density: tens of Tbit/km^2
Low power consumption and a large number of connections	Connection density: 10^6/km^2 Ultra-low power consumption and cost
Low latency and high reliability	Air interface latency: 1 ms End-to-end latency: at millisecond level Reliability: close to 100%

7.1.3.1   IoV

As far as China is concerned, the national car ownership has reached 217 million up to 2017, which is increased by 23.04 million with a growth rate of 11.85% compared with that of 2016. Moreover, the proportion of automobiles in motor vehicles increases constantly from 54.93 to 70.17% in the recent 5 years; automobiles have become the main part of motor vehicles. In terms of distribution, there are 53 cities in China whose car ownership is more than a million, of which 24 cities amount to 2 million and 7 cities possess more than 3 million, Beijing, Chengdu, Chongqing, Shanghai, Suzhou, Shenzhen, and Zhengzhou. In western areas, the motor vehicle ownership reaches 64.34 million with the fastest growth rate. In 2017, motor vehicle ownership in eastern, middle, and western areas were 155.44, 90.06, and 64.36 million, accounting for 50.17, 29.06, and 20.77% of total motor vehicles in China, respectively. Among them, the automobile ownership of western areas in recent five years rises by 19.63 million with a growth rate of 19.33%, which is higher than the 14.61 and 16.65% of eastern and middle areas [7].

As you may know, international internet tycoons are rushing to control the driving cabs. It is possible that the significance that they march into onboard system is

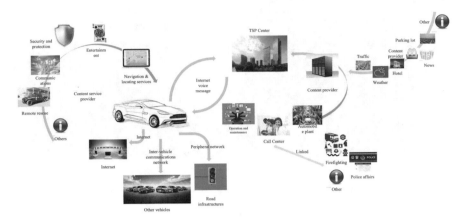

Fig. 7.5 Application of 5G in automobile industry

to reconstruct the ecosystem of the entire industry and establish a standard onboard operation platform, which is analogous to that once occurred to the smartphone. Some experts even predicted that the IoV will become the third internet entity, after PC-oriented internet and cell phone-oriented mobile Internet. A complete IoV involves a lot of links mainly including communications chip/module suppliers, external hardware suppliers, RFID and sensor suppliers, system integration suppliers, application equipment and software suppliers, telecom operators, service suppliers, automobile manufacturers, etc., as shown in Fig. 7.5. Thus, the automobile hardware market arising from the commercialization of 5G is also to be exploited.

With the continuous and rapid increase of the motor vehicle ownership, the number of drivers also substantially grows synchronously, driving the annual increment of recent five years to 24.67 million. In 2017, the number of national motor vehicle drivers reached 385 million, of which automobile drivers accounted for 342 million. 30.54 million drivers occupying 7.94% of the total drivers had less than 1 year of driving experience. On one hand, the surge of automobile ownership and the improper management of parking lots aggravate the low utilization rate of parking space, which raises the "parking problem" to be solved. On the other hand, the high requirements on motor vehicles and sharp increased inexperienced drivers are endangering the traffic safety. Therefore, the research, development, and upgrade of the driving assistance even self-driving technologies are imminent.

To solve the parking problem, the cloud based parking management system (as shown in Fig. 7.6) is designed focusing on the remote control and management of parking locks. It achieves centralized management and decentralized control for parking space, which benefits the owner to lease the parking space while it is idle, and effectively mitigates the supply and demand issue of parking space and increases the urban park utilization rate. This is one of the typical usage scenarios of the low power consumption and large number of connections [8]. As a pivot component of the cloud-based parking management platform, the remote control system for parking

Fig. 7.6 Architecture of the cloud-based parking management system

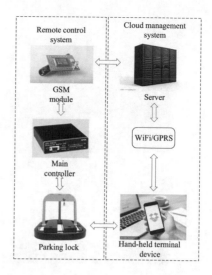

locks is mainly responsible for the collection of parking space status information and the control of parking space permission. The system includes hardware and software parts, where the hardware mainly refers to the design of the built-in hardware control system tailored to realize the remote control of parking locks, while the software part involves the development of cell phone client and software at the server side (communications programs, data storage programs, etc.).

On the condition of satisfying the performance requirements such as power consumption, latency, and throughput, to cope with the increase of motor vehicle ownership by leaps and bounds, the data processing scale of massive MIMO signal detection chips will be certainly raised. In this scenario, the reconfigurable architecture is more suitable compared with the customized ASIC architecture.

Regarding autopilot technologies, in May 2016, Florida, USA, a Model S engaged in autopilot mode at full speed crashed into a white tractor trailer cutting across the highway and caused the death of the driver. In March 2018, Tempe city, Arizona State, USA, a self-driving Uber struck a pedestrian who died after being sent to the hospital. Apparently, these two accidents are caused due to different reasons. The autopilot accident was caused due to the failure in identifying the vehicle while the Uber accident occurred because of pedestrian identification fault. To deal with the shortcomings of current autopilot technologies, Tesla recently announced that the new version of in-house navigation and maps engine, "light-years ahead" will preliminarily complete the upgrade of software. Also, the software algorithms should be updated unceasingly. Figure 7.7 shows the autopilot diagram of Tesla.

In high-speed driving mode, real-time data processing and information interaction are extremely important [9], which is one of the typical usage scenarios of the low delay and high reliability. Thus, the low delay and high throughput are the most pressing performance demands. ASIC does not only have the natural strength in energy efficiency but also have low chip manufacturing cost due to the large volume

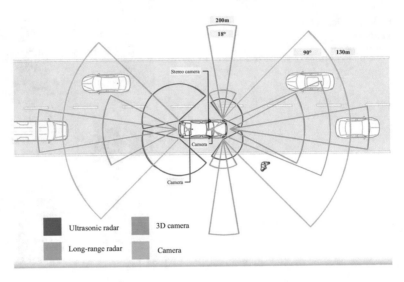

Fig. 7.7 Autopilot diagram of Tesla

of motor vehicle ownership (after mass production, one-time engineering cost can be amortized over all chips). Therefore, ASIC-based massive MIMO signal detection chips have an optimistic application prospect.

7.1.3.2 Cloud Computing

As network technologies are progressing, network size is increasing fast, and computer systems are growing complex, various novel systems and services spring up. Telecom operators and Internet application service suppliers are competing intensely with each other for attracting more users and achieving more profit. Recently, mobile Internet grows mature gradually, and numerous application service suppliers start to transform and develop over-the-top (OTT) services [6] to directly profit from users and advertisers without involving network operators. To tackle such challenge, although operators spend a lot on providing network services, they still do not find any effective solution, severely impacting their revenues. At the same time, application service suppliers attempt to break the technical barriers to obtain more network resources, resulting in issues such as "signaling storm" and the surge of terminal electricity energy consumption that dramatically harm the users' interests. The blind contest and the lack of cooperation platform lead to the constantly intensifying of conflicts among operators, users and service suppliers (Fig. 7.8).

Being the technical and development hotspot of the current Internet, cloud computing integrates infrastructures, application platforms, and application software into a complete network structure [10]. Based on the internet technologies, this system provides external services in self-service and on-demand manners; it is featured with

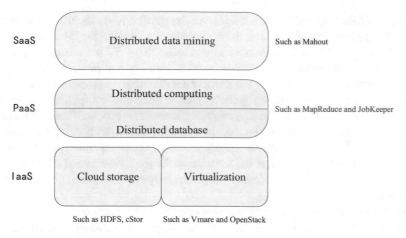

Fig. 7.8   Three-layer model of cloud computing

broad network access, virtualized resource pooling, rapid elasticity, measured service, and multi-tenant, posing as an active reference for operators to improve their network application capabilities. According to different service modes, cloud platform can divided into three service modes, infrastructure as a service (IaaS), platform as a service (PaaS), and software as a service (SaaS). Using the technologies such as virtualization and distributed computing, cloud computing incorporates various computer resources into an address pool via a computer network, which is a new type of on-demand service mode [6]. Mobile cloud computing (MCC) has the characteristics including weakened limit for terminal hardware, more convenient data storage, personalized services, and ubiquitous availability [11], which should be supported by large-scale and prompt data quantity processing at the server side. In summary, in this scenario, the ASIC-based massive MIMO has a broad application prospect.

7.2   Prospect of Mobile-Side Application

Mobile computing terminal, which is by definition referring to the computer equipment used during movement, mainly including the wireless onboard terminal and the wireless handheld terminal. Thanks to the rapid progress of broadband wireless access technology and mobile internet technology, people are eager to ubiquitously obtain information and services easily even during movement. As the access interface to wireless network, mobile terminals witness a flourishing tendency with many kinds of mobile equipment (smartphones and pads) springing up. Current mobile computing terminals cannot only accomplish voice chat, voice videos, and photographing but also enable rich functions such as Bluetooth, GPS location, and information processing, which plays an ever more important role in human society. At Mobile World

Congress 2018, "5G era" stood out as one of the spotlights. As the fifth generation of mobile communications network, 5G is capable of achieving the "internet of everything." Compared with 4G communications technology, 5G has a much higher data rate, and achieves significant improvements in stability and power consumption; it will significantly affect the mobile computing terminals. Different from previous generations of communications technologies, the mobile computing terminals of 5G cover a more extensive range, generating many new products such as wearable devices and home networking devices. In addition, mobile computing terminals are more humanized to satisfy users' requirements at faster information transmission rate. More importantly, 5G lays the foundation for the development of other related technologies because the fast data rate is universally required among big data, cloud computing, AI, and self-driving.

However, the development of mobile terminal still faces a series of challenges. Researches on the basic theories and key technologies have always been the concerns of researchers from either enterprises or academies. As one of the key technologies, massive MIMO can significantly improve the channel capacity and signal coverage range of mobile communications. Therefore, during the design of the massive MIMO detection processor, a better massive MIMO detection VLSI architecture means a lot for its high performance, low power consumption, low latency, flexibility, and scalability. In other words, seeking for an optimized MIMO detection architecture is vital to the development of the MIMO detection processor even the mobile terminal. Since the twenty-first century, relying on the proximity advantage, mobile terminals have already taken over the position where the competition is the intensest in marketing. As various technologies gradually mature, diversified mobile terminals step into the intelligent age featured with enriched functions, which evolve toward the integration of more functions. The development of global mobile communications terminal poses a forceful rising trend while the market has harsh performance requirements on the mobile terminal products. The performance and cost of mobile terminals mainly concentrate on the chip, in particular, baseband communications chip; thus, the primary link of terminal R&D focuses on the baseband chip [12]. The requirements of mobile terminals on baseband chips are mainly reflected in the following aspects.

(1) Low power consumption. As the most essential part of mobile terminals, baseband chip mainly synthesizes baseband signals to be transmitted and decodes the received signals. During the transmission, it encodes signals into baseband codes that can be transmitted, while it decodes the received baseband codes into audio signals during the receiving. In the smart terminal market, the data processing on the baseband chip of smart terminals is becoming increasingly heavy, therefore, the low power consumption design of the baseband chip is significant to the development of smart terminals [13].

(2) Low latency. A growing number of applications raise higher requirements on the path delay. In this case, baseband chip needs to process data in real time with a latency at millisecond level.

(3) Low cost. In the fifth generation of ultra-intensive network, the size of a micro base station will be very tiny with short distances between stations. As the deployment density is very high, the cost of micro base stations is very important to the operators. The deployment should cover both the indoor and outdoor scenarios using low-cost CMOS power amplifiers to access nodes ranging from several meters to 100 m.

(4) High capacity. Baseband chip needs to accomplish high capacity, energy efficiency, and spectrum efficiency.

As for the requirements of the baseband chip, massive MIMO detection VLSI architecture based on ASIC and the reconfigurable massive MIMO detection VLSI architecture may have a promising application prospect, from which aspects this section will be elaborated.

7.2.1 Application of ASIC-Based Detection Chips

In addition to supporting the mobile broadband development, 5G also supports numerous emerging application scenarios. Increasing numbers of applications promote higher requirements for data transmission, i.e., low latency and high throughput, which demands more for the design of massive MIMO detection chips. ASIC-based massive MIMO detection chips are endowed with the potential to meet the future application requirements in latency and throughput. The applications of ASIC-based massive MIMO detection are illustrated by taking the following future applications as examples.

7.2.1.1 VR and AR

VR and AR are revolutionary technologies that will radically change the content consumption of consumers and enterprise departments. VR is a shared and tactile virtual environment where several users are physically connected through a simulation tool to cooperate with each other via not only visual and but also tactile perception. Whereas, in AR, the real content and content generated by the computer is combined into the users' line of sight to be visualized. Compared with the static information augment today, future AR application mainly aims at the visualization of dynamic content. The tactile feedback is the proposition of the interactions with high fidelity in the VR. Specifically, perceiving objects in the VR through haptic results in the dependency of high precision of programs, which can be realized only when the latency between the user and VR is lowered to several milliseconds. The addition of extra information to the users' line of sight can boost the development of many assistance systems such as maintenance, driving assistance system, and education. With the tactile network, the content in the AR can be transferred from static to dynamic, enabling the virtual expansion of the user views in real time, and identifying and avoiding

| Usuario | Publicar | Salvar | Recuperar | Objetos | Textura | Cor | Som | Parar Func. | Rnglish |

| Remover |
| Fixar |
| Transladar |
| Transl. Y |
| Rotacionar |
| Escala |
| Escala2D |
| Encadear |
| Encad. Orig |
| Habilitar |
| ← | → |
| Pausar Enc. |

| Inserir Enc. | Editar Enc. | Colisao | Atrair | Repelir | Ocultar | Animar | Anim. Cir. | Anim. Lin. | Camera |

Fig. 7.9 Application of VR in remote education

Fig. 7.10 Subversion of VR in the traditional education

possible hazardous accidents. VR and AR technologies have extensive application prospects in the education industry by connecting the real world to the virtual one. Applying AR in the classroom deeply changes the traditional education mode, which enhances the teaching and learning effect (as shown in Figs. 7.9 and 7.10).

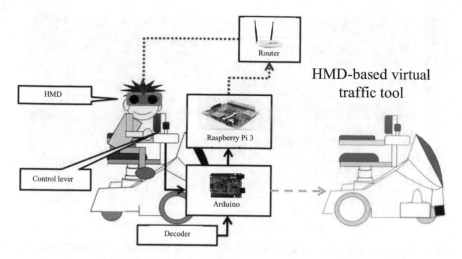

Fig. 7.11 AR-based driving assistance system. ©[2018] IEEE. Reprinted, with permission, from Ref. [14]

The perception capability of humans can be enhanced by using the AR-based driving assistance system (Fig. 7.11). First of all, the system adopts the virtual platoon control (VPC) to enable a real vehicle with passengers on to tightly follow a virtual one that is projected on the head-mounted display (HMD) manipulated from the objective view, which can ensure to drive without colliding with any obstacle [14].

The wireless transmission plays a crucial role in VR and AR. For example, the tactile network in VR and AR must process data in real time to meet users' demands. Under the circumstances where a large number of VR and AR terminals have data to be processed, massive MIMO detection needs to satisfy the requirements of high accuracy, low latency, and high throughput. Therefore, the ASIC-based massive MIMO is prospective particular for cases with the requirements of low latency and high processing rate.

7.2.1.2 Self-driving

Figure 7.12 outlines the driving assistance functions that will emerge in the near future. Most functions listed will involve radar sensors because they are relatively stable in different conditions such as rain, dust, and sunlight. However, there is no such a universal radar sensor that can satisfy all the functional requirements in the roadmap shown in Fig. 7.12. To meet the future demands, it is possible a good attempt to identify all key technologies required to apply in today's radar sensors. In the real application scenario, the radar sensors usually demand high angular and speed resolutions, high reliability, high throughput, low cost, and small size. As one of the key technologies of radar detection, massive MIMO technology makes a

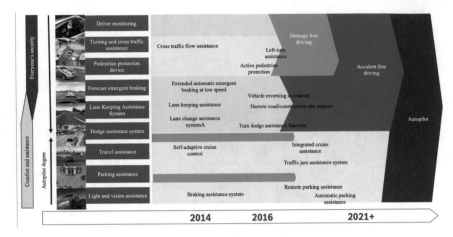

Fig. 7.12 Roadmap of driving assistance function. ©[2018] IEEE. Reprinted, with permission, from Ref. [15]

great improvement in angular resolution and data throughput. Moreover, the design that applies plane frequency-modulated continuous-wave (FMCW) MIMO array and TDMA concept will maintain the dominance of MIMO and enhance the antenna gains at the transmission end to improve the overall SNR [15].

In the massive MIMO detection system, the more optimal VLSI architecture can significantly improve the performance of the detection chip and lower the power consumption and latency of the system, which can better accomplish the real-time communications and achieve a higher safety in the self-driving field. This greatly benefits the reduction of traffic accidents and the improvement of the traffic congestion situation. In the current automobile security scenarios, the reaction time to avoid collisions is shorter than 10 ms, while the bidirectional data exchange of self-driving vehicles may require the latency be within 1 ms, which can be technically realized through tactile network and 1 ms end-to-end latency. Thus, full autopilot technology will definitely change traffic behaviors. In terms of the distance between vehicles, autopilot technology needs to detect the potential safety-critical conditions in advance, which can be supported by the future wireless communications system with high reliability and proactive predication [16]. With the increase of the self-driving terminals, data exchange is required among an increasingly number of users. For the self-driving terminal, how to cope with multiuser requirements and shield multiuser interference is a big challenge. The highly effective massive MIMO detection architecture can meet the high-speed processing requirements in the self-driving system to lower the latency. Meanwhile, massive MIMO detection architecture is capable of transmitting massive data and processing correspondingly to improve the system throughput. For the application scenarios with high interference and noise, the nonlinear massive MIMO detection architecture can enhance the detection precision while maintaining certain latency and throughput, which is very important for the security of self-driving. The ASIC-based massive MIMO detection architecture not only can meet the latency and throughput requirements but also poses certain advantages in power consumption. In addition, considering the high mobility of self-driving terminals, the massive MIMO detection architecture must adapt to distinct scenarios and requests.

The massive MIMO detector receives and restores information, which has great effects in improving the channel capacity and communications efficiency and ensuring instant communications of remote diagnosis. More importantly, a proper massive MIMO architecture can accelerate this process. Thus, designing a more optimal massive MIMO VLSI architecture has always been the research topic of many researchers. In 5G era, most communications systems of mobile terminals are inevitably associated with massive MIMO technology. The massive MIMO can fulfill not only high capacity and speed but also low power consumption and cost, which will contribute to the boost of further flourish of mobile terminals.

7.2.2 Application of Reconfigurable Detection Chips

In the future, more applications will emphasize high energy efficiency as well as flexibility and scalability to adapt to different algorithms, MIMO system scales, and detection performance requirements. To accommodate these features, the reconfigurable MIMO signal detectors have gradually become the hotspot in the academia in recent years. This is because the reconfigurable MIMO signal detectors can fully exploit and utilize the data parallelism in algorithms and dynamically reconfigure chip functions via configuration flow, which can achieve a certain tradeoff between efficiency and flexibility compared with GPP and ASIC. The following sections give examples to show the applications based on the reconfigurable massive MIMO detectors.

7.2.2.1 Intelligent Manufacturing (IM)

IM is a man–machine integration intelligent system composed of intelligent machines and human experts, which is capable of performing a series of intelligent activities, such as analysis, reasoning, judgment, conception, and decision-making (Fig. 7.13). With the cooperation between human and intelligent machines, the brainwork of human experts is enlarged, extended, and partly replaced. IM updates the concept of manufacturing automation and expands it to flexibility, intelligentization, and highly integration. Undoubtedly, intelligentization is the future development direction of manufacturing automation. AI technology should be widely used in almost each link of manufacturing. Expert system technology can be used in engineering design, process design, production scheduling, fault diagnosis, etc. Also, the advanced computational intelligence methodologies such as neural network (NN) and fuzzy control can be used in the product formulation, production scheduling, etc., to achieve IM. AI technology is especially suitable to solve extremely complex and uncertain

Fig. 7.13 IM-related technologies

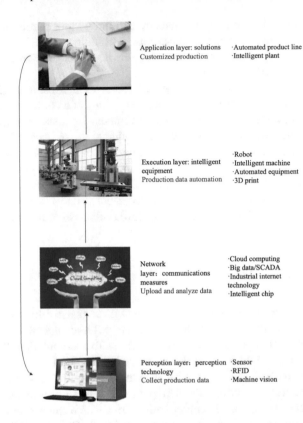

Application layer: solutions   ·Automated product line
Customized production   ·Intelligent plant

Execution layer: intelligent   ·Robot
equipment   ·Intelligent machine
Production data automation   ·Automated equipment
                                 ·3D print

Network   ·Cloud computing
layer: communications   ·Big data/SCADA
measures   ·Industrial internet
Upload and analyze data   technology
                                 ·Intelligent chip

Perception layer: perception   ·Sensor
technology   ·RFID
Collect production data   ·Machine vision

problems. In the previous three industrial revolutions, the traditional manufacturing system mainly focused on its five core elements to pursue constantly technical upgrade, which include materials (including characteristics and functions, etc.), methods (including technology, efficiency, productivity, etc.), machines (including precision, automation, production capacity, etc.), measurement (including sensor monitoring, etc.), and maintenance (including utilization rate, fault rate, O&M cost, etc.). Throughout the whole human industrialized process, these five elements have always been the essentials. The logic of the IM is as follows. The issue occurs first; it is then analyzed according to the model, and the model is adjusted based on the five core elements. Then, solve the issue. Finally, accumulate the experiences according to the solved issues and retrospect the source of the issue to avoid similar issues later. In essence, IM is the process of knowledge generation and inheritance.

IM must make the most of communications means at the network layer to control and operate all the intelligent equipment by using wireless communications. In turn, massive MIMO detection in IM equipment must meet the requirements of high stability, flexibility, and scalability. Therefore, how to realize the above requirements will be a challenge for massive MIMO detection chip. Nevertheless, the reconfigurable massive MIMO detector shows certain advantages in these aspects, which is featured with very high potential application values. Moreover, with the popularization of IM, a growing number of industrial intelligent equipment will leverage wireless transmission systems, which raises the issues of upgrade and compatibility for equipment systems. Thus, the precision requirement for the design of massive MIMO detection chips will be increased. Therefore, how to reduce the interference between equipment and the impact of other environmental noise on the signal transmission, and improve flexibility and scalability will be the primary research directions for the design of reconfigurable massive MIMO detectors.

7.2.2.2 Wireless Medical

Communications technology is a key technology in wireless medical [17], as shown in Fig. 7.14. The remote diagnosis, remote surgery, and telerehabilitation using wireless communications and information technologies can ignore the geographical distance and provide effective, reliable, and real-time health services for patients [18]. In addition, in the remote surgeries assisted by robots, to promptly and accurately provide audio and video information and tactile feedback, e-health has very strict requirements for the reliability of the wireless connection. Especially in the remote surgeries and diagnosis, reliability is extraordinarily important. Unreliable connections may lead to the delay of imaging, and low image resolution may limit the remote handling efficiency of doctors. Furthermore, the accurate remote medical can only be realized by tactile feedback. Once human and machine can interact in real time, this demand can be achieved. However, the deterministic real-time act demanded is not supported by the existing communications systems. Human wearable devices can provide medical monitoring for the seniors, athletes, and children. Remote medical system offers a complicated communications environment for patients and healthcare professionals by monitoring patients via computer or cell

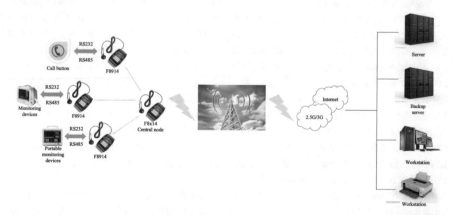

Fig. 7.14 Wireless medical and monitoring system

phone technologies. Owing to its low cost, lightweight, and low maintenance frequency, wearable devices have an extensive application prospect in the medical data collection for patients, the establishment of connections between translation devices, tracking, rescue, etc. [16].

In wireless medical, the reliability plays a vital role. To suit distinguished equipment and human characteristics, the hardware circuits with more flexible framework are required. Also, to adapt to the continuously developed and updated equipment requirements, the wireless baseband processing circuits should be scalable to lower the cost. Exactly, the reconfigurable massive MIMO detector will have a bright future regarding to these aspects. In addition, most massive MIMO detection algorithms are with deep parallel computing, and reconfigurable architecture shows its superiority in efficient processing of parallel computation [19]. Generally speaking, the higher the parallelism and the lower data dependency in the algorithm are, the more suitable it is to be accelerated using reconfigurable methods, which is also a reflection of the algorithm at the hardware level. Therefore, reconfigurable massive MIMO detector can effectively fulfill the computation with high parallelism.

7.3 Prospect of Applications of Edge Computing

As the continuous development of the socioeconomic level, people's demand on mobile internet has shown a clear diversification trend. As for capacity, the massive application demands boost the application and development of emerging technologies such as IoT, D2D, and M2M, which promotes the continuous upgrade of mobile internet equipment and intelligent mobile equipment. The numbers of users and intelligent communications equipment in mobile internet are exploding, which will reach the order of tens of billions or even hundreds of billions according to the forecast. Correspondingly, the data traffic of 5G mobile communications will reach an unprecedented level along with the growth of the communications equipment. Some new application scenarios, i.e., self-driving, smart grid, AR and VR, propose higher requirements on the latency, energy efficiency, number of devices that can be accommodated, and reliability for the communications system [20]. Currently, the emergence of the applications such as online gaming, cloud desktop, smart city, environmental monitoring, and intelligent agriculture, puts the real-time computing capacity of mobile terminals under a harsh test. On one hand, limited by the reality factors such as volume, power consumption, and weight, the processing capability of existing terminal devices is far from meeting the requirements of the aforementioned applications in low latency, high energy efficiency, and high reliability, which severely affects the user experience. In this case, the MCC stands out as one of the effective solutions at present. MCC allows the user equipment to partially or completely migrate the local computation tasks to the cloud server for execution, which solves the problem of resource shortage of mobile equipment and saves the energy consumption of the locally task execution. However, offloading tasks to the core cloud server not only consumes the backhaul link resources and generates additional latency overhead but also impacts the reliability; thus, the requirements of low

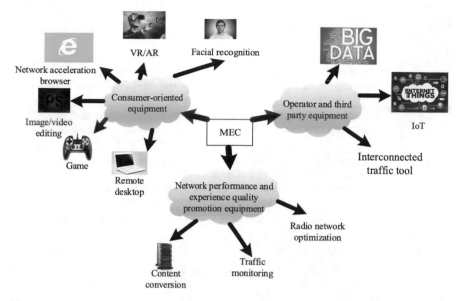

Fig. 7.15 Typical MEC application scenarios

latency and high reliability are cannot meet for new application scenarios. Therefore, the emerging mobile edging computing (MEC) becomes the key to address this issue. Typical application scenarios of MEC are shown in Fig. 7.15. The basic idea of MEC is to migrate the cloud computing platform to the edge of mobile access network and allow the user equipment to offload computation tasks to the nodes at the edge of the network, e.g., base stations and wireless access spots. Apart from meeting the scalability requirements of the computing capability for terminal devices, MEC also makes up the shortcoming of MCC in long latency. Hence, MEC will become a key technology to assist 5G services to fulfill the technical indicators such as ultra-low latency, ultra-high energy efficiency, and ultra-high reliability [20, 21].

7.3.1   Concept of Edge Computing

The concept of MEC was first proposed by the European Telecommunications Standards Institute (ETSI) in 2014, and was defined as a new platform that "*provides IT and cloud computing capabilities within* the edge of mobile access network, *the Radio Access Network (RAN) in close proximity to mobile subscribers*" [22]. MEC offers cloud computing capabilities within the RAN. MEC connects the user directly to the nearest cloud-enabled edge network, which avoids the direct mobile communication between the core network and end users. Deploying MEC at the base station enhances computation and avoids the performance bottleneck and possible system failures [23]. As shown in Table 7.2, the comparison between MEC and tra-

Table 7.2 Comparison between MEC and MCC

Comparison items	MEC	MCC
Server hardware	Small-sized data center requiring moderate resource [2, 9]	Large-scale data center (each possessing a lot of powerful servers) [10, 25]
Server location	Coexist with wireless gateway, Wi-Fi router, and LTE base station [2]	Installed in exclusive buildings with the scale comparable to several football courts [11, 26]
Deployment	Intensive deployment by telecom operators, MEC suppliers, enterprises, and family users participated and light configuration and plans required [2]	Deployed in a few places all around the world by IT corporations such as Google and Amazon with complicated configuration and plans required [10]
Distance to end user	Short (dozens of meters to several hundred meters) [26]	Long (probably across continents) [26]
Backhaul use	Infrequently used, alleviating congestion [12]	Frequently used, causing congestion [12]
System management	Layered control (centralized/distributed) [13]	Centralized control [13]
Supported latency	Less than tens of milliseconds [14, 26]	More than a 100 ms [15, 16]
Application	Compute-intensive applications with high requirements on latency, e.g., VR, self-driving, and online interactive games [2, 17]	Compute-intensive applications without high requirements on latency, e.g., online social, mobile commerce/healthcare/study [18, 19]

ditional MCC shows that there are significant differences between MEC and MCC systems in terms of computing server, distance to end users, and typical latency, etc. Compared with MCC, MEC has the advantages of achieving lower latency, saving energy for mobile devices, supporting context-aware computing, and enhancing privacy and security for mobile applications [24]. First, MEC can lower the task execution latency. By migrating the cloud computing platform to the edge of access network, it narrows the distance between the computing server and user equipment. Since the task offloading does not need to travel through the backhaul link or core network, the transmission latency overhead is reduced. In addition, the computation capability of the edge server is significantly superior to user equipment, which dramatically lowers the task computation latency. Second, MEC can greatly improve network energy efficiency. Although IoT equipment can be widely applied to various scenarios such as environmental monitoring, group awareness, and intelligent agriculture, most deployed IoT equipment is powered by batteries. As MEC shortens the distance between the edge server and mobile equipment, it significantly saves the energy consumed by task offloading and wireless transmission, extending the service life of IoT equipment. Research results show that for different AR equipment, MEC can extend the battery service life ranging from 30 to 50%. Finally, MEC can provide a higher service reliability. Due to the distributed deployment, small-scale nature, and the less concentration of valuable information, MEC servers are much less likely to become the target of a security attack in contrast with the big data center of MCC, being able to provide more reliable services. And, most MEC servers are private-owned cloudlets, which shall ease the concern of information leakage [24] and ensure higher security. In general, the technical characteristics of MEC are mainly embodied in proximity, low latency, high bandwidth, and location awareness.

Proximity: As MEC server is deployed proximal to the information source, edge computing is very suitable for capturing and analyzing the key information of big data. Moreover, edge computing can directly access user equipment; thus, specific business applications are easily derived.

Low latency: As MEC server is proximal to or directly operating on the terminal devices, the latency is greatly lowered. This makes the application feedback faster, improves user experience, and dramatically reduces the possibilities of congestion incurred in other parts of the network.

High bandwidth: As MEC server is proximal to the information source, it can complete simple data processing without uploading all data or information to the cloud, which reduces the transmission pressure of the core network, decreases network congestion, and enhances network transmission speed.

Location awareness: When the network edge is a part of the radio access network, no matter Wi-Fi or honeycomb, local services can identify the specific location of each connection equipment with a relatively little information.

Figure 7.16 shows the basic system architecture of MEC. Note that, MEC server is closer to the end user than the cloud server. In this case, although the computing capability of MEC servers is weaker than that of the cloud computing servers, they still can provide better QoS for end users. Apparently, unlike cloud computing, edge computing incorporates edge computing nodes into the network. In general, the structure of edge computing can be categorized into three aspects, i.e., front-end, near-end, and far-end. The front-end mainly refers to the terminal devices (e.g., sensors, actuators) deployed at the front-end of the edge computing structure. The front-end environment can provide more interaction and better responsiveness for the end users. Nonetheless, due to the limited computing capacity of the terminal devices, most requirements cannot be satisfied at the front-end environment. In these circumstances, the terminal devices must forward the resource requirements to the servers. The gateway deployed in the near-end environment will support most of the traffic flows in the network. The reason why edge computing can provide real-time services for some applications is that it endows the near-end equipment with powerful computing capabilities. Edge severs can have also numerous resource requirements, such as real-time data processing, data caching, and computation offloading. In edge computing, most of the data computation and storage will be migrated to this near-end environment. In doing so, the end users can achieve a much better performance

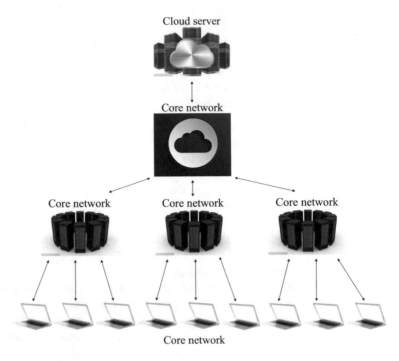

Fig. 7.16 Basic system architecture of MEC

on data computing and storage, with a small increase in the latency. As the cloud servers are deployed farther away from the terminal devices, the transmission latency is significant in the networks. Nonetheless, the cloud servers in the far-end environment can provide more computing power and more data storage. For example, the cloud servers can provide massive parallel data processing, big data mining, big data management, machine learning, etc. [27].

7.3.2   Application of Detection Chips in the Edge Computing

In the current network architecture, the high deployment position of core network results in a long transmission latency, failing to meet the business requirement of ultra-low latency. Additionally, businesses ended at the cloud are not completely effective while some regional businesses that do not end locally waste the bandwidth and increase latency. Therefore, latency and connection number indicators determine that the ending point of 5G businesses is not all on the cloud platform at the rear end of core network. Fortunately, MEC fits the demands [28]. Figure 7.17 shows how MEC enhances the integration of data center and 5G. From one aspect, MEC is deployed at the edge. The edge service operating on the terminal devices feeds back faster, which

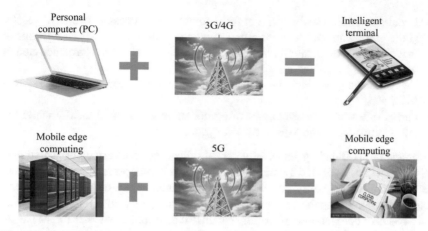

Fig. 7.17 MEC enhancing the integration of data center and 5G

resolves the latency issue. From another, MEC submerges computing content and capability, provides intelligent traffic scheduling, localizes services, caches content locally, and prevents part of regional services from the trouble of ending at cloud. As mobile network has to serve devices of different types and requirements, the cost is incredible if an exclusive network is established for each service. Network slicing technology allows operators to slice a hardware infrastructure into multiple end-to-end virtual networks. Each network slice is logically isolated from the equipment to the access network, to the transmission network to the core network, adapting to different requirements of various types of services, ensuring that from the core network to the access network including links such as terminals can allocate network resources dynamically, in real time and effectively to guarantee the performance of quality, latency, speed, and bandwidth. To a certain degree, the service awareness function of MEC is analogous to the network slicing technology. With low latency as one of the primary technical characteristics, MEC can support the most latency-sensitive services, which also means that MEC is the key technology for the slicing with ultra-low latency [29]. With the application of the MEC, the connotation of network slicing technology will be extended from purely slicing to slicing under different latency requirements to achieve multiple virtual end-to-end networks, which contributes to the development of 5G network slicing technology.

The key to achieving low latency and saving user equipment energy in MEC lies in the computation offloading, while a key to computing offloading is usually to decide whether to perform a computation offload. In general, there are three decisions with regards to computing offloading.

(1) Local execution, in which the entire computation process is executed locally at the user equipment without offloading computation to the MEC, e.g., due to that the MEC computation resources are unavailable or the performance cannot be improved by offloading.
(2) Full offloading, in which the entire computation is offloaded to be processed at MEC server.
(3) Partial offloading, in which part of computation is executed locally while the left is offloaded to the MEC server for processing.

Computation offloading, especially partial offloading, is a very complicated process which will be impacted by multiple factors such as user preference, wireless and backhaul connection quality, user equipment computation capability, or utilizability of cloud computing capability. Application model/category is also one of the significant aspects of computation offloading because it determines whether full or partial offloading fits, which computations can be offloaded, and how these computations can be offloaded [27]. MEC server is able to provide more powerful computation capabilities than user equipment, offloading computation to MEC server for processing can shorten the data processing time and save the energy of terminal devices consumed for data processing. However, we cannot ignore the fact that offloading data to be processed by the user equipment to the MEC server (uplink) needs to consume transmission time and energy so does it when the MEC server transmits the processed data to the user equipment (downlink). When the computation amount of an application is not very huge, especially when the processing capability of user equipment satisfies the requirements, the aforementioned data transmission (uplink and downlink) may waste time and energy, causing the performance loss. Thus, a reasonable mechanism is required to make the decision of whether to perform computation offloading. MEC technology has relatively high requirements on uplink and downlink data transmission, which are mainly reflected in the low latency, high throughput, and low power consumption in massive MIMO detection. ASIC-based massive MIMO detection chips show outstanding performances in these aspects and can be implemented at the MEC terminal, to reduce latency and power consumption, and improve throughput.

In recent years, there are a large number of research results targeting at the computation offloading for MEC systems. However, there are still many emerging issues need to be addressed including mobility management for MEC, green MEC, and security and privacy issues for MEC. Mobility is an inherent feature of many MEC applications such as AR-assisted museum visit to enhance visitor experience. In this kind of applications, the movements and trajectories of users offer the MEC server with location and personal preference information, which improves the processing efficiency of user computation requests. Furthermore, mobility also poses a great challenge to the realization of universally reliable computation (i.e., without interruptions or errors) for the following reasons. First of all, MEC is usually executed in a heterogeneous network composed of multiple macro and small base stations, and wireless access points. Thus, the user movements should be frequently switched among small coverage MEC servers, as shown in Fig. 7.18, which becomes

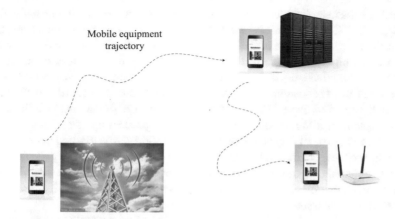

Mobile equipment trajectory

Fig. 7.18 MEC terminal management

more complex due to the diversified system configurations and associated strategies between users and servers. Subsequently, serious signal interference and pilot pollution can be generated while users move among different base stations, dramatically deteriorating the communications performance. Finally, frequent switch increases computation latency, which affects the user experience [24]. To meet the communications performance demands, higher detection precision is required during the signal restoring by the detector. Hence, more optimal detector architecture is in demand, which shall be addressed by nonlinear or even more complicated detection algorithms. Therefore, how to support different algorithms and sizes of mobile terminal, and algorithm scalability should all be considered for the development of massive MIMO detection chips. Reconfigurable massive MIMO detectors ensure detection performance and can reach certain energy efficiency at the same time. Most importantly, this detector enables high flexibility, reliability, scalability, etc.

The MEC server is a small data center, and each data center consumes less energy than a traditional cloud data center. However, its intensive deployment mode causes serious problems in the energy consumption of the whole system. Therefore, it is definitely a key to developing innovative technologies to achieve green MEC. Compared with green communications system, the computation resources of MEC server must be appropriately allocated to realize the required computation performance, making traditional green wireless technologies no longer suitable. In addition, as the past researches on green data communications network has not considered wireless resource management, they are not applicable to green MEC. Besides, the highly unpredictable computation workload pattern in MEC server poses another big challenge for the resource management in MEC systems, calling for advanced estimation and optimization techniques [24]. What's more, there are increasing demands for secure and privacy-preserving mobile services. While MEC enables new types of services, its unique features also bring new security and privacy issues. First of all, the innate heterogeneity of MEC systems makes the conventional trust and authentication mechanisms inapplicable. Second, the diversity of communication technologies that support MEC and the software nature of the networking management mechanisms bring new security threats. Besides, secure and private computation mechanisms become highly desirable as the edge servers may be an eavesdropper or an attacker. These motivate us to develop effective mechanisms [24]. We can also circumvent some power consumption and security related issues from hardware circuits. The reconfigurable massive MIMO detector is close to ASIC in energy efficiency, and can implement different algorithms and signals processing of different scales, demonstrating high flexibility and scalability. In addition, as the PEs and interconnect muddles inside the reconfigurable massive MIMO detector are relatively regular, it is difficult to obtain the algorithm information by observing the hardware architecture and circuit composition. This feature can improve the hardware security and avoid some MEC security issues.

Next, the practical application of the IoV is used as an example to demonstrate the advantages of MEC. The IoV has special requirements for the data processing. The first requirement is low latency, i.e., to achieve the early warning of collision when vehicles are moving at high-speed, the communications latency should be within several milliseconds. The second requirement is high reliability. For safe driving requirements, the IoV requires higher reliability compared with ordinary communications. Meanwhile, as vehicles are moving at high speed, signals must meet the high reliability requirements on the basis of being able to support high-speed motion. With the increase of networked vehicles, the data quantity of the IoV also grows and as a return, the requirements for latency and reliability are higher. After MEC technology is applied to the IoV, due to the location characteristics of MEC, the IoV data can be saved in places proximal to the vehicles to lower the latency, which is quite suitable for the service types with high latency requirements such as anti-collision and accident warning. Meanwhile, the IoV should ultimately be used to help in driving. The location information of vehicles changes rapidly when vehicles are moving at high-speed. Nevertheless, the MEC server can be placed on the vehicle to accurately sense the location change in real time, which improves the communications reliability. In addition, what the MEC server processes are the real-time IoV data with great values. The MEC server analyzes the data in real time and transmits the analysis results to other networked vehicles in the proximal area with ultra-low latency (usually in milliseconds) to facilitate the decision-making of other vehicles (drivers). This approach is more swift, autonomous, and reliable than other processing methods.

References

1. Björnson E, Larsson EG, Marzetta TL (2015) Massive MIMO: ten myths and one critical question. IEEE Commun Mag 54(2):114–123
2. Larsson EG, Edfors O, Tufvesson F et al (2014) Massive MIMO for next generation wireless systems. IEEE Commun Mag 52(2):186–195

3. Datsika E, Antonopoulos A, Zorba N et al (2017) Cross-network performance analysis of network coding aided cooperative outband D2D communications. IEEE Trans Wireless Commun 16(5):3176–3188

4. Yang M, Li Y, Jin D et al (2014) Software-defined and virtualized future mobile and wireless networks: a survey. Mobile Netw Appl 20(1):4–18

5. Vereecken W, Van Heddeghem W, Colle D et al (2010) Overall ICT footprint and green communication technologies. In: International symposium on communications, control and signal processing, pp 1–6

6. Zhang S (2016) Study on the technical proposal of the base station application capability expansion based on cloud computing technologies in LTE Network. Beijing University of Posts and Telecommunications

7. Wen BJ (2017) Resource environmental effect analysis based on the development planning of chinese new energy vehicles. China Min Mag 10:76–78

8. Zhu W, Gao D, Zhao W et al (2017) SDN-enabled hybrid emergency message transmission architecture in internet-of-vehicles. Enterp Inf Syst 2017:1–21

9. Gerla M, Lee EK, Pau G et al (2016) Internet of vehicles: from intelligent grid to autonomous cars and vehicular clouds. In: Internet of Things, pp 241–246

10. Garg SK, Versteeg S, Buyya R (2013) A framework for ranking of cloud computing services. Future Gener Comput Syst 29(4):1012–1023

11. Buyya R, Yeo CS, Venugopal S (2008) Market-oriented cloud computing: vision, hype, and reality for delivering IT services as computing utilities. 11(4):10–1016

12. Yin S (2016) Reasearch on the verification platform of communications baseband chips at module level. Xidian University

13. Li D (2016) Low power consumption design of CPU in the baseband chips. Xidian University

14. Kimura R, Matsunaga N, Okajima H et al (2017) Driving assistance system for welfare vehicle using virtual platoon control with augmented reality. In: Conference of the society of instrument and control engineers of Japan, pp 980–985

15. Hasch J (2015) Driving towards 2020: automotive radar technology trends. In: IEEE Mtt-S international conference on microwaves for intelligent mobility, pp 1–4

16. Simsek M, Aijaz A, Dohler M et al (2016) The 5G-enabled tactile internet: applications, requirements, and architecture

17. Khodashenas PS, Aznar J, Legarrea A et al (2016) 5G network challenges and realization insights. In: International conference on transparent optical networks, pp 1–4

18. Kang G (2012) Wireless eHealth (WeHealth)—from concept to practice. In: IEEE International conference on E-Health networking, applications and services, pp 375–378

19. Khalaf A, Abdoola R (2017) Wireless body sensor network and ECG Android application for eHealth. In: International conference on advances in biomedical engineering, pp 1–4

20. Tian H, Fan SS, Lv XC et al (2017) 5G-oriented mobile edging computing. J Beijing Univ Posts Telecommun 40(2):1–10

21. Yu YF, Ren CM, Ruan LF et al (2016) A brief analysis on the development of mobile edging computing. Commun Netw Technol 11:46–48

22. Liu J, Mao Y, Zhang J et al (2016) Delay-optimal computation task scheduling for mobile-edge computing systems. 1451–1455

23. Abbas N, Zhang Y, Taherkordi A et al (2017) Mobile edge computing: a survey. IEEE Internet of Things J PP(99):1

24. Mao Y, You C, Zhang J et al (2017) A survey on mobile edge computing: the communication perspective. IEEE Commun Surv Tutorials PP(99):1

25. Buyya R, Yeo CS, Venugopal S et al (2009) Cloud computing and emerging IT platforms: vision, hype, and reality for delivering computing as the 5th utility. Future Gener Comput Syst 25(6):599–616

26. Vecchiola C, Pandey S, Buyya R. High-performance cloud computing: a view of scientific applications. In: International symposium on pervasive systems, algorithms, and networks, pp 4–16

27. Yu W, Liang F, He X et al (2017) A survey on the edge computing for the internet of things. IEEE Access PP(99):1
28. Liu J, Mao Y, Zhang J et al (2016) Delay-optimal computation task scheduling for mobile-edge computing systems. 1451–1455
29. Corcoran P, Datta SK (2016) Mobile-edge computing and the Internet of Things for consumers: extending cloud computing and services to the edge of the network. IEEE Consum Electron Mag 5(4):73–74

Printed in the United States
By Bookmasters